Everything you always wanted to know about…

Chemistry

2nd edition

STERLING
Education

Our Commitment to the Environment

Sterling Test Prep is committed to protecting our planet's resources by supporting environmental organizations for conservation, ecological research, education, and preservation of vital natural resources. A portion of our profits is donated to help these organizations continue their critical missions.

Since 1992, the Ocean Conservancy advocates for a healthy ocean by supporting sustainable solutions based on science and cleanup efforts. Ocean Conservancy laid the groundwork for an international moratorium on commercial whaling, played an instrumental role in protecting fur seals from overhunting and banning the international trade of sea turtles.

Since 1988, the Rainforest Trust saves critical lands for conservation through land purchases and protected area designations. Rainforest Trust has helped protect more than 72 protected areas in over 16 countries, such as the Falkland Islands, Costa Rica, Peru.

Since 1980, Pacific Whale Foundation saves whales from extinction and protects oceans through science and advocacy. As an international organization, with ongoing research projects in Hawaii, Australia, and Ecuador, PWF participates in global efforts for threats to whales and other marine life.

Your purchase helps support environmental causes around the world.

Table of Contents

Table of Contents *(continued)*

Table of Contents *(continued)*

Chapter 1

Electronic Structure and the Periodic Table

- **Electronic Structure**
- **The Periodic Table: Variations of Chemical Properties with Group and Row**
- **The Periodic Table: Classification of Elements into Groups by Electronic Structure**

Electronic Structure

The *atom* is the smallest unit of an element that retains the characteristics of that element. An atom consists of several *subatomic particles*, including protons, neutrons, and electrons.

The *nucleus* is the densely packed region at the center of an atom that consists of protons and neutrons. The diameter of the nucleus is approximately ~10,000 times smaller than the overall diameter of the atom. Most of an atom's volume comes from its *electron cloud*, which is the outer region of an atom that surrounds the nucleus.

The *proton* is the positively charged particle located in the nucleus of the atom. Each proton has a charge of +1, and its mass is approximately equal to a neutron.

The *neutron* is an uncharged particle located in the nucleus of the atom.

The *electron* is a small, negatively charged particle located in the electron cloud. Each electron has a charge of –1, and its mass is about 2,000 times smaller than a proton or neutron. In a neutral atom, the number of electrons equals the number of protons.

The *electrostatic attraction* (i.e., due to opposite electrostatic charges) between the positive core protons and orbiting negative electrons holds electrons around the nucleus. The *repulsion* between neighboring electrons spreads them over the entire volume of the electron cloud.

An *ion* forms when an atom loses or gains electrons, causing the atom to have either a net negative or a net positive charge. The loss of electrons produces positively charged *cations*, and the gain of electrons produces negatively charged *anions*.

Cations and anions are represented by a superscript of a positive or negative sign after a chemical symbol.

The *atomic number* (Z) equals the number of protons in an atom. If the atom has no charge, then Z equals the number of electrons. On the periodic table, elements are arranged by their atomic numbers.

The *mass number* (A) is the total number of protons and neutrons (i.e., *nucleons*) of an atom.

Isotopes are atoms of the same element with the same number of protons (i.e., same atomic number Z) but a different number of neutrons. Therefore, isotopes have different mass numbers *A*. The isotope of an atom has virtually identical chemical properties as the atom because they have the same number of protons and therefore are the same element.

Most elements naturally occur as a mixture of two or more stable isotopes. For example, the element carbon (Z=6) includes the isotopes ^{12}carbon, ^{13}carbon, and ^{14}carbon. The 12, 13, and 14 are the mass numbers (A) of the respective isotopes. ^{12}Carbon has six neutrons, ^{13}carbon has seven neutrons, and ^{14}carbon has eight neutrons.

The *relative atomic mass* of an element (also known as *atomic weight*) is the weighted average of the masses of its stable isotopes.

Mass spectrometry is an experimental method used to determine the atomic masses of isotopes. In mass spectrometry, a sample is ionized by bombarding it with electrons, which causes it to break into charged fragments. The ions are separated by subjecting them to an electric or magnetic field. The amount of deflection that the ions experience is proportional to the ions' weight.

The unit of measurement used for atomic weight is the *atomic mass unit* (amu) or *dalton* (Da), which is approximately equivalent to the mass of one nucleon (a single proton and neutron). The dalton is based on the atomic mass of the carbon-12 isotope, meaning 1 amu is equal to 1/12 the mass of a ^{12}C atom, or 1.66×10^{-27} g.

^{12}Carbon is the only atomic species with an atomic mass that is exactly a whole number. The atomic masses of other elements are always remarkably close to whole numbers of the atomic mass units.

Atomic mass is expressed by the equation:

$$avg.\ atomic\ mass = (mass_1)\cdot(abundance_1) + (mass_2)\cdot(abundance_2) + \ldots$$

Hydrogen has two isotopes, and their masses and abundances are shown below:

Isotope	Mass	% Abundance
^1H	1.0078 amu	99.985
^2H	2.0140 amu	0.015000

The relative atomic mass of hydrogen (i.e., atomic weight) can be calculated:

$$(0.99985 \times 1.0078\ \text{amu}) + (0.00015000 \times 2.0140\ \text{amu})$$

$$1.0076 + 0.00030210 = 1.0079\ \text{amu for H}$$

Orbital structure of hydrogen atom, principal quantum number *n*, number of electrons per orbital

The *electron configuration* of an atom is the arrangement of electrons around the nucleus. Electron configurations describe electrons as each moving independently in a predefined orbital around the nucleus.

Nobel laureate Niels Bohr (1885-1962), a Danish physicist of the 20th century, was the first to apply quantum physics, using wave functions, to define the energy of the electrons to discrete values. A single-electron atom is an elementary particle that consists of one electron around its core nucleus of a proton and a neutron. The *Bohr model*, discussed in more detail later, focuses on the hydrogen atom and the single electron that orbits its nucleus. In quantum mechanics, the hydrogen electron exists in a spherical probability density cloud around its nucleus.

The *principal quantum number* (*n*) defines which shell the electron occupies and describes the *size* of the orbital and its distance from the nucleus. The principal quantum number can have only positive integer values of $n = 1, 2, 3...$

Electron shells, called principal energy levels, are labeled by their principal quantum numbers ($n = 1, 2, 3...$), but can be labeled alphabetically as n = K, L, M. . .

Higher n shells indicate larger orbitals further from the nucleus and higher energy levels. The maximum number of electrons per shell is given by $2n^2$.

For example, the second shell can hold up to $2(2)^2 = 8$ electrons.

Electrons usually occupy outer shells only after inner shells have been filled. However, this is not necessary as some outer shell electrons are promoted to a higher shell if it imparts stability to the atom.

Conventional notation for electronic structure

The *Aufbau principle* states that electrons fill their orbitals in the order of lowest energy to highest energy. Orbitals fill according to the diagonal lines shown.

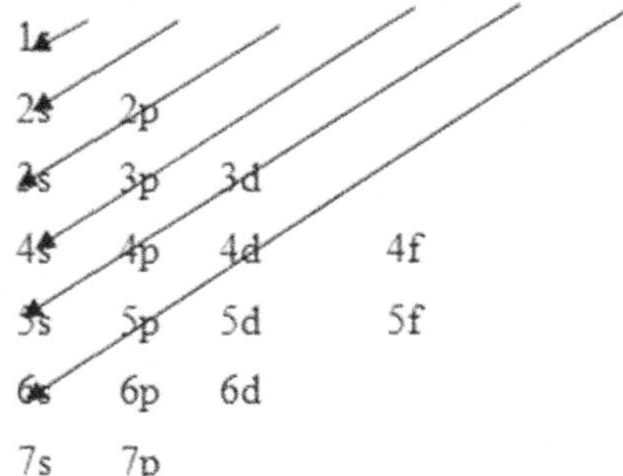

$1s^2\ 2s^2\ 2p^6\ 3s^2\ 3p^6\ 4s^2\ 3d^{10}\ 4p^6\ 5s^2\ 4d^{10}\ 5p^6\ 6s^2\ 4f^{14}\ 5d^{10}\ 6p^6\ 7s^2\ 5f^{14}\ 6d^{10}\ 7p^6$

In multi-electron atoms, the energy level for an electron is affected by both the *n* and *l* quantum numbers. The *l* quantum number is explained later but know that it is represented by the letters *s, p, d,* and *f.* The *l* quantum number describes a subshell. Subshells are comprised of atomic orbitals, and each orbital can hold two electrons.

The 4*s* orbital is lower in energy than the 3*d* orbitals. From the *Aufbau principle,* electrons populate the 4*s* orbital before the 3*d* orbital, even though the 4*s* orbital has a higher principal (*n*) value than the 3*d* orbital.

Hund's rule is another essential guideline for writing electron configurations. Hund's rule states that an electron must occupy every orbital in a sublevel (e.g., s, p, d, f) before a second electron can occupy an orbital in that sublevel. Hund's rule maximizes the number of electrons with the same electron magnetic spin (discussed below) to increases the stability of the atom. Double-occupied orbitals are much higher in energy and, therefore, less stable than orbitals with a single electron.

The *octet rule* refers to the tendency of atoms to gain, lose or share electrons to achieve a full orbital of eight electrons in its *valence* (i.e., outermost) orbital. The octet rule only applies to the *s* and *p* orbital electrons. Therefore, the octet rule is particularly useful when applied to the *representative elements.* Representative elements are not in the transition block (*d* orbitals) or the inner-transition metal block (*f* orbitals, also known as lanthanides and actinides). A complete octet is typically represented as an electron configuration ending in s^2p^6. When atoms have either an excess or deficiency in the number of valences (outermost) electrons, they react to satisfy their octets and form more stable compounds.

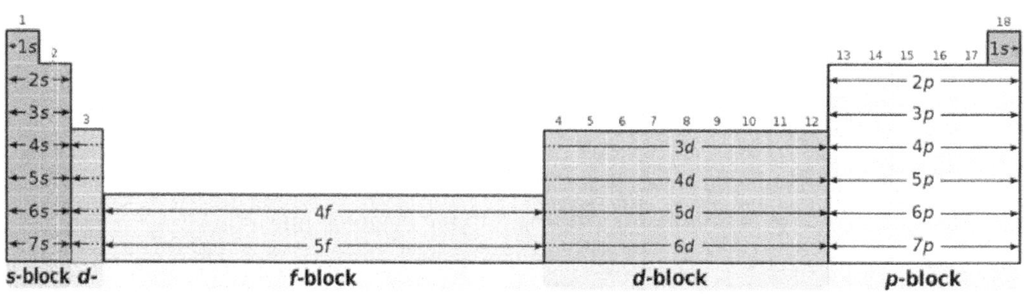

The vertical columns are groups or families. The representative elements are in the left two groups with s shells (metals) and the six right-hand side p orbitals (nonmetals).

Using the Aufbau Principle and Hund's Rule for writing electron configurations, the electron configuration for zinc (Zn) atom with 30 electrons is:

$$1s^2\ 2s^2\ 2p^6\ 3s^2\ 3p^6\ 4s^2\ 3d^{10}$$

- The first number indicates the principal energy level (n).
- The letter (s, p, d, f) indicates the subshell (which ranges from 0 to $n-1$).
- The superscript indicates the number of electrons in the subshell.

As the number of electrons in an atom increases across the periodic table, the length of written electron configurations can become exceedingly long and tedious.

There is a shorthand approach to writing electron configurations that references the electronic configuration of the noble gases (or inert gases). The noble gases have complete octets and are the elements located in the far-right column (group or family) of the periodic table (Group 18).

The electronic configurations of Na (atomic number or Z is 11):

1. Identify the noble gas that comes before the element in the periodic table (i.e., the noble gas in the previous row; also called *periods*).

   The noble gas before Na ($Z = 11$) is Ne ($Z = 10$)

   The noble gas before Cl ($Z = 17$) is Ne ($Z = 10$), not Ar ($Z = 18$)

2. Write the noble gas in square brackets:

   [Ne]

   Ne has the electron configuration of $1s^2\,2s^2\,2p^6$

3. Write the electron configuration for Na:

   Na: $1s^2\,2s^2\,2p^6\,3s^1$

4. Abbreviate the electron configuration for Na by referencing Ne:

   [Ne] $3s^1$

Example: the electronic configurations of Zn (atomic number or Z is 30):

Zinc: $1s^2\,2s^2\,2p^6\,3s^2\,3p^6\,4s^2\,3d^{10}$

The noble gas argon has a Z value = 18

The short-hand electronic configuration for Zn is:

[Ar] $4s^2\,3d^{10}$

When writing electron configurations for ions of an element, add or subtract the number of electrons gained or lost from the atom when filling the subshells (*s*. *p*, *d*, *f*).

For sodium (Na), the electron configuration is:

Na: $1s^2\,2s^2\,2p^6\,3s^1$

However, when sodium reacts to form the Na^+ cation, it loses one electron, and its electron configuration becomes:

Na^+: $1s^2\,2s^2\,2p^6$

Notice that the electron configuration for the sodium cation is identical to Ne because both have 10 electrons.

When writing electron configurations, there are a few exceptions to the guidelines above because of the *stability rule*. The stability rule states that a sublevel is more stable when it has a half-filled configuration (one electron in every orbital) or a full configuration (two electrons in every orbital).

If an atom is only one short from achieving a half-filled or full configuration, it takes an electron from a neighboring *s* orbital to increase its stability. This promoting of an electron from the *s* orbital is most commonly seen with the transition metals.

Chromium (Cr, 24 electrons) appears to have the electron configuration:

[Ar] $4s^2 3d^4$

According to the stability rule, the Cr electron configuration is more stable with a half-filled configuration:

[Ar] $4s^1 3d^5$

The 3*d* subshell took an electron from the neighboring 4*s* subshell. The 4*s* subshell now has a partially filled orbital.

Copper (Cu) has 29 electrons with the expected electron configuration:

[Ar] $4s^2 3d^9$

From the stability rule, the electron configuration changes to the more stable:

[Ar] $4s^1 3d^{10}$

The 3*d* subshell of Cu has the electron of $4s^1$, which is now partially filled.

An *orbital diagram* is another way to express an atom's electron configuration. Orbital diagrams are a visual representation that uses fishhook (single-headed) arrows to indicate electrons in orbitals, as shown on the left side below.

⇅ ⇅ ↿ ↿ ___ 1s 2s 2px 2py 2pz	$1s^2 2s^2 2p^2$
Single-headed fishhook arrow points up or down to represent an electron. Each line horizontal represents an orbital; the subshell and energy level are written below the line. Spaces separate the subshells *there are three horizontal lines for the three *p* orbitals (*x*, *y*, *z*)	Coefficient = energy level (1, 2) Letter = subshell (*s*, *p*) Exponent = # of electrons in subshell This method does not give as much information as orbital diagrams.

Bohr atom

It is vital to understand the scientific discoveries that contributed to the current model of the atom. At the beginning of the 19th century, chemists were able to show that pure compounds contained fixed and unvarying amounts of their constituent components.

In 1806, John Dalton (1766-1844) provided a significant step in explaining this model of an atom with his particle theory, known as *Dalton's Atomic Theory.*

Dalton's Atomic Theory consisted of these essential principles:

1. All matter is composed of microscopic particles, known as atoms.
2. Atoms of one element have the same shape, size, mass, and other properties. Atoms of an element differ in properties from atoms of other elements.
3. Atoms can neither be subdivided nor changed into another atom.
4. Atoms cannot be created nor destroyed.
5. The atom is the smallest unit of matter that undergoes a chemical reaction.

In addition to these principles, Dalton proposed the "*rule of greatest simplicity*," which suggested that atoms only combine in binary ratios (i.e., 1:1). This rule was problematic because Dalton assumed that the formula for water was HO instead of H_2O.

Dalton eventually suggested that the rule of greatest simplicity was not correct, and in 1810, he suggested that a water molecule has three atoms. Atoms combine in whole-number ratios (1:1, 1:2, 1:3, ...) following the "*law of multiple proportions*."

There are problems with Dalton's Atomic Theory. He claimed that all atoms of an element have the same mass. The discovery of isotopes, which are atoms of the same element that vary in mass and density (different number of neutrons), proves otherwise.

Additionally, it is now known that atoms can be subdivided into their constituent particles in nuclear processes; Dalton's postulate that atoms cannot be subdivided remains correct within the scope of chemical reactions.

In the early 1900s, German Nobel laureate physicists Max Planck (1858-1947) discovered that radiation is emitted in quantized amounts of energy, as opposed to a single continuous ray.

Based on Planck's discovery, Nobel laureate Niels Bohr (1885-1962) conducted experiments on hydrogen atoms and revised the model for the atom. Bohr suggested that electrons orbit the nucleus in fixed orbits with defined energies and sizes. Bohr's model is like how planets orbit the sun (except that the attraction within the atom is provided by electrostatic forces rather than gravity). When electrons transition between orbits, they emit or absorb energy equivalent to the difference in energy levels between the orbits.

Bohr identified each energy level using an integer *n*, which is now known as the principal quantum number. Bohr also determined that additional electrons always occupy the lowest available energy level.

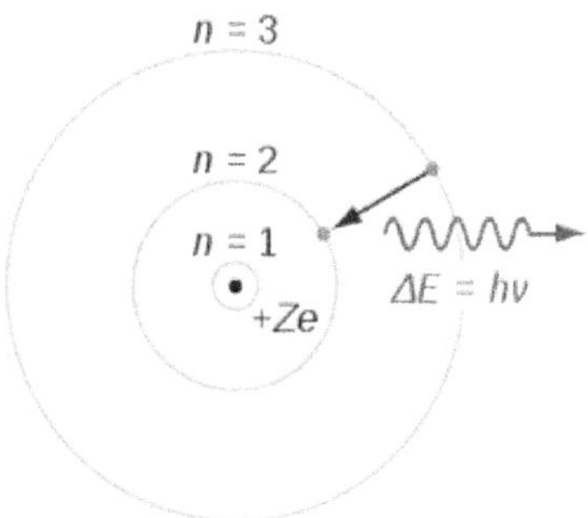

Bohr's model of electrons in discrete orbitals according to specific energy levels

Quantum numbers l, m_l and m_s and the number of electrons per orbital; Pauli Exclusion Principle

The *azimuthal quantum number* or *angular momentum quantum number* (l) is the second quantum number that defines the *shape* of an orbital. Except for the first level where $n = 1$, the principal energy levels have two or more possible subshells (*s, p, d, f*) with different energy levels.

For any n, the values for the azimuthal (angular momentum) quantum number l may range from 0 to $(n - 1)$. The number of subshells for any given energy level is equal to the value of n and is represented by the letters *s, p, d,* and *f.*

Orbital	Angular Momentum Quantum Number
s	$l = 0$
p	$l = 1$
d	$l = 2$
f	$l = 3$

The *magnetic quantum number* (m_l) is the third quantum number that describes the *orientation* of an orbital. The magnetic quantum number defines the number of orbitals within a subshell.

The possible values for m_l range from $-l$ to $+l$, including 0.

- An *s* subshell has $l = 0$, the m_l has only a single value (0), and thus it only has one orbital.

  A *s* subshell can hold 2 electrons.

- A *p* subshell has $l = 1$, the m_l can have the values of $-1, 0, +1$.

  A *p* subshell has three orbitals and can hold up to 6 electrons.

- A *d* subshell has $l = 2$, the $m_l = -2, -1, 0, +1, +2$.

  A *d* subshell has five orbitals and can hold up to 10 electrons.

- An *f* subshell has $l = 3$, the $m_l = -3, -2, -1, 0, +1, +2, +3$.

  An *f* subshell has seven orbitals and can hold up to 14 electrons.

Each successive subshell holds 4 more electrons than its predecessor.

The *spin quantum number* (m_s) is the fourth quantum number and describes the electron spin of an individual electron that occupies an orbital.

The spin quantum number distinguishes specific electrons within an orbital.

There are two opposing values for electron-spin: +½ and −½.

By convention, the first electron to occupy an orbital has a spin number of +½ while the second electron has a spin number of −½.

In orbital diagrams, the +½ electrons are represented by upward arrows, and the −½ electrons are indicated by downward arrows.

Energy Level	Number of subshells	Subshells	# of orbitals in that subshell	Total number of electrons in a subshell
1	1	*s*	1	2
2	2	*s*	1	2
		p	3	6
3	3	*s*	1	2
		p	3	6
		d	5	10
4	4	*s*	1	2
		p	3	6
		d	5	10
		f	7	14

Quantum numbers represent the locations of the electrons within the atom

In 1925, Nobel laureate Wolfgang Pauli (1900-1958) proposed the *Pauli Exclusion Principle,* which states that no two electrons in the same atom may have the same four quantum numbers (n, l, m_l, and m_s).

The maximum number of electrons in an orbital is two, and they must possess opposite spins (+½ or –½) because electrons of the same magnetic spin cannot both be in the same subshell of an orbital.

Common names and geometric shapes for orbitals: *s, p, d, f*

There are four types of orbitals whose shapes were predicted using wave mechanics (described below).

The four types of orbitals are *s*, *p*, *d*, and *f* from the initial letters of the words **s**harp, **p**rincipal, **d**iffuse, and **f**undamental. Each energy level has between 1 and 4 subshells. Each subshell is associated with an electron probability density of a shape.

Each subshell has a specific number of orbitals as given by the m_l value.

- *s orbitals* are spheres:

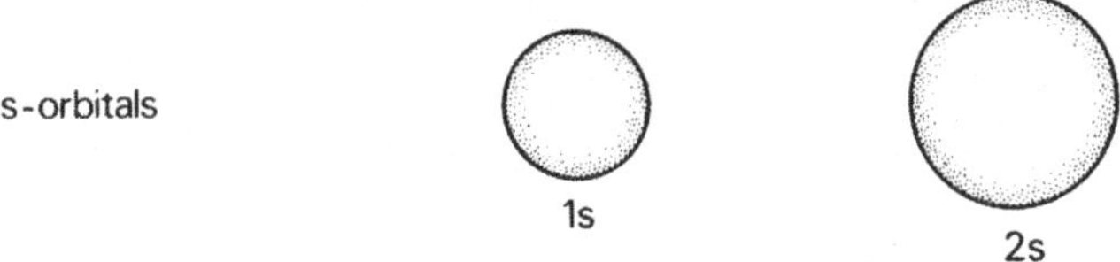

p orbitals have two lobes and are dumbbell shaped. There are three different p orbitals (three orientations: p_x, p_y, p_z):

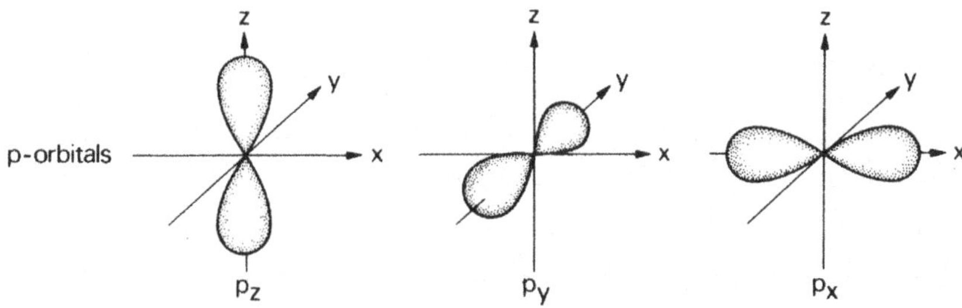

- *d orbitals have four lobes and are clover shaped. There are five different d orbitals (five orientations: d_{yz}, d_{z^2}, d_{xy}, d_{xz}, $d_{x^2-y^2}$):*

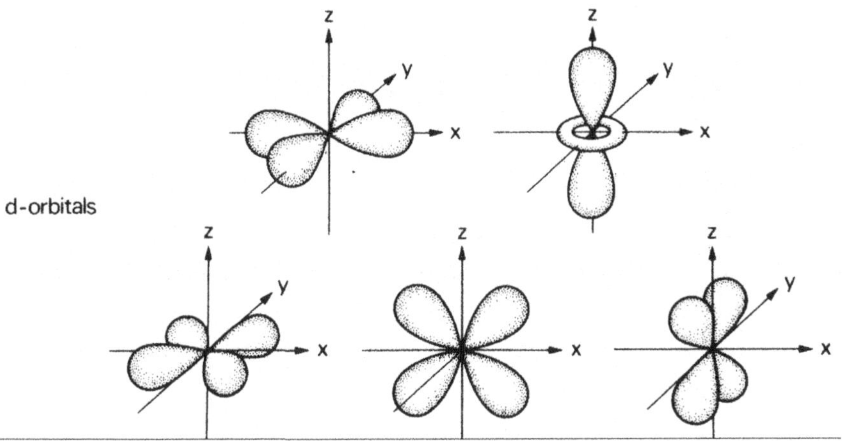

- *f orbitals are complicated, and the shape is a combination of the shapes of the other orbitals. There are seven f orbitals:*

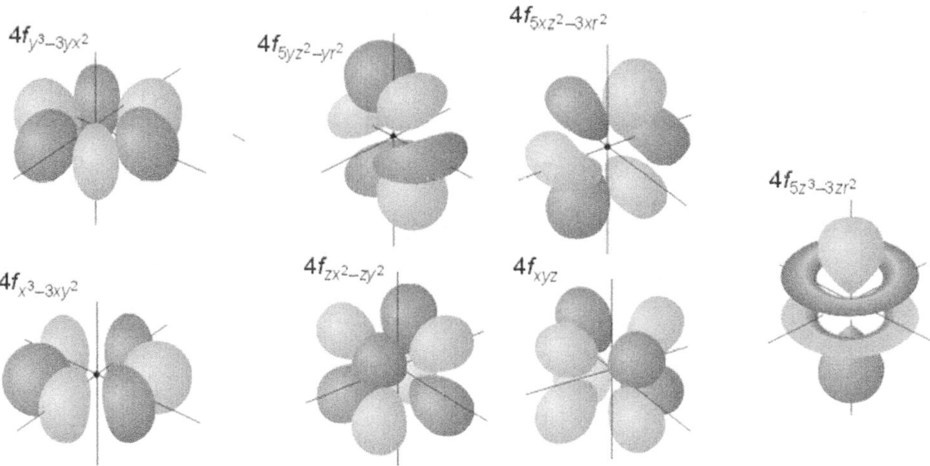

Heisenberg Uncertainty Principle

There were a series of remarkable discoveries that led to quantum mechanics and enabled a model to understand electron behavior.

In 1924, Nobel laureate Louis de Broglie (1892-1987) published his work on *wave-particle duality*. He proposed that electrons could be classified as both *particles* and *waves* simultaneously. Louis de Broglie proposed that electrons are particles with mass and velocity, but that they also possess *wave properties* such as wavelength and frequency.

The de Broglie equation demonstrates that all objects exhibit wave-behavior. However, the wave-behavior of everyday objects becomes negligible as the mass increases. Subatomic particles have tiny masses:

$$\lambda = h / mv$$

where λ is the wavelength, m is the mass, h is Planck's constant (6.626×10^{-34} J·s), and v is the velocity.

In 1926, Nobel laureate Erwin Schrödinger (1887-1961) derived a wave equation that models the movement of electrons based on their wave properties. In quantum mechanics, the solution to Schrödinger's wave equation yields a *wave function*, represented by the uppercase Greek letter psi (Ψ), which, by itself, does not indicate much about the atom. However, using the square of the absolute value of psi, $|\Psi|^2$ gives the *probability density* of an electron's position.

The wave function $|\Psi(x)|^2$ gives the probability of locating an electron at position x. Orbitals are electron "*clouds*" because they represent the probability of an electron's specific location in the atom. Since Schrödinger's equation showed that electrons behave like waves, it implies that electrons do not travel in clearly defined paths and that it is not possible to determine an electron's exact position and speed.

German physicist Nobel laureate Werner Heisenberg (1901-1976) developed a theory based on the premise that there is a theoretical limit to how small the uncertainty in the measurements can be. The *Heisenberg's Uncertainty Principle* and states that both the position and momentum of an electron cannot be measured simultaneously at any point in time. The measurement of position or momentum distorts the value for the other.

The *Heisenberg's Uncertainty Principle* is expressed as:

$$\Delta x \times \Delta p \geq \frac{h}{4\pi}$$

where Δx is the *uncertainty in position*, Δp is the *uncertainty in momentum*, and h is Planck's constant of 6.626×10^{-34} J·s.

Paramagnetism and diamagnetism

Electrons are constantly spinning in a fixed direction, which generates magnetic fields. According to the Pauli Exclusion Principle, if one electron is spinning in a clockwise direction, the other electron must be spinning in a counterclockwise direction. These opposing spins result in the orbital having no net spin.

Since electrons occupying the same orbital always have opposite values for their spin quantum numbers (m_s), they always have different sets of quantum numbers.

Atoms can be classified based on their electron-induced magnetic behavior as ferromagnetic, diamagnetic, or paramagnetic.

1. *Diamagnetic* materials generate an induced magnetic field in a direction opposite to an externally applied magnetic field. The applied magnetic field repels these materials.

   All electrons in diamagnetic atoms are paired.

2. *Paramagnetic* materials, when in the presence of a magnetic field, generate internally-induced magnetic fields in the same direction as the external field. A paramagnetic electron is an unpaired electron. An atom is paramagnetic if it has at least one unpaired

paramagnetic electron in any orbital regardless of the number of paired electrons.

Paramagnetic atoms are slightly attracted to a magnetic field and cannot retain magnetization without an external magnetic field.

3. *Ferromagnetic* materials generate permanent magnetic moments without needing an applied external magnetic field. Certain transition metals, such as iron and nickel, have ferromagnetic properties.

   Electrons of ferromagnetic substances can spontaneously align themselves in the same direction, which reinforces each electron's magnetic properties.

Photoelectric effect

In 1905, Albert Einstein (1879-1955) proposed the *photoelectric effect* that explained how incoming visible light interacts with electrons when light is reflected off the surfaces of metals. Discrete packets of light energy, known as *photons*, are transferred to electrons on a metal surface as kinetic energy.

When electrons possess sufficient energy to escape, *emission* of electrons from the metal may occur. The emission of electrons only occurs if the incident light possesses energy greater than or equal to the *work function* ϕ_w of the electron.

Energy (E) of light waves is calculated by the formula:

$$E = hv$$

where h is Planck's constant and v is the frequency of the light wave (Hz)

Einstein's model for the photoelectric effect accurately predicts that the energy of a photon is proportional to its frequency and not its intensity.

Ground state, excited states

Every electron in an atom occupies a fixed orbital. When an electron occupies its default orbital, the electron is in its *ground state*. A ground-state electron is at its lowest energy level and is the most stable. If an electron absorbs energy, it has greater potential energy and occupies a higher energy level. An electron in a higher energy level is known as an *excited state* electron. An electron can remain in its excited state for a limited time before it drops back to its ground state and emits the excess energy as visible electromagnetic radiation.

Absorption and emission spectra

Electrons can only move between fixed orbitals, which means that the wavelength of light that is released or absorbed when they transition between energy levels is limited to a specific set of visible wavelengths known as the *line spectrum.*

A line spectrum is produced by gas under low pressure. Solids, liquids, and compressed gases produce *continuous spectra,* which contain all the colors of visible light.

Different elements have different numbers of electrons, which means that each element has a unique set of *atomic emission spectra* represented by discrete lines along a frequency scale. Due to the uniqueness of this chemical property, *the atomic emission spectra can* be used to identify elements in a mixture.

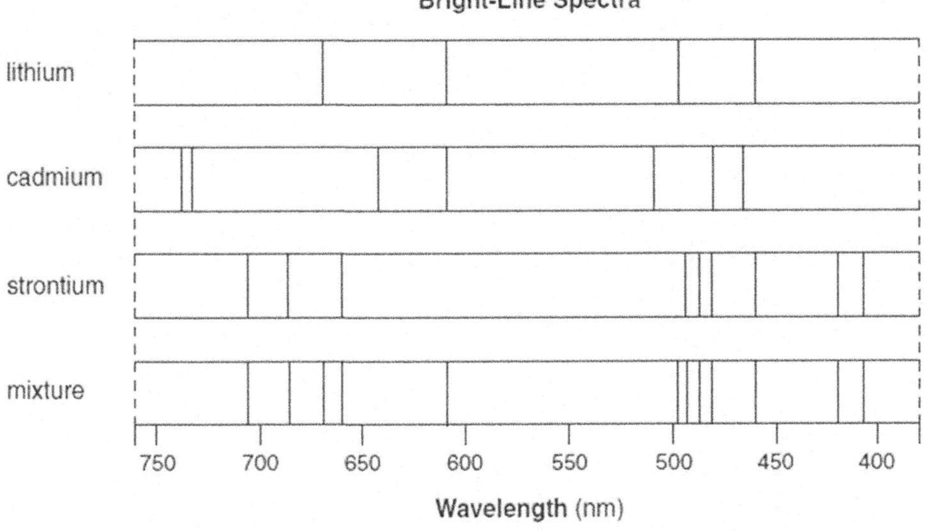

Each element produces a unique atomic emission spectrum that permits the classification of elements within a mixture

The *atomic absorption spectra* indicate the amount of energy absorbed when electrons are promoted to orbitals with higher energy levels.

An element's absorption and emission spectrum correlate because the net difference between energy levels will be the same, regardless of the direction of the electron's movement.

The atomic emission spectrum of hydrogen has been widely observed. Its spectrum consists of five series of lines, and each series is named after its discoverer. Lyman ($n = 1$), Balmer ($n = 2$), Paschen ($n = 3$), Brackett ($n = 4$), Pfund ($n = 5$) and Humphreys ($n = 6$). The most important emission series are the *Lyman, Balmer,* and *Paschen* series, which correspond to *ultraviolet, visible,* and *infrared* radiation, respectively.

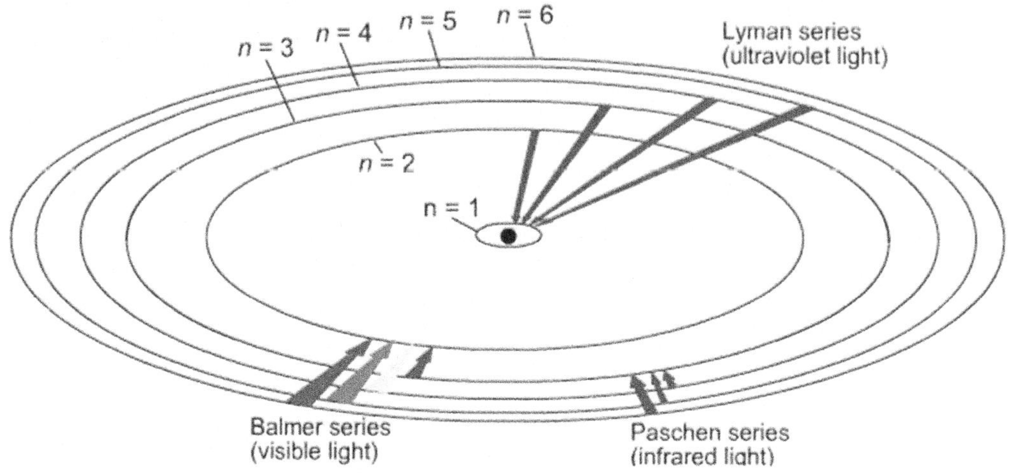

Atomic emission spectrum for a hydrogen atom

Swedish physicist Johannes Rydberg (1854-1919) proposed an equation to describe the relationship between the wavelengths of spectral lines of different elements.

The *Rydberg formula* for hydrogen, which calculates the energy emission of hydrogen electrons, is expressed as:

$$E = hc / \lambda$$

$$E = -R_h\left[\frac{1}{(n_i)^2} - \frac{1}{(n_f)^2}\right]$$

where R_h is Rydberg's constant (2.179×10^{-18} J), n_i is the principal quantum number of the initial orbital, n_f is the principal quantum number of the final orbital, c is the speed of light (3.0×10^8 m/s), λ is the wavelength and h is Planck's constant (6.626×10^{-34} J·s).

The Periodic Table: Variations of Chemical Properties with Group and Row

After the discovery of several new elements in the 18th and 19th centuries, scientists noticed that elements could be categorized into groups with similar properties. In 1869, the first significant development for this observation occurred when Russian chemist Dimitri Mendeleev (1834-1907) and German chemist Julius Lothar Meyer (1830-1895) both independently published variations for the periodic table.

Mendeleev arranged the elements by their atomic masses, which resulted in the similarity of physical and chemical properties between elements in the same column or group. Mendeleev's table was revolutionary because it accurately predicted the position of undiscovered elements. Meyer arranged his table based on increasing atomic volume.

Although Mendeleev's table exhibited patterns of similarities between elements in the same column or group, there were some exceptions. For example, argon exists in elemental form as an unreactive gas (noble gas or inert gas), but it was categorized with the highly reactive sodium and lithium metals.

In 1913, British physicist Henry Moseley (1887-1915) improved Mendeleev's periodic table by proposing that chemical properties of elements are related to their atomic number (Z, number of protons) instead of their atomic weight (A, number of protons and neutrons). Moseley's discovery became the foundation of the modern periodic table.

In the modern periodic table, elements are arranged in rows and columns. The horizontal rows are *periods*, which are identified by numbers (1-7). The vertical columns are *groups* or *families* (1-18).

There are several different standards for group nomenclature:

- In the older system, each group was assigned a Roman numeral (I-VIII) based on the number of their valence electrons.
- The groups were also assigned a letter (A or B) based on their location on the table. However, there were two different standards on the A/B designation. The American standard assigned A to the main (representative) group of elements on the sides and B to the transition group in the center. The European standard assigned A to the left side and B to the right side of the table.
- For consistency, a new universal naming standard was devised from left to right; the groups are given consecutive integers (1-18).

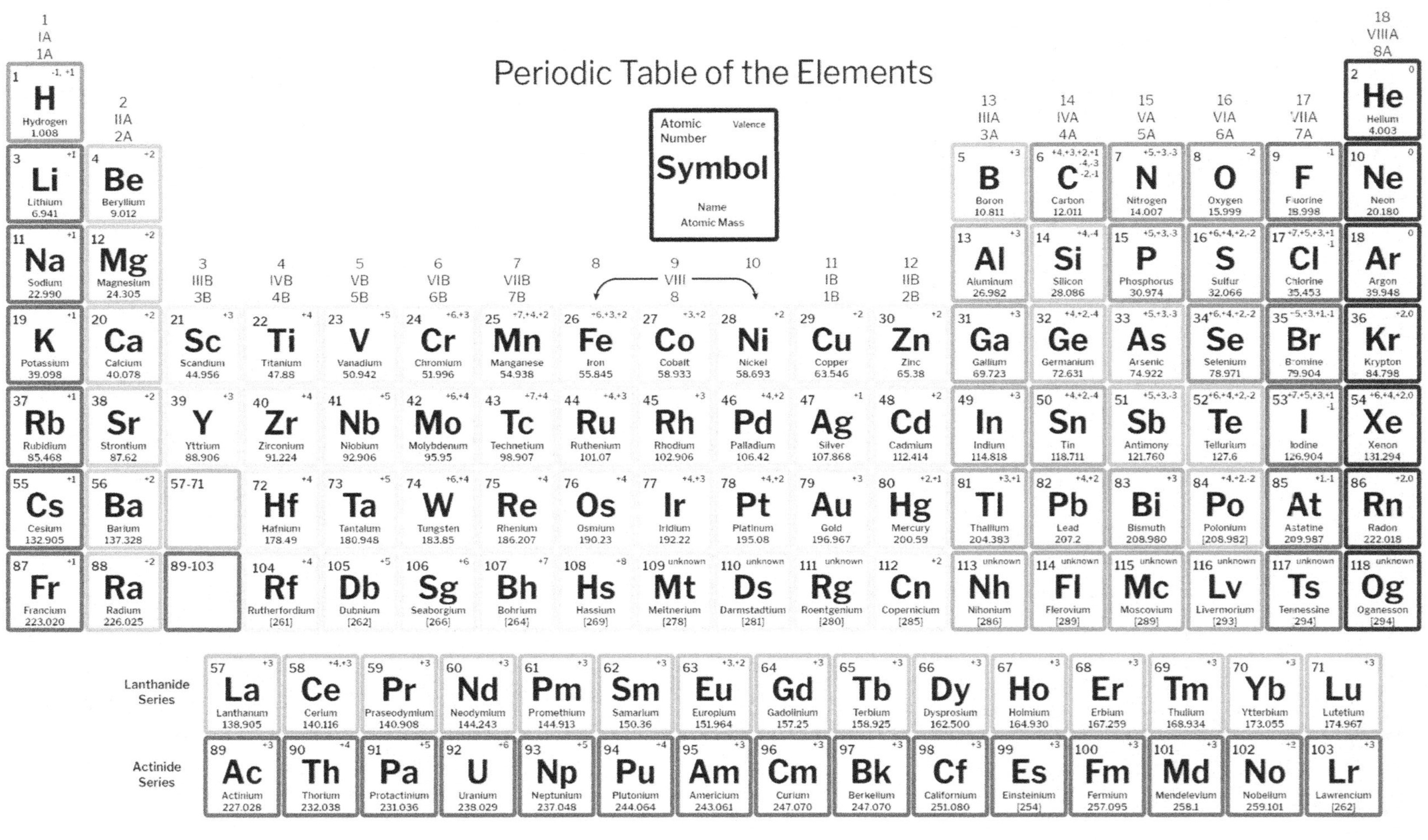
Periodic Table of the Elements
Atomic Number
Valence
Symbol
Name
Atomic Mass
1 IA 1A
2 IIA 2A
3 IIIB 3B
4 IVB 4B
5 VB 5B
6 VIB 6B
7 VIIB 7B
8
9 VIII 8
10
11 IB 1B
12 IIB 2B
13 IIIA 3A
14 IVA 4A
15 VA 5A
16 VIA 6A
17 VIIA 7A
18 VIIIA 8A
1 H Hydrogen 1.008 -1, +1
2 He Helium 4.003 0
3 Li Lithium 6.941 +1
4 Be Beryllium 9.012 +2
5 B Boron 10.811 +3
6 C Carbon 12.011 +4,+3,+2,+1 -4,-3 -2,-1
7 N Nitrogen 14.007 +5,+3,-3
8 O Oxygen 15.999 -2
9 F Fluorine 18.998 -1
10 Ne Neon 20.180 0
11 Na Sodium 22.990 +1
12 Mg Magnesium 24.305 +2
13 Al Aluminum 26.982 +3
14 Si Silicon 28.086 +4,-4
15 P Phosphorus 30.974 +5,+3,-3
16 S Sulfur 32.066 +6,+4,+2,-2
17 Cl Chlorine 35.453 +7,+5,+3,+1 -1
18 Ar Argon 39.948 0
19 K Potassium 39.098 +1
20 Ca Calcium 40.078 +2
21 Sc Scandium 44.956 +3
22 Ti Titanium 47.88 +4
23 V Vanadium 50.942 +5
24 Cr Chromium 51.996 +6,+3
25 Mn Manganese 54.938 +7,+4,+2
26 Fe Iron 55.845 +6,+3,+2
27 Co Cobalt 58.933 +3,+2
28 Ni Nickel 58.693 +2
29 Cu Copper 63.546 +2
30 Zn Zinc 65.38 +2
31 Ga Gallium 69.723 +3
32 Ge Germanium 72.631 +4,+2,-4
33 As Arsenic 74.922 +5,+3,-3
34 Se Selenium 78.971 +6,+4,+2,-2
35 Br Bromine 79.904 -5,+3,+1,-1
36 Kr Krypton 84.798 +2,0
37 Rb Rubidium 85.468 +1
38 Sr Strontium 87.62 +2
39 Y Yttrium 88.906 +3
40 Zr Zirconium 91.224 +4
41 Nb Niobium 92.906 +5
42 Mo Molybdenum 95.95 +6,+4
43 Tc Technetium 98.907 +7,+4
44 Ru Ruthenium 101.07 +4,+3
45 Rh Rhodium 102.906 +3
46 Pd Palladium 106.42 +4,+2
47 Ag Silver 107.868 +1
48 Cd Cadmium 112.414 +2
49 In Indium 114.818 +3
50 Sn Tin 118.711 +4,+2,-4
51 Sb Antimony 121.760 +5,+3,-3
52 Te Tellurium 127.6 +6,+4,+2,-2
53 I Iodine 126.904 +7,+5,+3,+1 -1
54 Xe Xenon 131.294 +6,+4,+2,0
55 Cs Cesium 132.905 +1
56 Ba Barium 137.328 +2
57-71
72 Hf Hafnium 178.49 +4
73 Ta Tantalum 180.948 +5
74 W Tungsten 183.85 +6,+4
75 Re Rhenium 186.207 +4
76 Os Osmium 190.23 +4
77 Ir Iridium 192.22 +4,+3
78 Pt Platinum 195.08 +4,+2
79 Au Gold 196.967 +3
80 Hg Mercury 200.59 +2,+1
81 Tl Thallium 204.383 +3,+1
82 Pb Lead 207.2 +4,+2
83 Bi Bismuth 208.980 +3
84 Po Polonium [208.982] +4,+2,-2
85 At Astatine 209.987 +1,-1
86 Rn Radon 222.018 +2,0
87 Fr Francium 223.020 +1
88 Ra Radium 226.025 +2
89-103
104 Rf Rutherfordium [261] +4
105 Db Dubnium [262] +5
106 Sg Seaborgium [266] +6
107 Bh Bohrium [264] +7
108 Hs Hassium [269] +8
109 Mt Meitnerium [278] unknown
110 Ds Darmstadtium [281] unknown
111 Rg Roentgenium [280] unknown
112 Cn Copernicium [285] +2
113 Nh Nihonium [286] unknown
114 Fl Flerovium [289] unknown
115 Mc Moscovium [289] unknown
116 Lv Livermorium [293] unknown
117 Ts Tennessine 294] unknown
118 Og Oganesson [294] unknown
Lanthanide Series
57 La Lanthanum 138.905 +3
58 Ce Cerium 140.116 +4,+3
59 Pr Praseodymium 140.908 +3
60 Nd Neodymium 144.243 +3
61 Pm Promethium 144.913 +3
62 Sm Samarium 150.36 +3
63 Eu Europium 151.964 +3,+2
64 Gd Gadolinium 157.25 +3
65 Tb Terbium 158.925 +3
66 Dy Dysprosium 162.500 +3
67 Ho Holmium 164.930 +3
68 Er Erbium 167.259 +3
69 Tm Thulium 168.934 +3
70 Yb Ytterbium 173.055 +3
71 Lu Lutetium 174.967 +3
Actinide Series
89 Ac Actinium 227.028 +3
90 Th Thorium 232.038 +4
91 Pa Protactinium 231.036 +5
92 U Uranium 238.029 +6
93 Np Neptunium 237.048 +5
94 Pu Plutonium 244.064 +4
95 Am Americium 243.061 +3
96 Cm Curium 247.070 +3
97 Bk Berkelium 247.070 +3
98 Cf Californium 251.080 +3
99 Es Einsteinium [254] +3
100 Fm Fermium 257.095 +3
101 Md Mendelevium 258.1 +3
102 No Nobelium 259.101 +2
103 Lr Lawrencium [262] +3

Elements are ordered by increasing atomic number, so there are recurring properties within the rows (periods) and columns (groups or families). Several physical and chemical properties of elements can be predicted from their position in the periodic table. These periodic trends include the number of valence (outermost) electrons, ionization energy, electron affinity, atomic or ionic radii, and electronegativity.

Valence electrons

Valence electrons are the outermost shell electrons and are responsible for the formation of chemical bonds. On the periodic table, the group numbers represent the number of valence electrons for elements in the column. However, this rule only applies to the main group elements and does not apply to the transition metals.

Two factors contribute to valence electrons experiencing weaker electrostatic attraction from the protons in the nucleus. The first is the distance of the electron *n* shell from the nucleus. The increased distance of higher principal quantum numbers for the electron's orbital increased the distance from the nucleus and leads to decreased attraction.

Coulomb's Law expresses electrostatic force:

$$F = kq_1q_2 / r^2$$

where F is the electrostatic force, k is Coulomb's constant (9×10^9 N·m²/C²), q_1 and q_2 are the respective charges of the two particles, and r is the distance between the particles.

The second factor is the presence of *screening electrons*, which are inner electrons that shield the attractive force experienced by the valence electrons.

Effective nuclear charge

The *nuclear charge* is the charge of the positive protons in the nucleus, and it describes the extent to which electrons are pulled toward the nucleus.

The *effective nuclear charge* (Z_{eff}) is the net nuclear force experienced by valence electrons after accounting for electron shielding by inner orbital electrons.

The *effective nuclear charge* is calculated by the equation:

$$Z_{eff} = Z - S$$

where Z is the number of protons in the nucleus (i.e., the atomic number), and S is the average number of screening electrons between the nucleus and the valence electron.

The effective nuclear charge experienced by an electron is proportional to its stability: the more stable the electron, the higher the effective nuclear charge, and the more energy is required to remove the electron; ionization energy

Effective nuclear charge increases across a period, due to increasing nuclear charge (increasing number of protons) but no accompanying increase in shielding effect.

There is no general trend down a group because both the nuclear charge (Z) and shielding effect increase as more electron shells are added down the group.

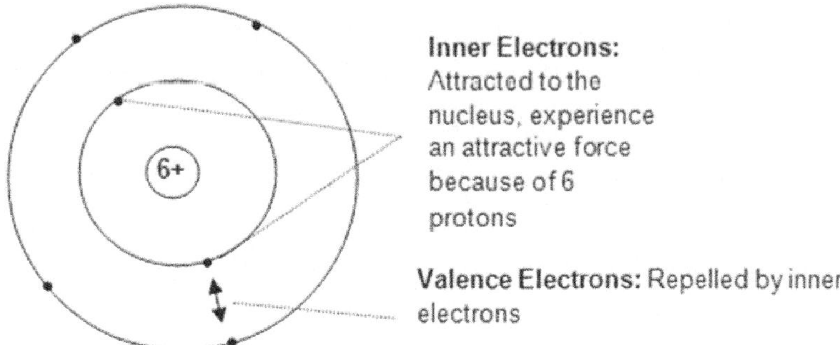

Effective Nuclear Charge: valence electrons only "feel" an attractive force of 4 protons. The two inner electrons cancel the +2 of the nuclear charge due to shielding.

First and second ionization energy

The *ionization energy* (SI Units kJ/mol) is the energy required to remove an electron from a neutral atom of an element or an ion in the gaseous state. The ionization of an electron requires energy and is an endothermic process (i.e., it absorbs heat).

While electrons are removed from the atom, the number of protons in the nucleus is constant. Therefore, the remaining electrons experience a stronger electrostatic force, as the constant force from the nucleus is shared among fewer remaining electrons.

Ionization energies increase for each successive removal of an electron, due to the increase in electrostatic force. Thus, the first electron is the easiest to remove.

The *first ionization energy* is the energy needed to remove an outermost electron.

The *second ionization energy* is the energy needed to remove a second electron.

Low ionization energy indicates that an electron is easily lost. This property is most commonly found in metals, with the Group I elements (alkali metal) having the lowest IE values. Conversely, high IE indicates that an electron is not easily removed. High ionization energy is mostly associated with nonmetals because of their natural tendency to gain electrons to form anions instead of losing electrons. The highest values are found for the inert gases (Group 8 or VIII) because they have octets, which make them extremely stable.

A more substantial positive nuclear charge results in a stronger electrostatic attraction and higher ionization energy for the valence electrons. In contrast, a larger atomic radius means that the outermost electrons are further from the nucleus and thus have a weaker electrostatic attraction, increased shielding, and lower ionization energy.

If an electron is in a lower shell (closer to the nucleus), it has lower energy and is more difficult to dislodge. Thus, it has higher ionization energy and requires more energy for the electron to be abstracted from the atom.

Half-filled shells are exceptions to the trend for ionization energy because these configurations are more stable than filled shell configurations. Half-filled shells occur when an atom's valence *p* or *d* orbital is half-filled (one electron in every subshell).

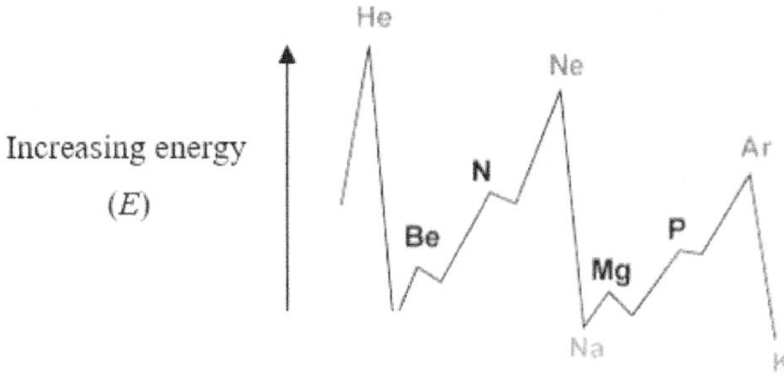

A complete octet (filled s and p subshells) increases the stability of the electrons

Ionization energies generally increase from left to right across a given period due to the increased effective nuclear charge of the atom. At the same time, the shielding effect due to the number of inner electron shells remains constant.

Ionization energies generally decrease from top to bottom down a group, due to the increasing distance from the nucleus corresponding to the increased principal quantum number *n* and the shielding effect of the electrons in the inner shells.

The ionization energy decreases down a group, because of the increasing distance for the valence electrons to the nucleus and the increase in shielding by more shells of inner electrons. The highest peaks on the graph are noble gases (stable due to filled octets), while the lowest valleys are alkali metals with single valence electrons in the valence shell. Local maxima occur for filled subshells and half-filled *p* subshells.

Electron Affinity

Electron affinity is the energy released when an electron adds to a neutral atom. As the name implies, elements with high electron affinity are strongly attracted to electrons and are associated with exothermic reactions (i.e., reactions that release heat). Electron affinity for nonmetals is indicative of their tendency to gain electrons and form negatively charged anions (Cl^- or Br^-).

The first electron affinity is the energy released when one electron adds; it is *exothermic* and releases heat.

The second electron affinity is *endothermic* because adding an electron to an orbital with excess electrons results in repulsion; thus, energy must be added to overcome the electrostatic repulsions among electrons and stabilize the additional electron.

The elements that prefer electrons most are the nonmetals, which have the highest electron affinities. The elements that do not prefer extra electrons are the metals, which have low electron affinities.

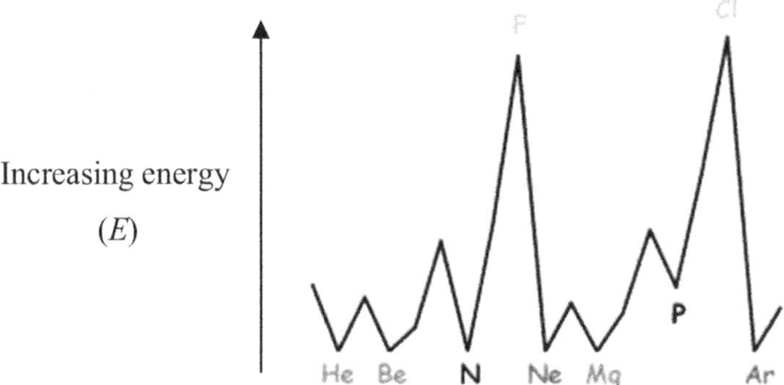

Electron affinity is highest for F, and Cl which results in complete p subshells

On the electron affinity graph, the highest peaks are for the halogens (F. Cl. Br, I, At), and the lowest peaks are for noble gases (e.g., Ne. Ar, Kr).

Local minima in energy occur for filled subshells and half-filled *p* subshells; this is observed for the noble gases of neon (Ne) and argon (Ar).

Like ionization energy, electron affinity is a periodic trend that generally increases from left to right and decreases from top to bottom (except the noble gases).

Electronegativity

In 1932, American chemist Nobel laureate Linus Pauling (1901-1994) derived a set of elemental values based on ionization energies (tendency to lose electrons) and electron affinities (tendency to gain electrons).

Electronegativity is the element's tendency to gain bonding electrons. Electronegativity is the measure of the attraction of an atom for the electrons in a bond with another atom. On the Pauling scale, electronegativity values range from a low of 0.7 (Fr) to a high of 4.0 (F).

From the periodic table below, electronegativity values generally increase from left to right and decrease from top to bottom on the periodic table. The most electronegative elements are at the top right corner of the periodic table. The least electronegative (i.e., most electropositive) elements are in the bottom left corner of the periodic table. Therefore, nonmetals are typically more electronegative than metals. There are no electronegativity values given for the noble gases because they usually do not bond with other elements.

In general, these electronegativity trends can be explained similarly to ionization energy and electron affinity: as the atomic number increases, the positive nuclear charge increases, and the electrons within the same energy level (same shielding effect) are more strongly attracted to the positive nucleus.

H																
2.1																
Li	Be											B	C	N	D	F
1.0	1.5											1.5	2.5	3.0	3.5	4.0
Na	Mg											Al	Si	P	S	Cl
0.9	1.2											1.5	1.8	2.1	3.5	3.0
K	Ca	Sc	Ti	V	Cr	Mn	Fe	Co	Ni	Cu	Zn	Ga	Ge	As	Se	Br
0.8	1.0	1.3	1.5	1.6	1.6	1.5	1.8	1.9	1.8	1.9	1.6	1.6	1.8	2.0	2.4	2.8
Rb	Sr	Y	Zr	Nb	Mo	Tc	Ru	Rh	Pd	Ag	Cd	In	Sn	Sb	Te	I
0.8	1.0	1.2	1.4	1.6	1.8	1.9	2.2	2.2	2.2	1.9	1.7	1.7	1.8	1.9	2.1	2.5
Cs	Ba		Hf	Ta	W	Re	Os	Ir	Pt	Au	Hg	Tl	Pb	Bi	Po	At
0.7	0.9		1.3	1.5	1.7	1.9	2.2	2.2	2.2	2.4	1.9	1.8	1.9	1.9	2.0	2.2
Fr	Ra															
0.7	0.9															

Pauling's electronegativity values assigned to elements on the periodic table

A *covalent bond* is the relatively equal sharing of electrons between two elements. If the electronegativity of the two atoms is the same (or similar), they share the electrons equally (or almost equally).

A *polar covalent bond* results if there is a sufficient difference in electronegativity between the atoms (see the table below for value ranges). In a polar covalent bond, the more electronegative element has a greater tendency to attract electrons. This atom gets a larger share of the electron density, giving it a partial negative (δ^-) charge. The less electronegative element in a polar covalent bond has a weaker tendency to attract electrons. The atom gets a smaller share of the electron density, giving it a partial positive (δ^+) charge.

An *ionic bond* occurs if the electronegativity difference is significant (see the table below). Ionic bonds transfer of an electron from the electropositive element to the electronegative element. Ionic bonds generally occur between a metal (left two columns; groups 1 and 2) and a nonmetal (right side; generally, groups 16 and 17) element.

The difference in electronegativity between atoms participating in a bond is the *dipole moment* and measured in units of *debye* (*D*). Based on the difference of electronegativity values, bonds are classified as covalent, polar covalent, or ionic.

Type of Bond	**Dipole Moment (*D*)**
Covalent	0 – 0.6
Polar covalent	0.6 – 1.6
Ionic bonds	> 1.6

Electron shells and the sizes of atoms

Electron shells are defined by the principal quantum number *n*. With an increasing atomic number, elements have more electrons and, therefore, more electron shells.

From top to bottom down a group, the shielding effect increases because the closer shells are between the nucleus and the outermost shell.

The addition of an extra shell increases the atom size because of a new orbit (i.e., higher shells have a more considerable distance from the nucleus than lower shells).

For elements left to right across a row on the periodic table, the principal quantum number *n* is the same; therefore, the same shell fills with the additional electrons. As a shell fills, the effective nuclear charge increases because of the increasing number of protons, while additional electrons fill the same orbit and no increase in shielding. With the increasing effective nuclear charge, the electrostatic attraction between the nucleus and the electrons increases, so the atom becomes more compact.

Atomic radius is ½ the distance between the two atomic nuclei and measures the sizes of atoms since electron clouds are too ill-defined to measure precisely.

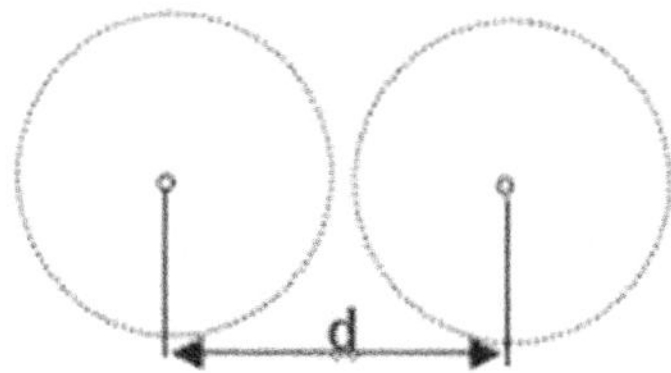

The atomic radii measure the distance between the atomic nuclei of two atoms

The trends in relative atomic radii in periods 2 and 3 are illustrated below.

On a relative scale, if a nucleus were the size of a ping-pong ball, the atom's diameter would be 25 miles. Therefore, an increase in the size of the nucleus does not substantially affect the overall size of an atom.

From top to bottom down a group on the periodic table, the atoms become larger because electrons add to new orbitals that are farther from the nucleus.

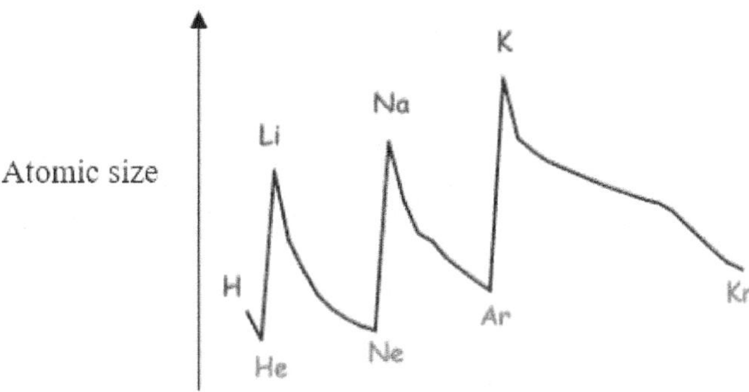

Atomic radii trend increases as the principal number increases for additional orbitals

The peaks on the energy *vs*. atomic number graph represent atoms with a single electron in the valence shell, while the valleys are for atoms with a filled valence shell.

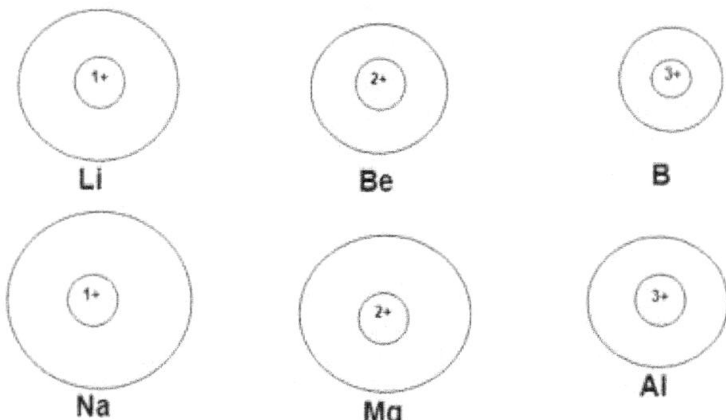

The relative size of the electron shells is indicated across a group and down periods. The nucleus is shown at the center, and the corresponding 1s orbital for the ions (Li and Na are 1+; Be and Mg are 2+; B and Al are 3+). Across the group, the effective nuclear charge increases with the addition of protons (increasing Z value) in the nucleus.

The size of an atom increases down a column due to an increasing number of shells (*n* number) and decreases across a row due to the increased nuclear attraction (higher number of nuclear core protons).

The Periodic Table: Classification of Elements into Groups by Electronic Structure; Physical and Chemical Properties of Elements

The vertical columns (groups or families) of the periodic table have recurring trends. For this reason, some of the groups are given unique names, sometimes referred to as common names or *unsystematic names* (e.g., alkali metals). Elements in the same group have the same number of valence electrons and tend to show patterns in physical properties such as ionization energy, atomic radius, and electronegativity.

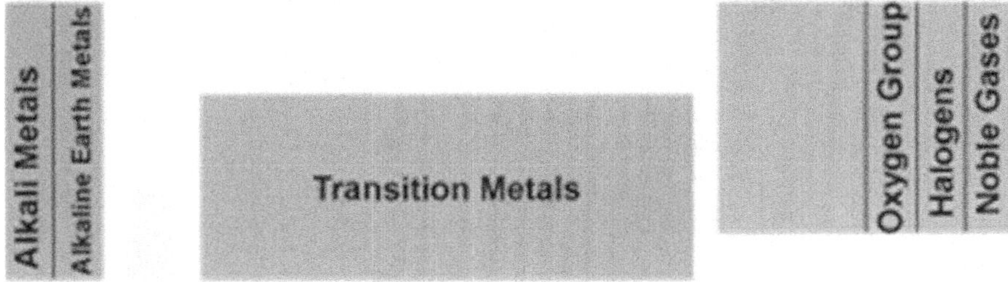

Sample common names for groups on the periodic table

The periodic table can also be divided into blocks, which are named based on the subshell (*s*, *p*, *d*, *f* orbitals) in which the valence (outermost) electron is located.

In some of these blocks, the horizontal trends are especially important.

The four blocks are the *s*-block, the *p*-block, the *d*-block, and the *f*-block. These blocks are seen in the figure below, indicating the principal quantum number *n* and the azimuthal quantum number *l* for the subshells *s*, *p*, *d*, and *f* are shown on the figure. The *f*-block (i.e., lanthanides and actinides) is generally displayed separately below the periodic table.

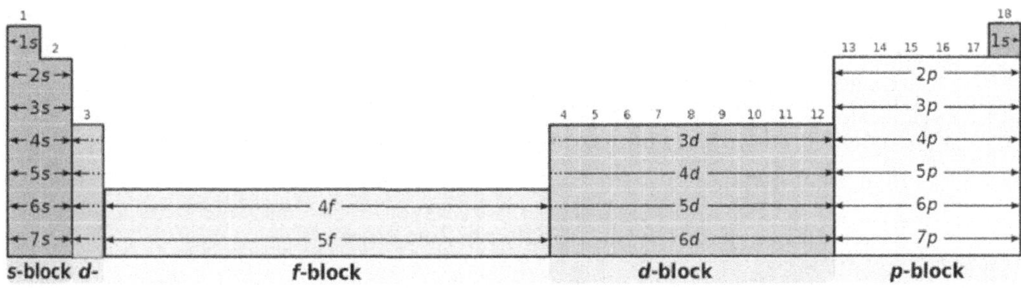

Representative elements

The representative (or main-group) elements (Groups 1, 2 and 13-18) comprise the *s* (1-2) and *p*-blocks (13-18) of the periodic table, and their properties are as follows:

- No free flowing (or loosely bound) outer *d* electrons, so the number of valence electrons is given by the group number
- Valence shell fills from the left (1 electron) to the right (8 electrons)
- Collectively, the essential elements for life on Earth, comprising 80% of the Earth's surface

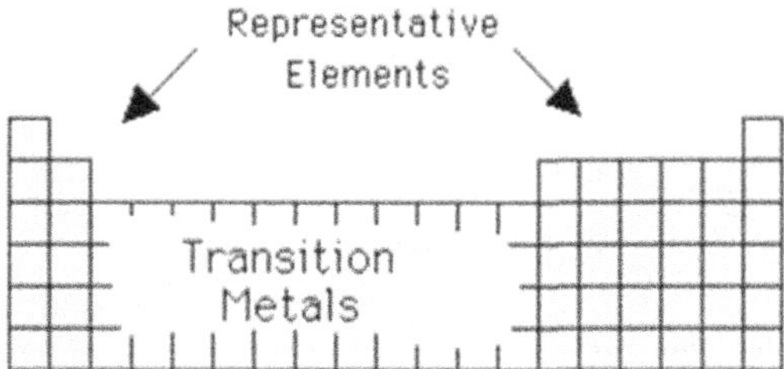

Representative elements have s and p orbitals, while transitions metals have d orbitals

Alkali metals

The alkali metals (Group 1 or formerly as1A) include the elements Li, Na, K, Rb, Cs and Fr, and the properties of the alkali metals are as follows:

- Single valence electron that is easily ionized. Therefore, they have low ionization energy and are very reactive
- React to lose one electron and form a cation with +1 oxidation state
- More reactive from top to bottom because of increasing atomic radii (Cs and Fr most reactive)
- Malleable, ductile, good conductor like all metals, but usually softer than other metals
- React with oxygen to form strongly alkaline oxides
- React with water to form hydroxides and releases hydrogen
- React with acids to form salts and releases hydrogen
- Commonly found in nature as compounds with the halogens

Alkaline earth metals

The known alkaline earth metals (Group 2 or formerly as 2A) include Be, Mg, Ca, Sr, Ba, and Ra, and the properties of the alkaline earth metals are as follows:

- 2 valence electrons, so relatively low ionization energy and reactive
- React to lose two electrons and form cation with +2 oxidation state
- Increasingly reactive from top to bottom because of the increasing atomic radii
- React with oxygen to form strongly alkaline oxides
- React with water to form hydroxides and releases hydrogen
- React with acids to form salts and releases hydrogen
- Due to their reactivity, they are generally not found in nature

Oxygen group

The oxygen group, also known as the chalcogens (Group 16 or formerly as VIA) includes O, S, Se, Te, Po, and Lv and the properties of the chalcogens are as follows:

- Oxygen, sulfur, and selenium are nonmetals, while tellurium and polonium are metalloids
- Lighter chalcogens (O, S) are typically non-toxic and critical to life, while heavier chalcogens are toxic
- Form ions with multiple oxidation numbers; sulfur's oxidation states are between –2 to +6
- Exist as polyatomic molecules in the elemental state (e.g., O_2, S_8)

Halogens

The halogens (Group 17 or formerly as 7A) include F, Cl, Br, I and At, and the properties of halogens are as follows:

- 7 valence electrons (2 from *s* subshell and 5 from *p* subshell) - high electron affinity, very reactive (most reactive of all nonmetals)
- Most electronegative of all the elements
- React to gain one electron and achieve full valence shell; forms anions with –1 oxidation state
- Less reactive from top to bottom because of decreasing atomic radii
- React with alkali metals and alkaline earth metals to form salts (crystalline ionic solids)
- Form diatomic molecules in the elemental state (e.g., Cl_2, Br_2, I_2)

Noble gases - physical and chemical characteristics

The noble gases (Group 18 or formerly as 8A) include He, Ne, Ar, Kr, Xe, Rn and Og and the properties of the noble gases are as follows:

- Full valence shell of 8 have high ionization energy coupled with a low electron affinity
- Inert (i.e., very unreactive); only heavier elements of the group react with few very electronegative elements, such as fluorine
- Exist in nature as monatomic gases (e.g., He, Ne, Ar)

Transition metals

The transition metals (Groups 3-11 or formerly as 3B-1B) have unique chemical properties due to their loosely bound outermost *d* orbital electrons, and the properties of the transition metals are as follows:

- High conductivity and malleability due to loosely bound, free-flowing outer *d* electrons.
- When bonded with other ions to form metal complexes, the *d* orbitals become non-degenerate (different in energy).

  Electron transitions between non-degenerate *d* orbitals give transition metal complexes vivid colors.
- Lose electrons from more than one shell to form cations of different oxidation states, indicated using Roman numerals (e.g., Iron (II) Fe^{2+} and Iron (III) Fe^{3+})

The periodic table is represented with two detached rows of elements located under the table (the *f*-block). Those elements are collectively known as *inner transition elements* or the *lanthanide and actinide series*. They are in the 6th and 7th periods starting at the 5*d* block of the periodic table.

The inner transition elements were once classified as rare-earth elements, due to the scarcity with which they naturally form.

However, even the rarest of all the inner transition metals were found to be still more common than the platinum-group metals (which are transition metals).

Most inner-transition metals, including plutonium and uranium, only exist in nature in radioactive form and decay rapidly.

Metals

The elements of the periodic table are divided into metals, nonmetals, and metalloids, according to their shared physical and chemical properties.

Highly ordered, closely-packed atoms characterize metals.

The metals comprise almost three-fourths of all known Earth elements, such as aluminum, iron, calcium, and magnesium.

Many of the chemical and physical properties of metals are a result of their metallic bonds, where the valence electrons of *s* and *p* orbitals delocalize and form an aggregate of electrons that surround the nuclei of the interacting metal atoms.

The high degree of freedom with which their outermost electrons move between metal atoms gives rise to many of the following properties of metals:

- Metallic luster - can shine or reflect light
- Malleable - can be hammered or rolled into thin sheets
- Ductile - can be drawn into wire
- Hardness ranges from hard (iron, chromium) to soft (sodium, lead, copper)
- Conduct heat and electricity
- Crystalline solids at room temperature, except mercury (Hg) which is the only liquid metal
- High melting point
- Chemical reactivity varies greatly. Some are very unreactive (e.g., Au, Pt) while others are very reactive and will burst into flames upon contact with water (e.g., Na, K)

Nonmetals

The nonmetals are substances that do not exhibit any of the chemical or physical traits most associated with metal elements because they bond with each other to form covalently bonded compounds.

The nonmetals are in the upper right-hand corner of the periodic table, and exhibit the following properties:

- Mostly polyatomic elements, except the monoatomic noble gases
- At room temperature, nonmetal elements can be found in all phases: gas (H_2, O_2, N_2, F_2, Cl_2), solid (I_2, Se_8, S_8, P_4) and liquid (Br_2)

- Brittle – they pulverize when struck
- Insulators (or extremely poor conductors) of electricity and heat, because electrons do not possess the high degree of movement as those in metal atoms
- Chemical reactivity ranges from inert (noble gases) to reactive (F_2, O_2, H_2). Nonmetals react with metals to form ionic compounds
- Some nonmetals have allotropes: different forms of the element in the same phase, such as carbon (diamond and graphite)

Differences between metals and nonmetals

Chemical Properties	
Metals	**Nonmetals**
Likes to lose electrons to gain a positive (+) oxidation state (good reducing agent)	Likes to gain electrons to form a negative (–) oxidation state (good oxidizing agent)
Lower electronegativity – partially positive in a covalent bond with nonmetal	Higher electronegativity – partially negative in a covalent bond with the metal
Forms basic oxides	Forms acidic oxides
Physical Properties	
Good conductor of heat and electricity	Poor conductor of heat and electricity
Malleable, ductile, luster and solid at room temperature (except Hg)	Solid, liquid, or gas at room temperature. Brittle if solid and without luster.

Metalloids (e.g., B, Si, As) exhibit properties of both metals and nonmetals. The metalloids are in a “stair-step” line of the periodic table that divides metals and nonmetals. Some metalloids are lustrous like metals, while others are brittle like nonmetals.

Metalloids are also unique in that they do not typically conduct electricity at room temperatures but conduct electricity when heated to higher temperatures.

Practice Questions

1. The attraction of the nucleus on the outermost electron in an atom tends to:

A. decrease from right to left and bottom to top on the periodic table
B. decrease from left to right and bottom to top on the periodic table
C. decrease from left to right and top to bottom of the periodic table
D. increase from right to left and top to bottom on the periodic table
E. decrease from right to left and top to bottom on the periodic table

2. Which species shown below has 24 electrons?

A. $^{52}_{24}Cr$ **B.** $^{55}_{25}Mn$ **C.** $^{24}_{12}Mg$ **D.** $^{45}_{21}Sc$ **E.** $^{51}_{23}V$

3. Which characteristics describe the mass, charge, and location of a proton, respectively?

A. approximate mass 5×10^{-4} amu; charge +2; inside nucleus
B. approximate mass 1 amu; charge 0; inside nucleus
C. approximate mass 1 amu; charge +1; inside nucleus
D. approximate mass 5×10^{-4} amu; charge –1; outside nucleus
E. approximate mass 1 amu; charge +2; outside nucleus

4. Which of the following subshell notations for electron occupancy is NOT possible?

A. $4f^{11}$ **B.** $2p^1$ **C.** $5s^3$ **D.** $4p^5$ **E.** $4d^3$

5. The *f* subshell can hold how many total electrons?

A. 4 **B.** 8 **C.** 12 **D.** 14 **E.** 18

6. Which of the following is the best description of the Bohr atom?

A. Sphere with a heavy, dense nucleus containing electrons
B. Sphere with a heavy, dense nucleus encircled by electrons in orbits
C. Indivisible, indestructible particle
D. Homogeneous sphere of protons, neutrons, and electrons
E. Sphere with a sparse nucleus surrounded by dense electron orbitals

7. Which of the following compounds does NOT exist?

A. H^-, because H forms only positive ions
B. PbO_2, because the charge on a Pb ion is only +2
C. SF_6, because F does not have an empty *d* orbital to form an expanded octet
D. $^-CH_3$, because C lacks *d* orbitals
E. OCl_6, because O does not have *d* orbitals to form an expanded octet

8. The elements in Groups 1, 17, and 18 (or formerly as IA, VIIA, and VIIIA) are called, respectively:

A. alkaline earth metals, transition metals, and halogens
B. alkali metals, halogens, and noble gases
C. alkali metals, alkali earth metals, and halogens
D. alkaline earth metals, halogens, and alkali metals
E. halogens, alkali earth metals, and noble gases

9. Which of the following is a general characteristic of a metallic element?

A. Reacts with nonmetals
B. Reacts with metals
C. Dull, brittle solid
D. Low melting point
E. None of the above

10. Which of the following elements would be shiny and flexible?

A. bromine (Br)
B. selenium (Se)
C. helium (He)
D. ruthenium (Ru)
E. silicon (Si)

11. The masses on the periodic table are expressed in what units?

A. picogram
B. nanogram
C. gram
D. microgram
E. amu

12. Isotopes of an element have the same number of [ ] and different numbers of [ ].

A. protons, electrons
B. neutrons, electrons
C. neutrons, protons
D. protons, neutrons
E. electrons, protons

13. Paramagnetism, the ability to be pulled into a magnetic field, is demonstrated by:

A. any substance containing unpaired electrons
B. nonmetal elements that have unpaired *p* orbital electrons
C. transition elements that have unpaired *d* orbital electrons
D. nonmetal elements that have paired *p* orbital electrons
E. any substance containing paired electrons

14. Which element has the electron configuration $1s^22s^22p^63s^23p^64s^23d^{10}4p^65s^24d^1$?

A. Y **B.** La **C.** Si **D.** Sc **E.** Zr

15. Dmitri Mendeleev's chart of elements:

A. predicted the behavior of unidentified elements
B. developed the basis of our modern periodic table
C. predicted the existence of elements undiscovered at his time
D. placed elements with the same number of valence electrons in the same period
E. all of the above

Detailed Explanations

1. E is correct.

The attraction of the nucleus on the outermost electrons determines the ionization energy, which increases towards the right and increases up on the periodic table.

2. A is correct.

The number of protons is the atomic number (*Z*) of the element and is shown as a subscript.

Atoms of an element do not necessarily have the same mass (*A*), because they may have different numbers of neutrons (i.e., isotopes of the element).

For neutral elements, the number of electrons equals the number of protons (i.e., subscript).

Atoms of an element may also have different numbers of electrons, forming charged ions.

For neutral atoms, the number of electrons equals the number of protons.

3. C is correct.

Protons are the positively charged particles located inside the nucleus of an atom.

Like neutrons (also in the nucleus), protons have a mass of approximately 1 amu.

Protons have a +1 charge, while neutrons have a charge of 0 (i.e., neutral).

4. C is correct.

Each orbital can hold two electrons.

The maximum number of electrons in each shell:

The *s* subshell has 1 spherical orbital and can accommodate 2 electrons

The *p* subshell has 3 dumbbell-shaped orbitals and can accommodate 6 electrons

The *d* subshell has 5 lobe-shaped orbitals and can accommodate 10 electrons

The *f* subshell has 7 orbitals and can accommodate 14 electrons

Therefore, the *s* subshell can accommodate only 2 electrons.

5. D is correct.

The number of orbitals in a subshell is different from the maximum number of electrons in the subshell.

Each orbital can hold two electrons.

The maximum number of electrons in each shell:

The *s* subshell has 1 spherical orbital and can accommodate 2 electrons

The *p* subshell has 3 dumbbell-shaped orbitals and can accommodate 6 electrons

The *d* subshell has 5 lobe-shaped orbitals and can accommodate 10 electrons

The *f* subshell has 7 orbitals and can accommodate 14 electrons

The capacity of an *f* subshell is 7 orbitals × 2 electrons/orbital = 14 electrons.

6. B is correct.

The Rutherford-Bohr model of the hydrogen atom is often referred to as the Bohr atom, which depicts a sphere with a heavy, dense nucleus encircled by electrons in orbits, held together by the electrostatic forces between the positively charged nucleus and the negatively charged electrons.

7. E is correct.

O contains only *s* and *p* orbitals and is not capable of forming 6 bonds due to the lack of *d* orbitals.

Hydrogen forms a negative ion in NaH (sodium hydride), $LiAlH_4$ (lithium aluminum hydride), or $NaBH_4$ (sodium borohydride).

Each of these hydrides (i.e., H^-) is a relatively strong base with a pK_a higher than 32.

Lead (Pb), although not a transition metal, does have two common valences: +2 and +4.

Fluorine in SF_6 does not have an expanded octet because each F is bonded only to S.

The sulfur has an expanded octet, but it is in the 3rd row of the periodic table and has empty *d* orbitals.

$^-CH_3$ (i.e., carbanion; carbon with a negative charge) does not have *d* orbitals but exists with sp^2 hybridization (i.e., trigonal planar molecule).

8. B is correct.

A group (or family) is a vertical column, and elements within groups share similar properties.

Group 1 is the alkali metals, which include lithium (Li), potassium (K), sodium (Na), rubidium (Rb), cesium (Cs), and francium (Fr).

Group 17 are the halogens, which include fluorine (F), chlorine (Cl), bromine (Br), iodine (I), astatine (At), and tennessine (Ts). Halogens gain one electron to become a –1 anion, and the resulting ion has a complete octet of valence electrons.

Group 18 is the noble gases that include helium (He), neon (Ne), argon (Ar), krypton (Kr) xenon (Xe), radon (Rn), and Oganesson (Og).

The known alkaline earth metals (Group 2) include beryllium (Be), magnesium (Mg), calcium (Ca), strontium (Sr), barium (Ba), and radium (Ra).

9. A is correct.

Most elements on the periodic table (over 100 elements) are metals. There are about four times more metals than nonmetals.

Alkali metals (Group 1) include lithium (Li), potassium (K), sodium (Na), rubidium (Rb), cesium (Cs), and francium (Fr).

The known alkaline earth metals (Group 2) include beryllium (Be), magnesium (Mg), calcium (Ca), strontium (Sr), barium (Ba), and radium (Ra).

Metals form positive ions by losing electrons during chemical reactions; thus, metals are electropositive elements.

Metals are good conductors of both heat and electricity because the molecules in metals are tightly packed together.

Metals are not brittle or fragile (i.e., they do not break easily); to the contrary, they are malleable and can be pressed or hammered without breaking.

Metals are opaque and not transparent because free moving electrons surround all of the atoms in the metal; therefore, any light that strikes the metal will hit these electrons, which will absorb and re-emit it, and the light is not able to pass through.

Metals, except mercury, are solids under normal conditions.

Potassium has the lowest melting point of the solid metals at 146 °F.

10. D is correct.

Transition metals occur in groups (vertical columns) 3–12 of the period table.

The transition metals are in periods (horizontal rows) 4–7. The transition metals include silver, iron, and copper.

11. E is correct.

The masses on the periodic table are the atomic masses of the elements.

The atomic mass is a weighted average of the various isotopes of the element (the mass of each isotope multiplied by its relative abundance).

Atomic mass is sometimes called atomic weight, but the term *atomic mass* is more accurate. This mass is expressed in atomic mass units (amu), also called Daltons (D).

1 amu (or u) is equal to 1 g/mol.

An amu is one-twelfth the mass of a neutral carbon-12 atom.

12. D is correct.

Elements are defined by the number of protons (i.e., atomic number).

The isotopes are neutral atoms: # electrons = # protons.

Isotopes are variants of an element that differ in the number of neutrons.

All isotopes of the element have the same number of protons and occupy the same position on the periodic table.

The number of protons within the atom's nucleus is the atomic number (Z) and is equal to the number of electrons in the neutral (non-ionized) atom.

Each atomic number identifies a specific element, but not the isotope; an atom of a given element may have a wide range in its number of neutrons.

The number of both protons and neutrons (i.e., nucleons) in the nucleus is the atom's mass number (A), and each isotope of an element has a different mass number.

13. A is correct.

Paramagnetic elements are substances that move into a magnetic field and have one or more unpaired electrons.

Most transition metals and their compounds in oxidation states involving incomplete inner electron subshells are paramagnetic.

An applied magnetic field repels diamagnetic elements because they create an induced magnetic field in a direction opposite to an externally applied magnetic field.

14. A is correct.

Identify an element using the periodic table by using its atomic number. The atomic number is equal to the number of protons or electrons.

The total number of electrons can be determined by adding all the electrons in the provided electron configuration:

$$2 + 2 + 6 + 2 + 6 + 2 + 10 + 6 + 2 + 1 = 39$$

Element #39 in the periodic table is yttrium (Y).

15. E is correct.

In 1869, Russian chemist Dmitri Mendeleev created the prototype version of the periodic table of elements, in which he placed elements with the same number of valence electrons in the same group of family (i.e., vertical columns).

Mendeleev's table was the basis of the modern-day periodic table. However, our modern-day table has a different system where elements with the same number of valence electrons are in the same group (or family) rather than periods (i.e., horizontal row).

Mendeleev predicted the existence of undiscovered elements, along with their behavior.

Notes

Chapter 2

Chemical Bonding

The Ionic Bond: Electrostatic Forces Between Ions

- **Electrostatic Energy ∝ q_1q_2 / r**
- **Electrostatic Energy ∝ Lattice Energy**
- **Electrostatic Force ∝ q_1q_2 / r^2**

The Covalent Bond: Electron Pair Sharing

- **Sigma (σ) and Pi (π) Bonds**
- **Lewis Electron Dot Formulas**
- **Partial Ionic Character**

Electrostatic Forces Between Ions

An *ionic bond* forms when electrons transfer from one atom to another. This electron transfer results in oppositely charged species that attract each other via electrostatic interaction.

When atoms gain or lose electrons and become charged, they are known as ions.

The table below summarizes some typical ions that are formed by elements in the periodic table.

Notice that nonmetals (on the right side of the periodic table) have a greater tendency to gain electrons and form negatively charged anions. Metals (on the left side of the periodic table) exhibit weaker nuclear forces on valence electrons and tend to lose electrons and form positively charged cations.

Ionic bonds form between a metal cation and a nonmetal anion so that both atoms obtain a full *valence shell.*

If a metal has low ionization energy and a nonmetal has a high electron affinity (concepts described in the previous chapter), ionic bonding is even more likely.

A familiar ionic compound is sodium chloride (NaCl), or table salt.

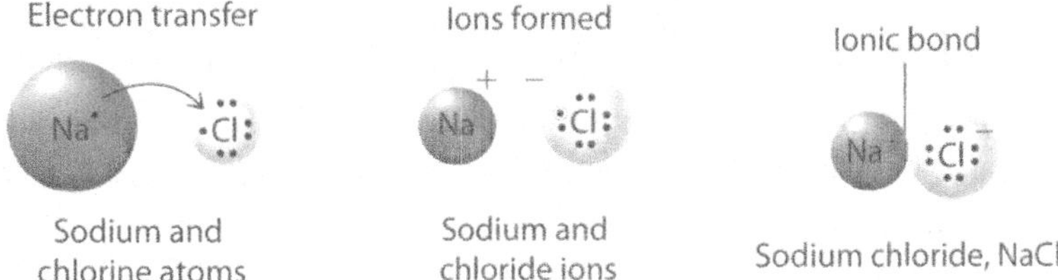

The cations and anions in an ionic compound are like the North and South poles of a magnet, which attract each other. Similarly, two cations (or anions) repel.

Therefore, the molecules are arranged into rigid crystalline structures.

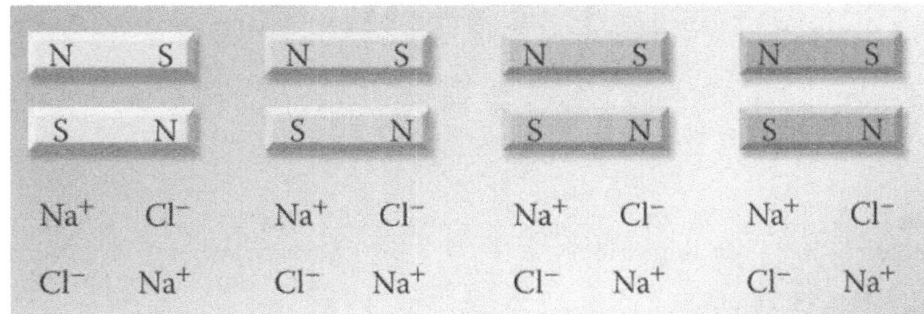

Crystalline structure of Na^+ and Cl^- with an alternating positive and negative charge

Ionic compounds, due to the strong, polarized, intermolecular forces generally possess high melting and boiling points and are, therefore, solid at room temperature.

Ionic crystals are three-dimensional arrangements of ions in an ionic compound, sometimes called a crystal lattice (e.g., solid crystalline structure of sodium chloride).

Sodium chloride has a high melting point (800 °C) and dissolves in water to give a conducting solution. Like all ionic compounds, sodium chloride is a strong electrolyte, so it dissociates completely in water.

Solid ionic compounds are non-conductive; however, molten compounds (i.e., compounds that have been liquefied by heat) or compounds dissolved in water conduct electricity.

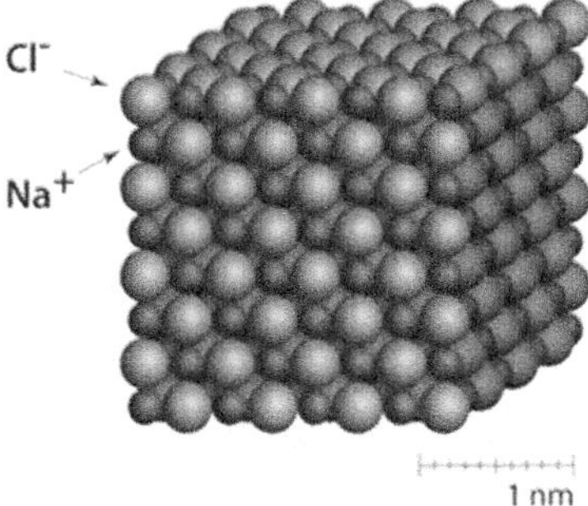

Crystalline Lattice Structure of Table Salt (NaCl)

Like in all types of bonding, the valence (outermost) electrons are the electrons involved in ionic bonding. Transfer of the lone 3*s* electron of a sodium atom to the half-filled 3*p* orbital of a chlorine atom generates a sodium cation and a chloride anion, forming the compound sodium chloride.

Initially, the sodium and chlorine atoms each had a net charge of 0. After the transfer of one electron took place, they had a charge of +1 and –1, respectively, due to the imbalance between protons and electrons in the atom.

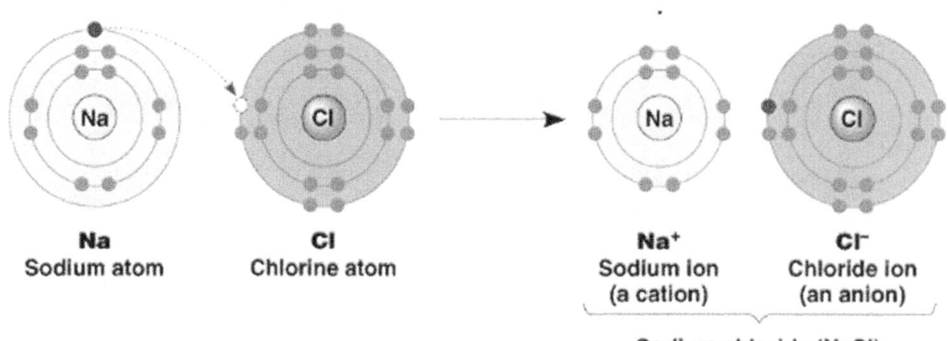

Ionic bonds form by the transfer of a valence electron

Electrostatic Energy ∝ q_1q_2 / r

The strength of an ionic bond can be determined by the amount of *electrostatic potential energy* between the two opposing charges.

The electrostatic potential energy is expressed by:

$$U_E = kq_1q_2 / r$$

where U_E is electrostatic potential energy, k is Coulomb's constant (9×10^9 m/F), q_1 and q_2 are the two charges, and r is the distance between the charges.

Electrostatic energy is negative in ionic bonds because q_1 and q_2 have opposing charges. If q_1 and q_2 did not have opposing charges (i.e., both positive or both negative), then they repel each other, and no ionic bond forms. However, the negative sign is often dropped, and only the magnitude of the electrostatic energy is used.

The higher the magnitude of the electrostatic potential energy, the stronger the ionic bond. The formation of strong ionic bonds are promoted by high charge magnitudes (q values) that are close together (small r value). Ions that form strong ionic bonds have a high charge density; that is, the charge-to-size ratio is high. A high charge density of a cation can distort the electron cloud of the anion to promote electron sharing.

Electrostatic Energy ∝ Lattice Energy

Lattice energy is the energy required to break an ionic bond.

Conversely, it can also be the energy given off when oppositely charged; gaseous ions come together to form an ionic crystalline solid.

This energy released by the reaction correlates to the ionic bond strength, which means that it is also proportional to electrostatic energy.

Electrostatic Force ∝ q_1q_2 / r^2

The electrostatic force that holds charged particles together is defined in terms of the electrostatic energy.

Coulomb's Law defines the electrostatic force in terms of electrostatic charges and distance.

$$F = kq_1q_2 / r^2$$

where F is the electrostatic force, k is Coulomb's constant (9×10^9 m/F), q_1 and q_2 are the two charges, and r is the distance between the charges.

From the equation, the electrostatic force is directly proportional to the product of opposite charges that attract (negative F) or the same charges that repel (positive F).

The electrostatic force is inversely proportional to the square of the distance between them. Therefore, larger charge magnitudes that are closer together exhibit a greater electrostatic force.

Coulomb's Law is analogous to the *universal law of gravitation*:

$$F = Gm_1m_2 / r^2$$

where G (gravitational constant) is analogous to k, and m (mass) is analogous to q.

The major difference is that G is tiny compared to coulomb's constant k (for stationary electrically charged particles) because the gravitational force is weak compared to the much stronger electrostatic force between charged particles.

There is another way to write the formula for an electrostatic force that is specific to ionic compounds.

The equation below describes the force of attraction between the cation n^+ and the anion n^- at a distance d apart.

$$F = R(n^+ \times e)\cdot(n^- \times e) / d^2$$

where R is coulomb's constant (usually written as k)

$n^+ \times e$ = charge of cation in coulombs = positive charge (n^+) × coulombs per electron (e)

$n^- \times e$ = charge of anion in coulombs = negative charge (n^-) × coulombs per electron (e)

Electron Pair Sharing

A *covalent bond* forms when two atoms share an electron pair.

These electron pairs are shared pairs or *bonding pairs*, and the formation of covalent bonds results in the overlap of their electron orbitals, as shown below.

Covalent bonds typically form between nonmetal elements.

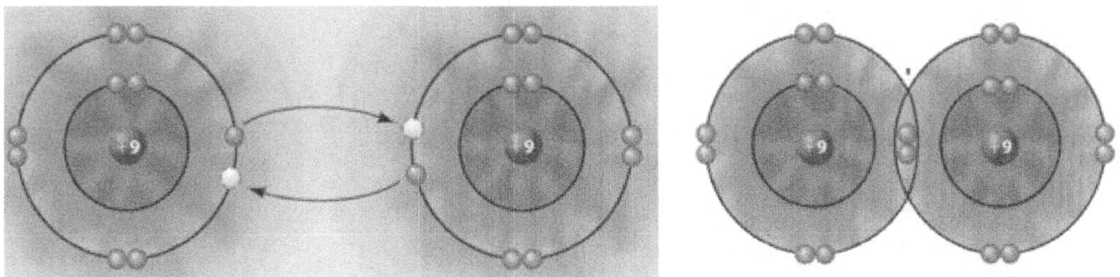

A covalent bond formed between two fluorine atoms. The image on the left shows the valence electron from each fluorine atom that becomes shared to form the covalent bond formed for the F_2 molecule

Most atoms with the same or similar electronegativity share electrons to achieve a full valence shell of eight valence electrons (i.e., the octet rule with 2*s* and 6*p* electrons). Illustrations of covalent bonding that follows the octet rule are displayed below.

Below are simple atomic notations, where dots designate the valence electrons.

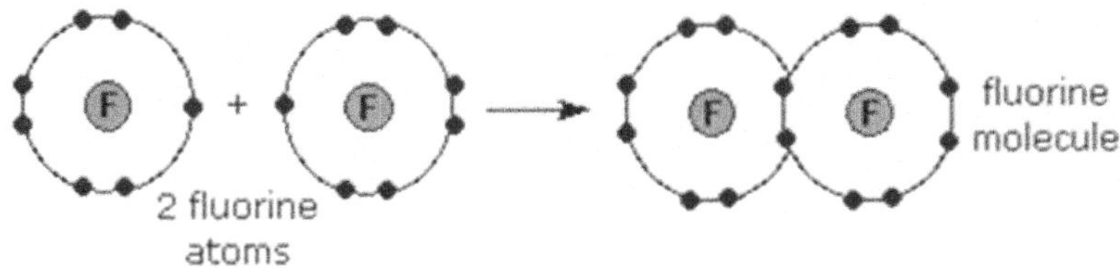

Covalent bond forms between 2 fluorine atoms that share a valence electron

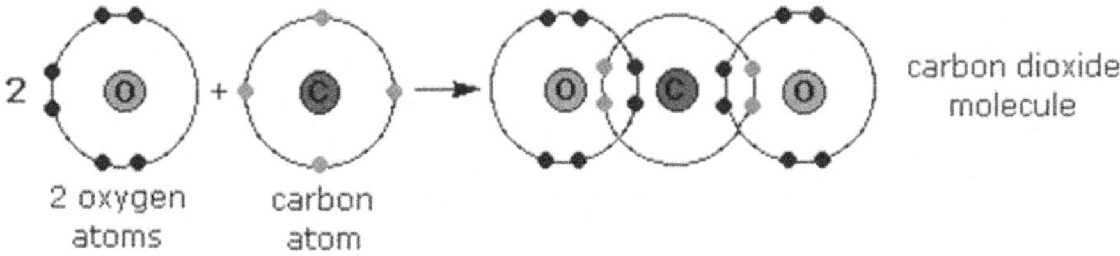

Covalent bond forms between 2 oxygen atoms and one carbon atom that share two pairs of valence electrons from the carbon atom to form carbon dioxide (CO_2)

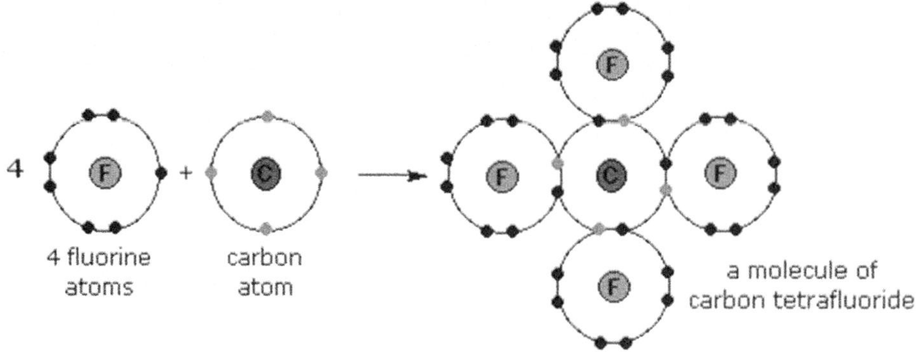

Covalent bond forms between 4 fluorine atoms and one carbon atom. Each atom contributes one valence electron to form carbon tetrafluoride (CF_4)

The hydrogen atom is an exception to the octet rule because it is at its lowest energy when it has two 1*s* electrons in its valence shell.

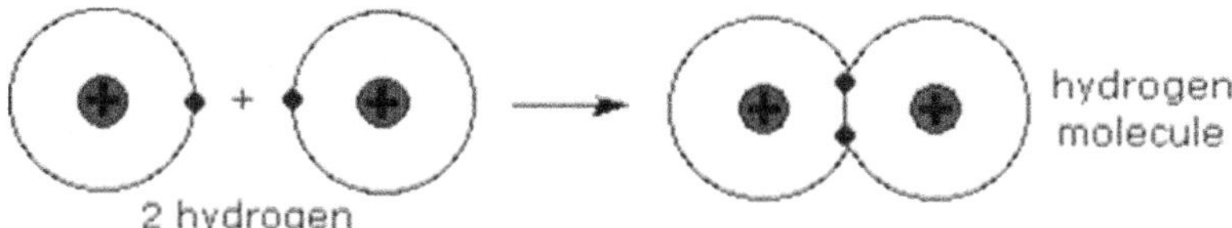

In the images below, the atoms form covalent bonds to obtain the ideal number of electrons. The *non-bonding lone pairs* are electrons that are not involved in bonding.

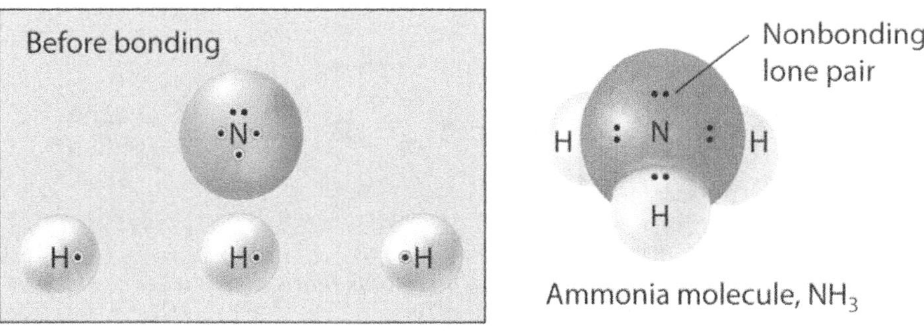

Nitrogen has 5 valence electrons. Nitrogen shares one valence electron with each of three hydrogen atoms. Each hydrogen contributes 1 valence electron. With three covalent bonds, nitrogen has a complete octet and hydrogen satisfies a complete 1s subshell

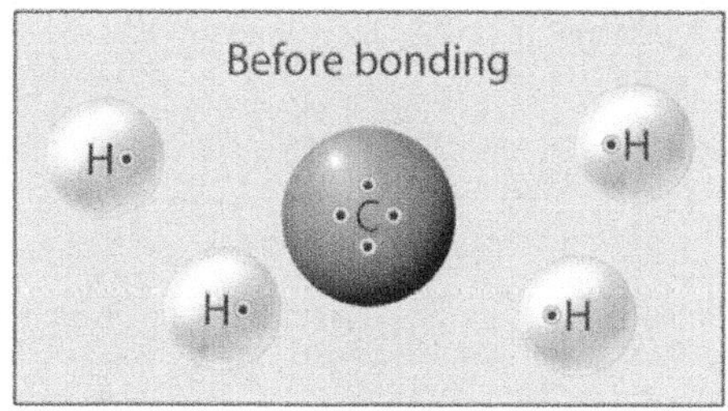

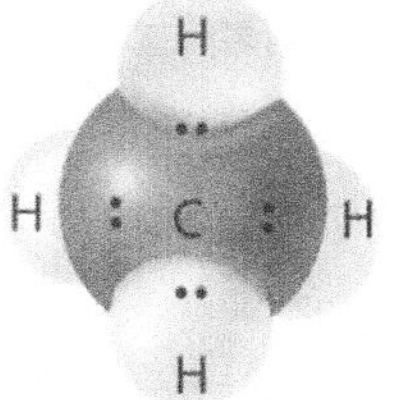

Methane molecule, CH_4

Carbon has 4 valence electrons and shares one valence electron with each of four hydrogen atoms. Each hydrogen contributes 1 valence electron. With four covalent bonds, carbon has a complete octet and hydrogen satisfies a complete 1s subshell

Covalent bonds can also involve more than a single pair of electrons, which results in double- and triple-bonded nonmetal atoms.

In double bonds, two pairs of electrons (four electrons) are shared.

In triple bonds, three pairs of electrons (six electrons) are shared.

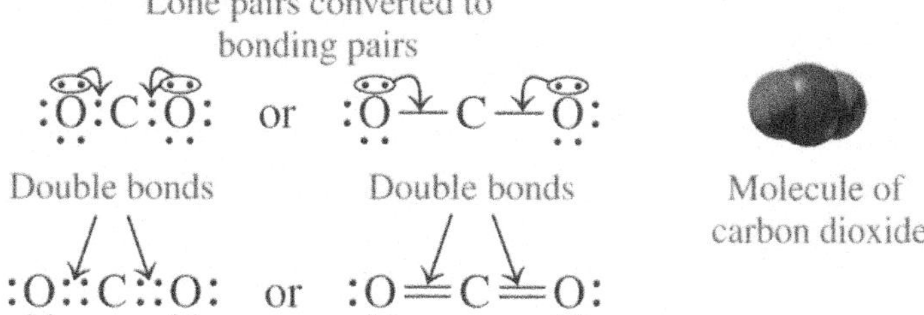

Carbon dioxide forms from two double bonds on the central carbon. Each double bond results from the oxygen and the carbon contributing a pair of electrons

·N:N· ⟶ :N⋮N: :N≡N:

Three shared pairs *Triple bond* *N_2 molecule*

A covalent bond resulting from one atom donating an electron pair to another atom is a *coordinate covalent bond.*

Both electrons in a coordinate covalent bond only come from one donor atom.

An example of a molecule with a coordinate covalent bond is ozone, O_3:

:O::O: + O: ⟶ :O::O:O:

nonbonding electron pairs *coordinate covalent bond: O=O–O*

Notes

Sigma and *Pi* Bonds

The electron density of a chemical bond can be divided into different regions.

Sigma (σ) *bonds* form when one orbital from each atom overlap and create a new orbital. Typically, a *sigma* bond is a single bond and is the first bond involved in double and triple bonds. The sigma bonds lie along the imaginary line joining their two nuclei, called the *internuclear axis*. In σ bonds, the electron density is between the nuclei of the two atoms and is symmetrical about the internuclear axis. Atoms on both sides of a σ single bond are free to rotate with the bond as their axis.

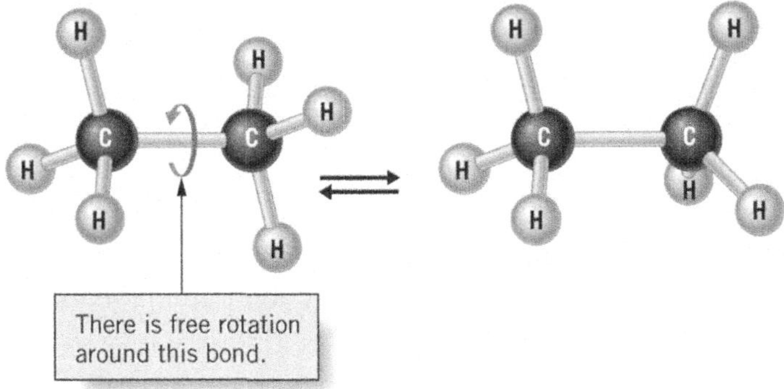

In the image below, the σ bonds of *s*–*s* and *p*–*p* orbitals are shown.

Hybrid orbitals also exist and are discussed in another section.

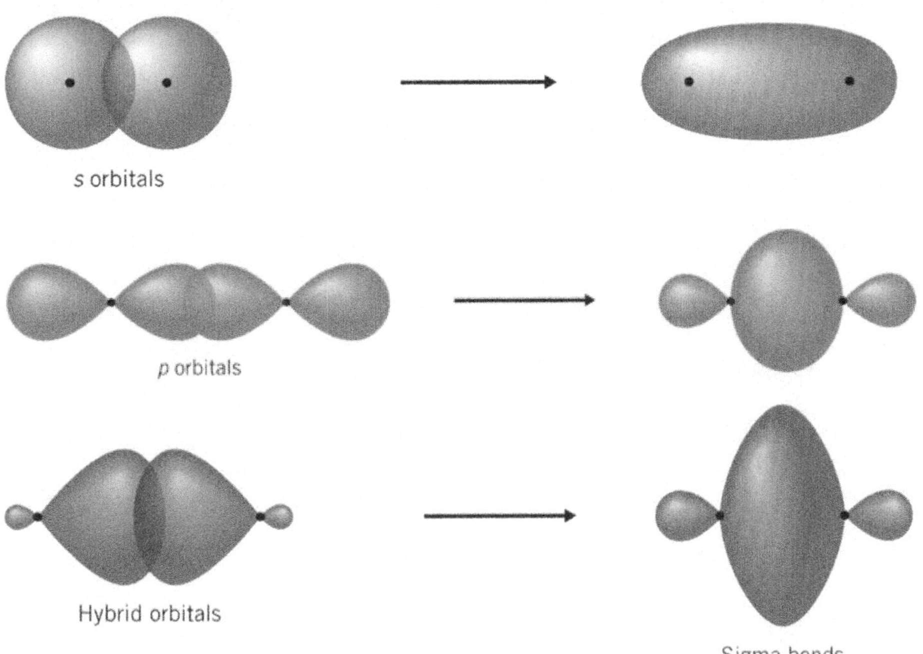

Pi (π) bonds form when unhybridized *p* orbitals on the top and bottom of the internuclear axis overlap. A single π bond consists of regions above and below the intermolecular axis. There is no electron density along the intermolecular axis.

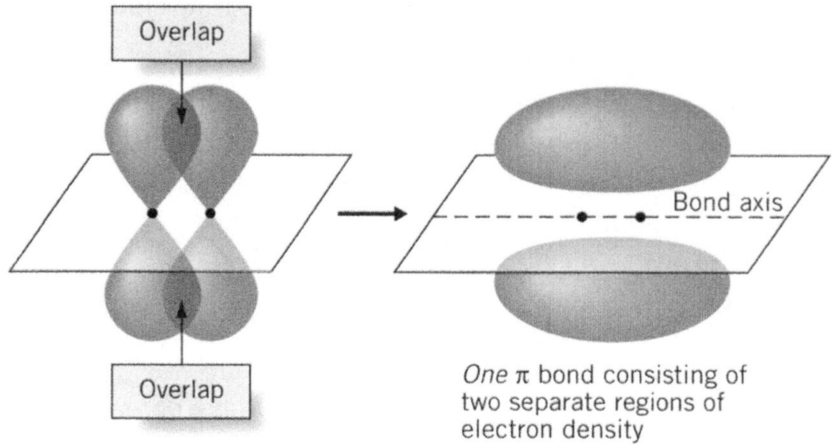

One π bond consisting of two separate regions of electron density

When electron density above and below the σ bonds overlap, a multiple bond (i.e., double or triple) forms. Thus, π bonds can only form after a σ bond has connected two atoms.

A double bond: one σ bond and one π bond.

A triple bond: one σ bond and two π bonds.

For example, nitrogen gas (N_2) is connected by a triple bond (N≡N). The first bond is a σ bond, and the remaining two bonds are π bonds. The π bond is distinct in that it introduces a particular property of rigidity to molecules; π bonds do not allow for the rotation of atoms. Molecules that involve π bonds require more energy to break.

Hybrid orbitals

Hybrid orbitals are produced by the hybridization (mixing) of existing electron orbitals to create geometries that can facilitate better bonding.

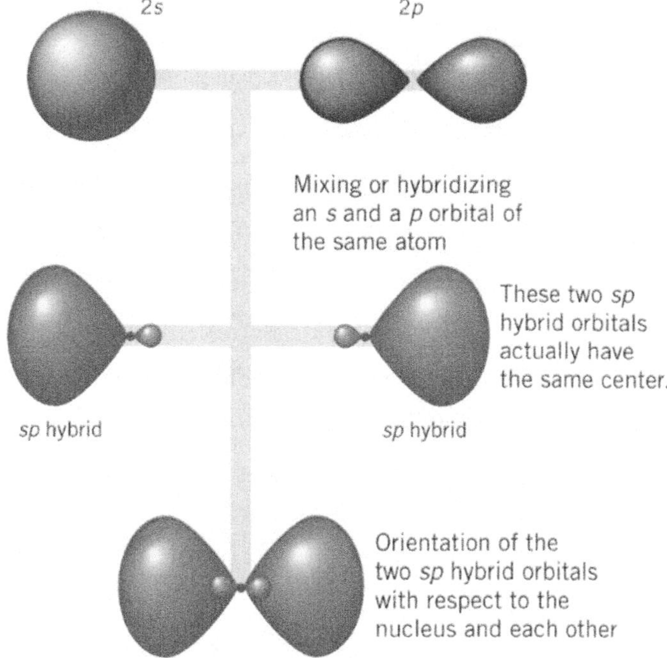

The possible hybridizations of the carbon atom are as follows.

- sp^3: a hybrid between one s with three p orbitals.

  Tetrahedral geometry. Contains single bonds only.

- sp^2: a hybrid between one s with two p orbitals.

  Trigonal planar geometry. Contains a double bond.

- sp: a hybrid between one s with one p orbital.

  Linear geometry. Contains a triple bond.

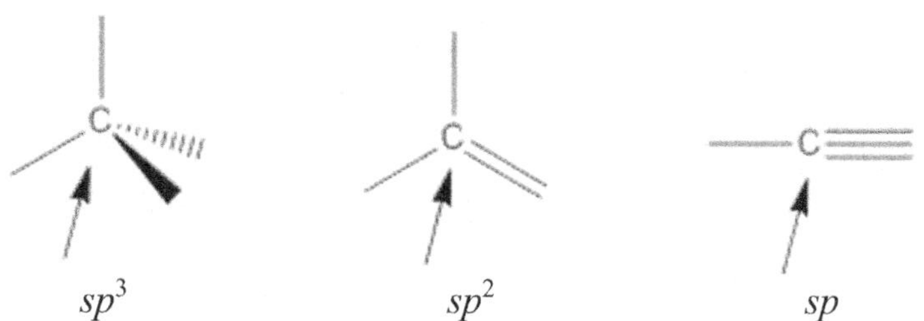

Hybridization of the carbon atom for 4 single bonds, 1 double or 1 triple bond

Valence shell electron-pair repulsion (VSEPR) theory, predictions of shapes of molecules

Valence Shell Electron Pair Repulsion (VSEPR) theory states that electron pairs surrounding an atom repel each other, and the largest possible bond angles separate the bonds and lone pairs around a central atom.

In VSEPR, *electron domains* refer to the electron pairs in the molecule.

Every molecule has bonding and non-bonding domains.

When atoms bond to create molecules, the atoms arrange themselves as far apart as possible to minimize same-charge repulsion between electrons.

VSEPR theory helps to predict the molecular geometry based on the number of electron pairs in the molecule.

The *electron pair domain geometry* (EDG) indicates the arrangement of bonding and non-bonding electron pairs around the central atom.

The *molecular shape geometry* (MG) indicates the arrangement of atoms around the central atom after accounting for electron repulsion.

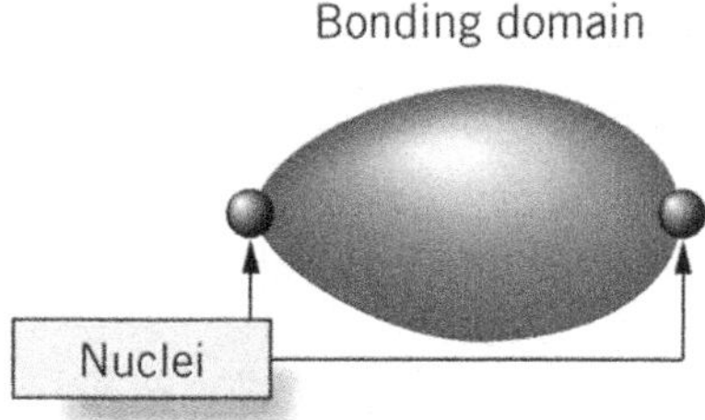

Nonbonding domain

Nucleus

Bonding domains are directed between two adjacent nuclei, while the antibonding domains are directed away from adjacent nuclei

Below are the configurations (i.e., specific shape) that maximize the distance between the bonding electron pairs and the associated substituents attached via the bond.

Molecules with *two* attached atoms (or substituents) are *linear* (a straight line, with one atom on each side of the central atom). Bond angle = 180°.

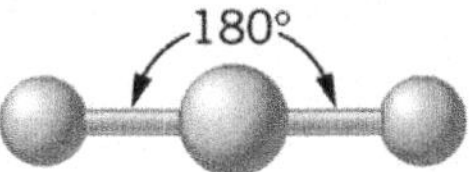

Molecules with *three* attached atoms are *trigonal planar*. Bond angle = 120°.

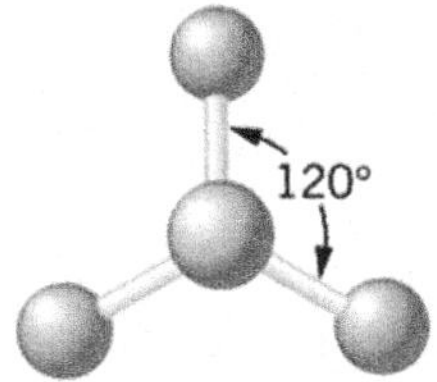

Molecules with *four* attached atoms are a *tetrahedron*. Bond angle = 109.5°.

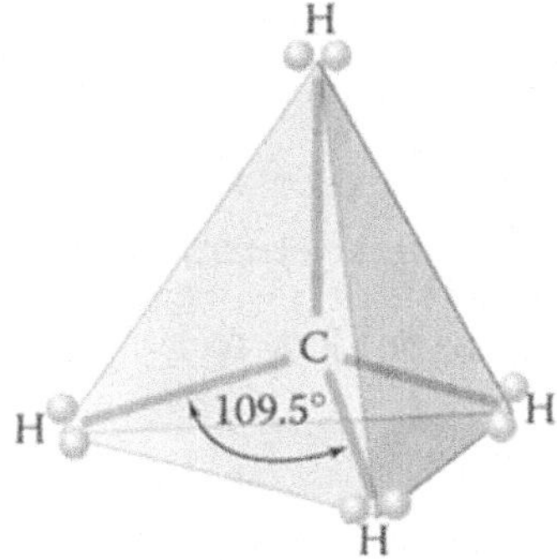

Molecules with *five* attached atoms are *trigonal bipyramid*. Bond angles = 120° and 90°.

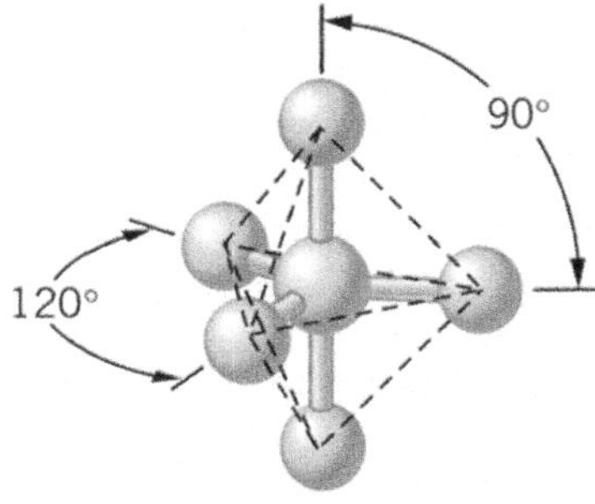

Molecules with *six* attached atoms are an *octahedron*. Bond angle = 90°.

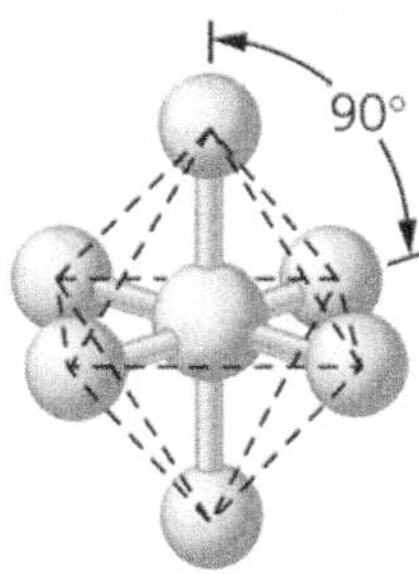

The molecular shapes above are accurate for molecules that contain only bonding domains (i.e., their central atoms do not have any non-bonding domains or lone pairs of electrons).

When the central atom has one or more non-bonding domains, the lone pair occupies the space where a bond usually forms. These non-bonding electron domains are relatively dispersed in space and exert electrostatic repulsions that cause bonding electrons to bend away.

Lone pairs need to be closer to the atom whose orbital they occupy than the bonding pairs shared along the internuclear axis between the two bonding atoms.

Thus, lone pairs are not restricted to the internucleus axis occupy more of the available space within the orbital (i.e., more electrostatic repulsion).

An example is SF_4, which has one lone pair:

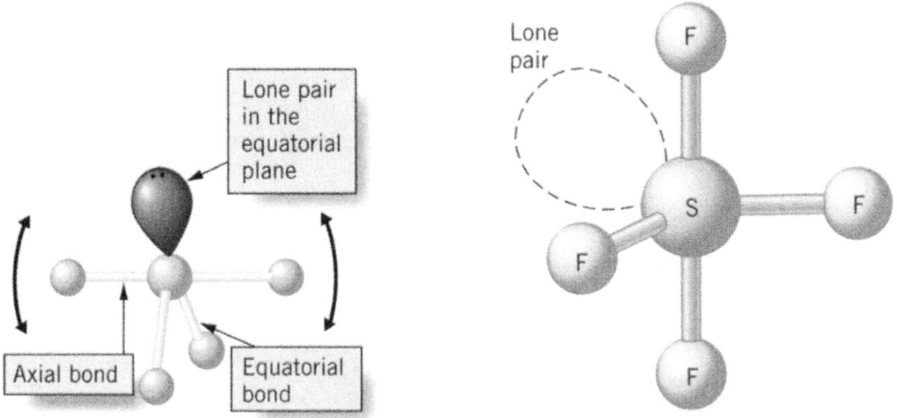

Electrons in sigma bonds occupy a restricted space while nonbonding lone pairs are more dispersed and therefore contribute electrostatic repulsion to the bonding electrons

Hybridization	Electron geometry	Electron groups	Bonding groups	Lone pairs	Approx. bond angles	Molecular geometry
sp	Linear	2	2	0	180°	Linear
sp^2	Trigonal planar	3	3	0	120°	Trigonal planar
		3	2	1	< 120°	Bent
sp^3	Tetrahedral	4	4	0	109.5°	Tetrahedral
		4	3	1	< 109.5°	Trigonal pyramidal
		4	2	2	<< 109.5°	Bent
sp^3d	Trigonal bipyramidal	5	5	0	120° (equatorial) 90° (axial)	Trigonal bipyramidal
		5	4	1	< 120° (equatorial) < 90° (axial)	Seesaw
		5	3	2	< 90°	T-shaped
		5	2	3	180°	Linear
sp^3d^2	Octahedral	6	6	0	90°	Octahedral
		6	5	1	< 90°	Square pyramidal
		6	4	2	90°	Square planar

A general strategy to determine the geometrical shape of a molecule:

1. Start by counting all the domains, both bonding, and non-bonding, around the central atom. The number of domains determines the

general shape of the molecule (one of the shapes discussed above). For every non-bonding domain, remove a surrounding atom.

Example: a molecule with four total domains has a tetrahedral shape.

If one of the domains is non-bonding, one of the surrounding atoms is removed, leaving a central atom with three surrounding atoms remaining; the molecule is in the shape of a trigonal pyramid.

If there are two non-bonding domains, remove two surrounding atoms; the molecule is in a bent shape.

2. For molecules with five domains, start by removing atoms above and below the pyramid before the 3 atoms surrounding it.

   The molecule is more stable when the atoms adopt a configuration with the least repulsion.

   The angles between the top or bottom atoms and the three side atoms are 90°.

   The angles between the side atoms are 120°.

   Thus, the molecule gets rid of the more repulsive atoms first (i.e., the ones with the smaller bond angle).

   The same approach applies to molecules with six domains. The domains above and below are removed before the domains surrounding the central atom.

Non-bonding domains affect the angle of bonds in the molecule due to their occupying more significant space (increased electrostatic repulsion) than bonded electrons.

The following diagram summarizes the shapes of atoms based on the number of bonding and non-bonding electron pairs (also called nonbonded electron pairs):

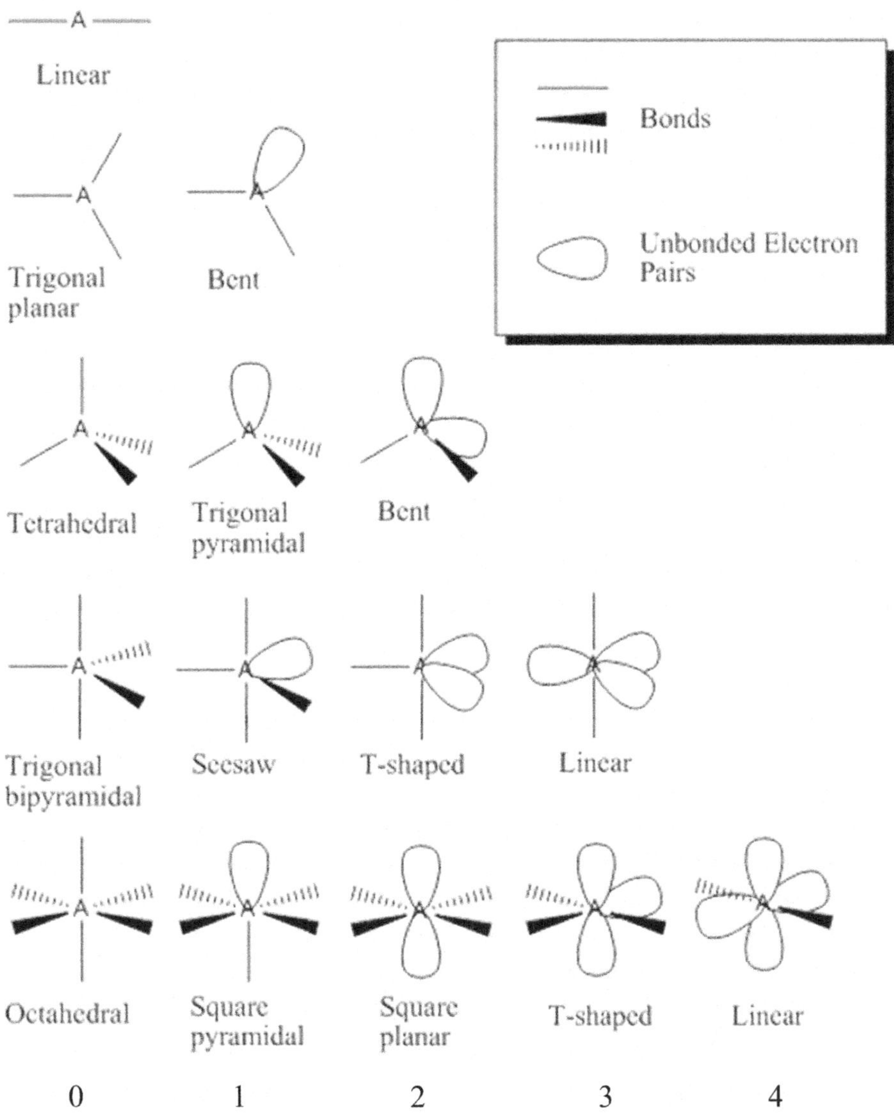

Numbers along the bottom indicate the # of nonbonded electron pairs

The oxygen of water (H_2O, shown below) has two bonding domains and two non-bonding domains. The water molecule experiences electron repulsion from the two lone pairs on its bonds. The water molecule adopts a bent shape (see the previous chart) to minimize the van der Waals repulsion of the negatively charged electrons.

Water's bent shape comes from a tetrahedral configuration, with two of its vertices removed. All bonds in a tetrahedron are 109.5° wide, which means that bonds in a bent molecule should have 109.5° between them.

However, because of the repulsion between lone electron pairs and the bonding electrons, the actual bond angle is slightly decreased to approximately 105°.

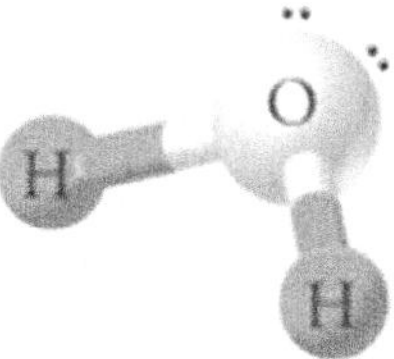

The water molecule (H_2O) is bent due to the two lone pairs of electrons on oxygen

Geometric shapes using VSEPR have double bonds counted as a single domain.

Formaldehyde (CH_2O) has one double and two single bonds, which adds to three bonding domains. The molecular geometry of formaldehyde is trigonal planar (i.e., flat shape).

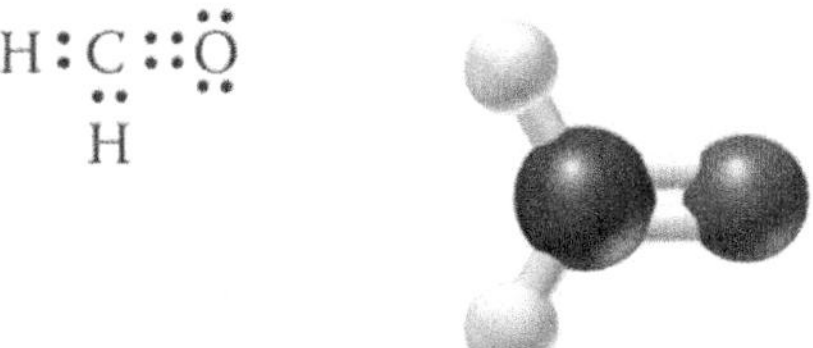

Trigonal plane geometry for formaldehyde (CH_2O) with one double and two single bonds

Two major theories explain how molecules bond: valence bond theory and molecular orbital theory. Both theories are used to predict the structure of a molecule.

The *Valence Bond Theory* (VB) is primarily based on valence electrons. Valence bond theory states that each atom has its orbitals and that orbitals overlap to form bonds. The extent of overlap of atomic orbitals is related to bond strength; the more significant the overlap, the stronger is the bond.

The VB theory can explain the bonding of F2. F_2 bonds form because of atomic valence orbitals overlap (in this case, 2*p* orbitals), one from each fluorine.

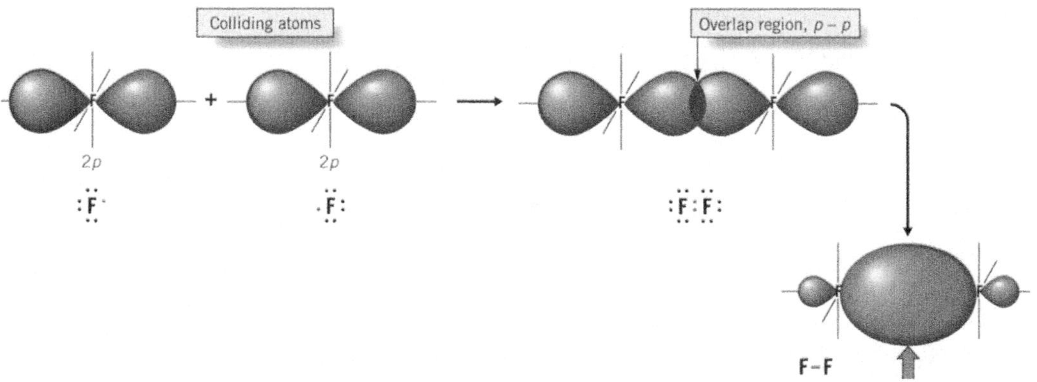

The overlap of p–p orbitals form the F_2 bond

The orbital overlap is not limited to identical orbitals. Bonding in the HF molecule involves the overlap between the 1*s* orbital of H and the 2*p* orbital of F.

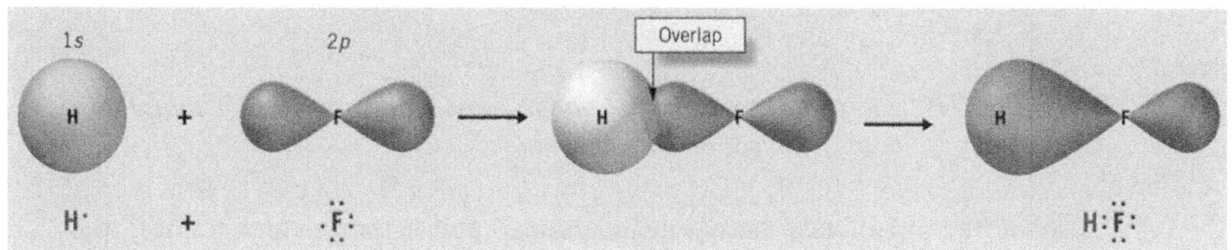

The bonding of H_2S can also be described according to VB theory. Sulfur has one filled orbital (contains two electrons) and two partially filled orbitals (contains one electron).

The partially filled *p* orbitals overlap with the *s* orbital from hydrogen.

The predicted 90° bond angle is close to the experimental value of 92°.

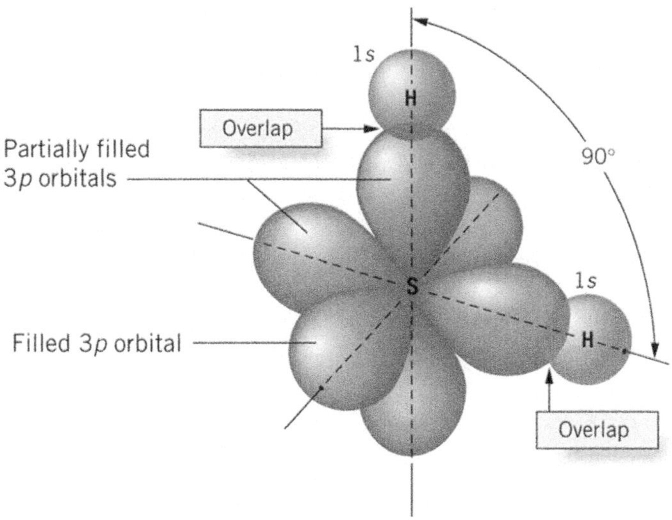

VB theory describes the bonding on many molecules, but it cannot explain all existing molecular geometries adequately.

Mcthanc (CH_4) has a tetrahedral shape, seen in the image below.

Carbon's electron configuration ($1s^22s^22p^2$) indicates that all its electrons are paired, except for two electrons that occupy *p* orbitals.

VB theory predicts that carbon can only bond with two hydrogen molecules, resulting in a CH_2 molecule and that the bond angle is close to 90° (like H_2S above).

However, observations of methane reveal that there are four equal bonds, adopts thc shape of a tetrahedral, and the bond angles are all 109.5°.

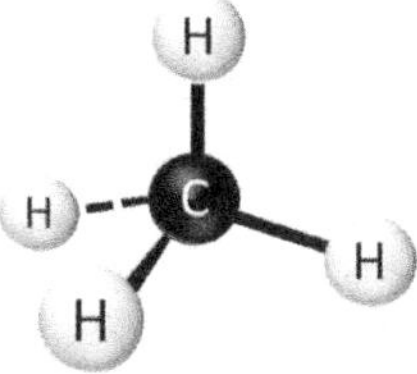

Methane (CH_4) adopts a tetrahedral shape to maximize repulsion from the 4 hydrogen substituents

To resolve the difference between theoretical and experimental values, the concept of *hybridization* was introduced to improve upon the VB theory.

Hybridization refers to the mixing of atomic orbitals to allow the formation of bonds that have realistic bond angles.

Atomic orbitals are areas with the highest probabilities of finding electrons, and hybridization is a rearrangement of orbital areas.

Hybrid orbitals have new shapes, new directional properties, and combined properties of the constituent orbitals.

The symbols for hybrid orbitals combine the symbols of the initial orbitals used to form them.

The sum of exponents in hybrid orbital notation must add up to the number of atomic orbitals used.

- One *s* and one *p* orbital form two *sp* hybrid orbitals
- One *s* and two *p* orbitals form three sp^2 hybrid orbitals

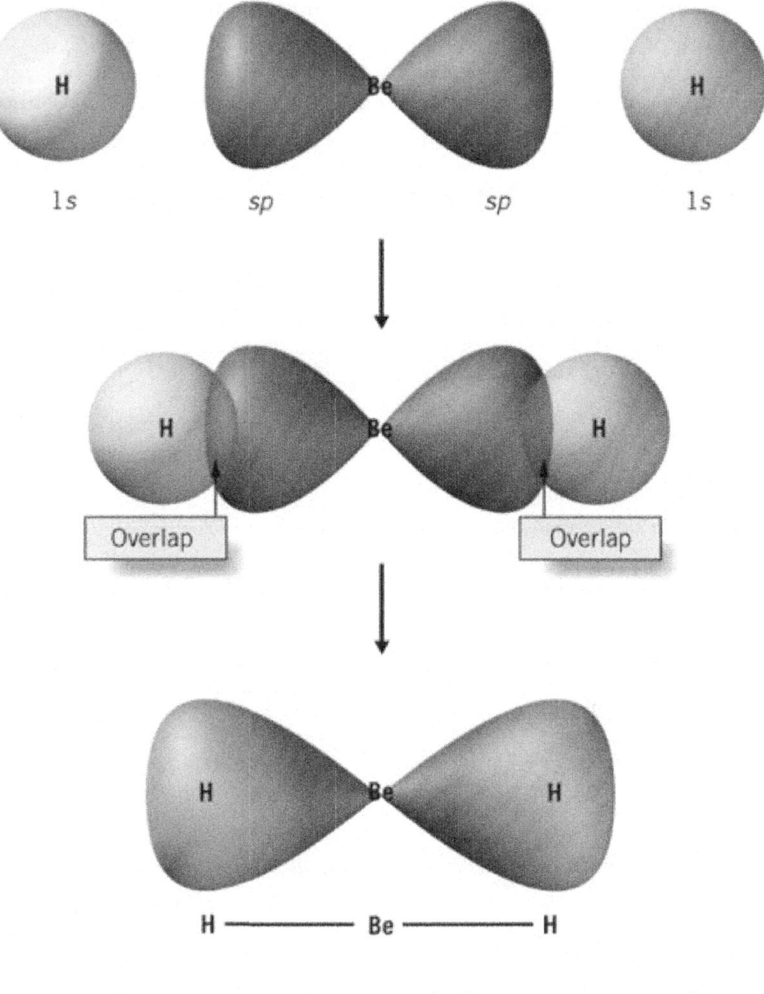

Bonding of BeH_2 Be: $1s^2 2s^2$ H: $1s^1$

Beryllium (Be) has two electrons in the 2*s* orbital. To bond with two hydrogens, it needs to split the electrons into two orbitals. The Be atom uses the next available orbital (2*p*) and combines it with 2*s* orbital, which results in two *sp* orbitals.

Each of those *sp* orbitals overlaps with hydrogen's 1*s* orbital, creating a covalent bond. VSEPR predicts that this molecule would have a linear shape, which is confirmed by experimental observations.

Because of the octet rule, most atoms can only form up to four bonds.

This means that four orbitals are needed (1 *s* and 3 *p* orbitals).

For molecules that have more than four bonds/lone pairs around them, *d* orbitals are included in the hybridization to accommodate the extra bonds; this is known as *expanded octet hybridization*.

Bond angle experiments suggest that NH_3 and H_2O both use sp^3 hybrid orbitals in bonding (with a total of four orbitals) even though each only has two bonds.

This evidence for sp^3 hybrid orbitals led to the conclusion that hybrid orbitals are not exclusively used for bonding; they also accommodate the non-bonding electron pairs.

Both bonding and non-bonding pairs contribute to the geometry of a molecule.

For example, the orbital diagram of hybridized nitrogen in NH_3:

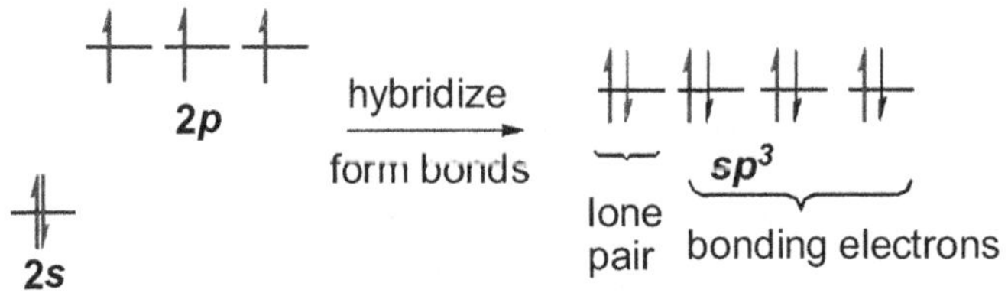

2p
hybridize
form bonds
sp³
lone pair
bonding electrons
2s

Hybridization accommodates both 3 bonding and 1 nonbonding electron pairs

This diagram shows the hybridization of oxygen in H_2O:

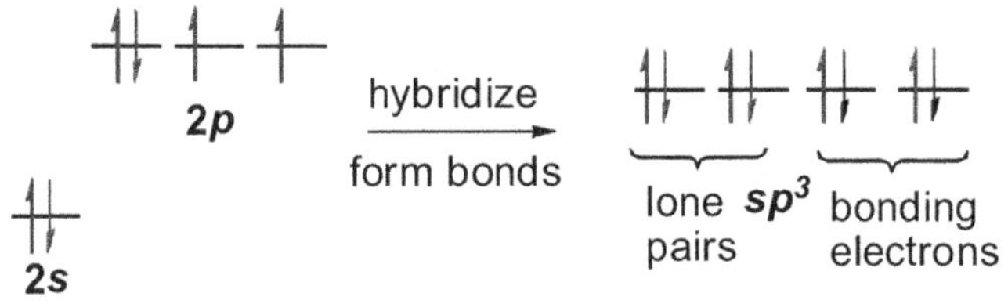

2p
hybridize
form bonds
lone pairs
sp³
bonding electrons
2s

Hybridization accommodates both 2 bonding and 2 nonbonding electron pairs

Molecular Orbital Theory (MO), the second dominant theory, views molecules as a collection of positively charged nuclei having a set of molecular orbitals that are filled with electrons (like filling atomic orbitals with electrons). The molecular orbital theory does not focus on the way that individual atoms come together to form molecules.

The MO theory is more difficult to visualize than the VB theory.

However, MO theory can be advantageous because it allows for the accurate prediction of magnetic properties of molecules, along with other characteristics.

The energies of molecular orbitals are determined by combining the electron waves of atomic orbitals.

For H_2 (shown below), two 1*s* wave functions, one from each atom, combine to make two MO wave functions. These two MO produce constructive interference of the waves. For H_2, the energy of the bonding MO is lower than the energy of the atomic orbitals, and the molecule is collectively more stable than the individual atoms (atoms join to form the molecule).

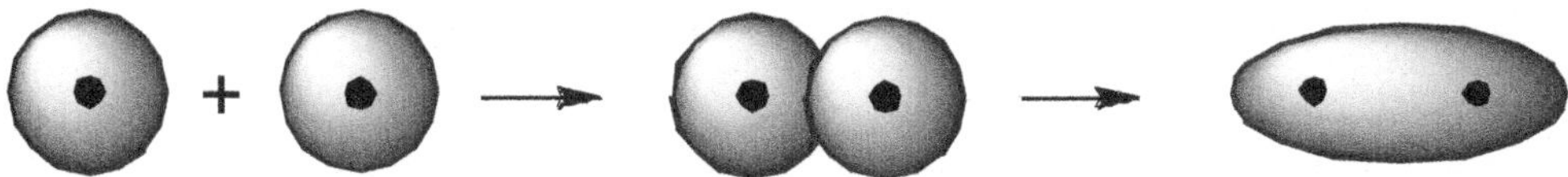

As there are σ orbitals, there are corresponding σ^* antibonding MOs. The other possible combination of two 1*s* orbitals involves destructive interference of the 1*s* waves.

In the example below, the energy of the antibonding MO is higher than the energy of parent atomic orbitals. The molecule does not form, and the atoms remain separate.

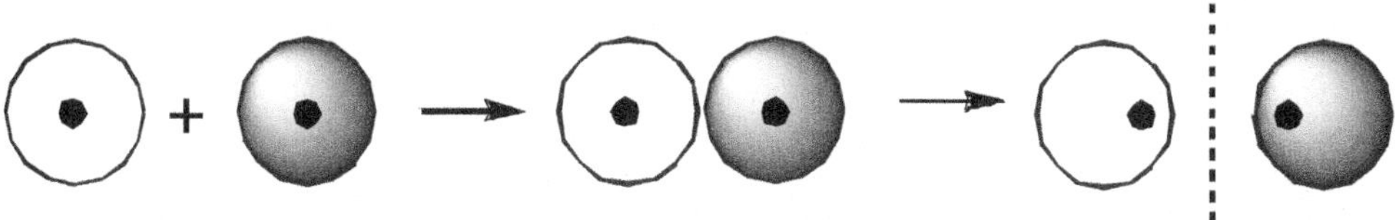

Antibonding MOs are high in energy and do not contribute to bond formation

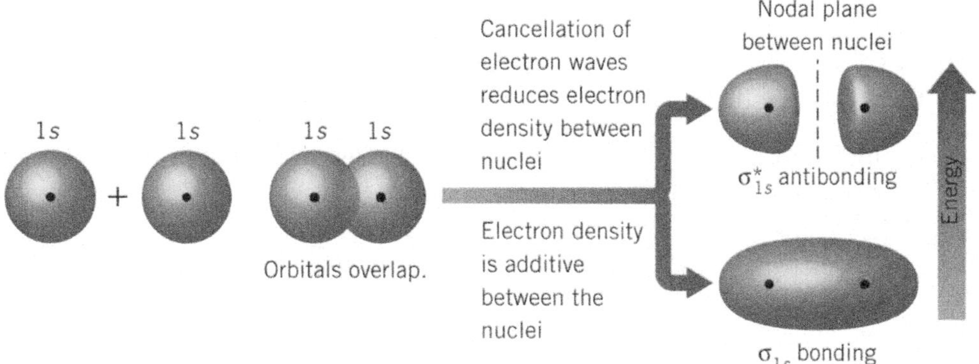

Bonding and antibonding MOs are shown for the relative energy of each

In the bonding MO, electron density builds up between nuclei; the electrons tend to stabilize the molecule.

In the antibonding MO, the cancellation of electron waves reduces the electron density between the nuclei.

The electrons located in antibonding MOs tend to destabilize the molecule.

MO energy diagrams represent the interactions between the atomic orbitals. A MO energy diagram displays the orbitals arranged vertically, from highest to lowest energy.

The atomic orbitals for the atoms are listed on the left and right sides of the diagram. The molecular orbitals are listed in a column down the center of the diagram.

The MO energy diagram for H_2

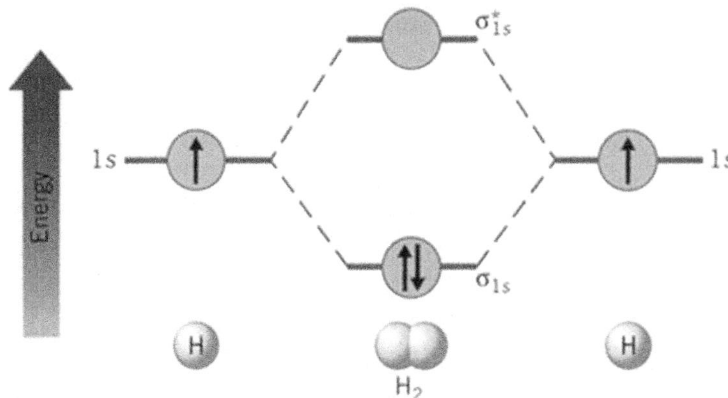

Atomic orbitals are designated as 1s and have single arrows pointing up while the lower energy bonding molecular orbital has 2 electrons of opposite spin. The high energy antibonding MO is vacant. Overall, the molecule forms because the bonded molecule is more stable than the separate atoms (relative stability with energy on the y-axis)

Three rules for filling MO energy diagrams:

1. Electrons fill the lowest-energy orbitals that are available (Aufbau's principle).

2. No more than two electrons, with opposite spins, can occupy any orbital (Pauli exclusion principle).

3. Electrons with unpaired spins spread out as much as possible over orbitals of the same energy (Hund's rule).

Bond order indicates the number of electron pairs shared between two atoms (i.e., the number of bonds between two atoms).

A bond order of one corresponds to a single bond.

For example, the H_2 bond order and the C−H bond order are both one, while the N≡N bond order in diatomic nitrogen is three.

The bond order is calculated by:

$$\text{Bond order} = \frac{(\text{number of bonding } e^-) - (\text{number of antibonding } e^-)}{2 \text{ electrons/bond}}$$

Neither the valence bonding (VB) nor the molecular orbital (MO) theory is entirely correct because neither theory explains all aspects of bonding. Each theory has its strengths and deficiencies.

VB theory is based on simple Lewis structures (discussed later) and the related geometric shapes.

MO theory correctly predicts the unpaired electrons in O_2, while the Lewis structures do not. MO theory is complicated because even simple molecules have complex energy level diagrams, and molecules with three or more atoms require extensive calculations.

However, VB theory, uses three-dimensional structures based on electron domains, without extensive calculations.

Simple hybrid orbitals are invoked where experimental evidence shows the need, and the integer bond orders are often correct.

Structural formulas for molecules involving H, C, N, O, F, S, P, Si, Cl

Lewis structures (explained in the next section), are frequently used to represent atoms and molecules visually. In general, the Lewis structures for elements in the same column (group or family) on the periodic table are similar. For example, sulfur can be substituted for oxygen in Lewis structures of oxygen.

Below are the structural formulas for some important molecules.

- Hydrogen Lewis structures

  Hydrogen Proton: H^+

  Hydride ion: H^-

- Boron – Group 13 Lewis structures

Since group 13 elements only have three valence electrons, often they only have six total electrons in a molecule (an exception to the octet rule).

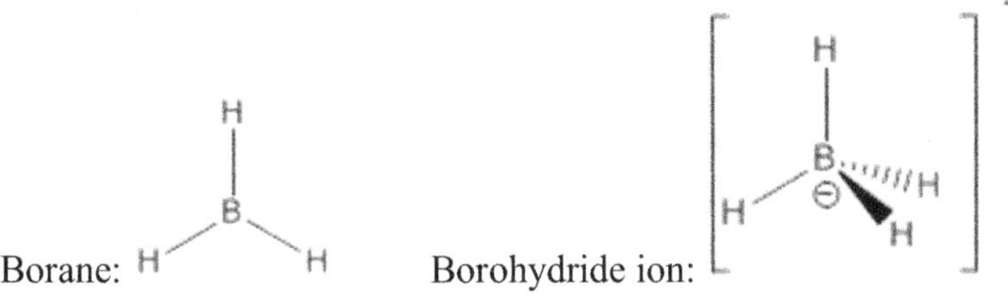

Borohydride is expressed as BH_4^- with the formal charge on boron. It is shown with brackets and a negative charge indicated on the top right of the bracket.

- Carbon – Group 14 Lewis structures

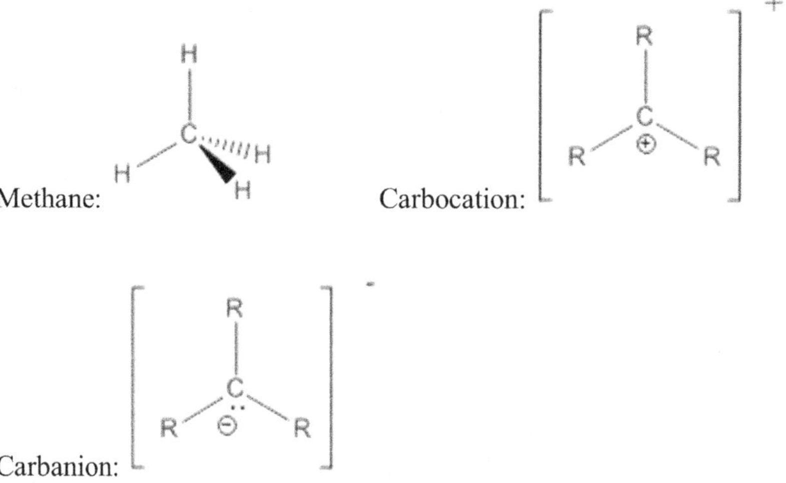

Note: R in the figures are either carbon or hydrogen.

- Nitrogen – Group 15 Lewis structures

Some group 15 elements have more than eight electrons after bonding, which is an exception from the octet rule (e.g., PCl_5). The presence of *d* orbitals permits the atom to accommodate these additional (more than 4) bonds.

Amine / Ammonia:

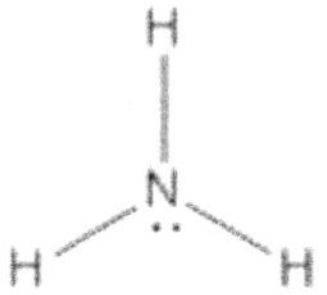

Ammonium:

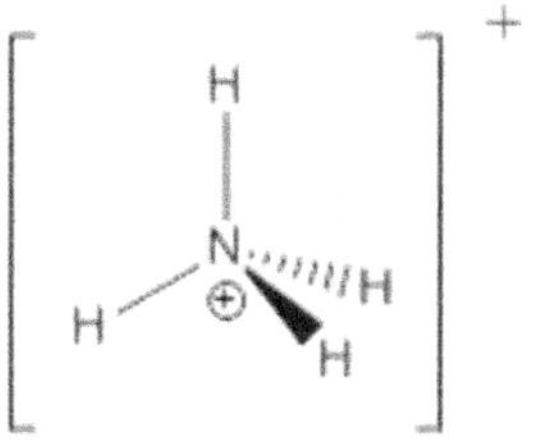

Imine:

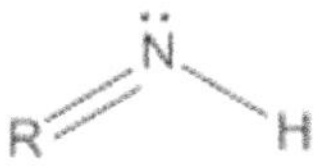

- Oxygen – Group 16 Lewis structures

Some group 16 elements have more than eight electrons after bonding, which is an exception from the octet rule (e.g., SF_6).

Molecular oxygen: $\ddot{\underset{\cdot\cdot}{O}}=\ddot{\underset{\cdot\cdot}{O}}$

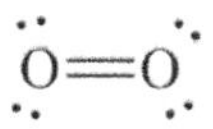

Water, alcohol, and ethers: R–Ö–R

Ozone:

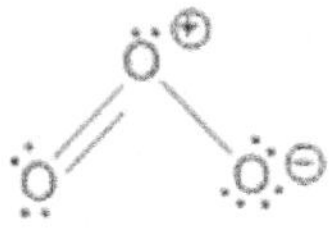

- Halogen – Group 17 Lewis structures

Hydrogen fluoride: H–F:

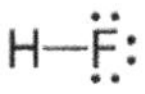

Chloromethane: H–C(H)(H)–Cl:

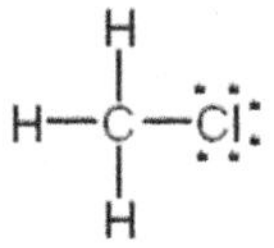

Bromide ion: $[:\ddot{\underset{\cdot\cdot}{Br}}:]^-$

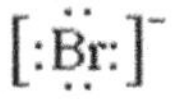

Notes

Lewis Electron Dot Formulas

Lewis dot formulas, or *Lewis structures*, are notations used to describe bonds and electrons in a molecule. For Lewis structures, each dot shown represents one electron, and each line represents one bond (two electrons); two dots represent a lone pair.

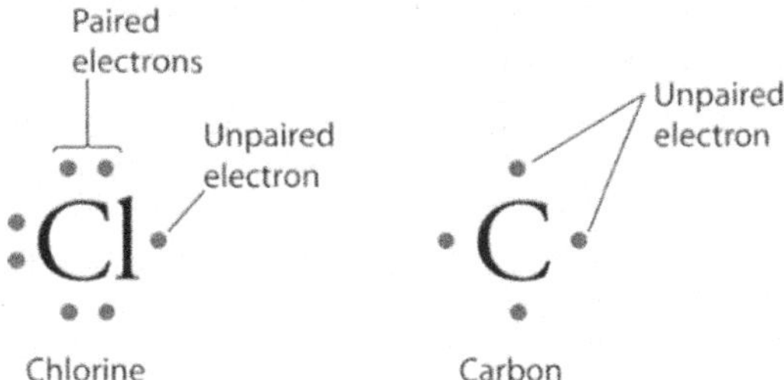

Lone pairs are shown as two dots while an unpaired electron has no partner

Cl:P:Cl   or   Cl—P—Cl
     Cl | Cl

Lone pairs are usually omitted from the molecule's final structure:

:F̤̈—F̤̈:   F–F

Recall the octet rule: most atoms need a total of eight valence electrons to achieve stable electron configurations. Electrons in a bond are shared and can be used to satisfy the octet for both bonded atoms.

This principle of electron configuration includes coordinate covalent bonds, where both electrons in a bond are contributed by only one atom. However, the electron pair also counts towards the other atom's octet rule.

Exceptions for the octet rule include atoms with more or less than eight electrons:

the boron column (they form three bonds and have a sextet),

large elements with periods of 3 or higher. These elements have *d* orbitals (e.g., the dectet P in PO_4^{3-} and the duodectet S in SO_4^{2-}),

radicals (compounds with an odd number of total electrons that possess a single, unpaired electron).

Some general rules for Lewis structures:

- From the octet rule, most atoms have up to eight electrons.
- When starting with atoms that have up to four valence electrons, separate them evenly around the molecule.
- If the atom has five or more valence electrons, divide them into four sections with a maximum of two electrons on each side. This is not a requirement but drawing them in this fashion makes it easier to determine if an atom has achieved stability (eight total electrons).

When sketching Lewis structures, use different symbols for electrons from different sources.

Consider the Lewis structure of PCl_3 below; it is unclear which electrons around P came from itself or Cl.

Knowing that P has five valence electrons, deduce that the remaining three must come from Cl.

However, when the bonds get more complicated, such as SO_3, it can be confusing. Use clear designations to differentiate electrons from different sources.

In the $BeCl_2$ example below, "x" represents electrons from Be, and dots represent electrons from Cl.

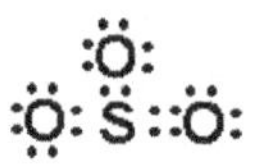

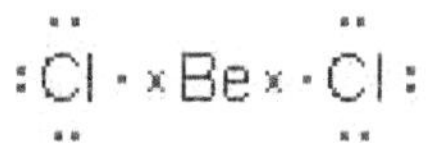

:Cl:
:Cl:P:Cl:

:O:
:O:S::O:

:Cl·xBex·Cl:

- Place square brackets around complete Lewis structures and indicate the charge on the top of the right bracket.

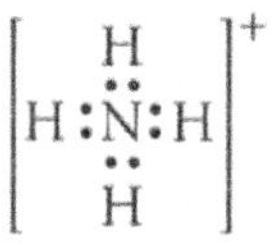

[H:N:H with H above and below]$^+$

Guide to drawing electron dot formulas:

1) Determine the arrangement of the atoms.
2) Determine the total number of valence electrons.
3) Attach each bonded atom to the central atom with a pair of electrons.
4) Place the remaining electrons using single or multiple bonds to complete the octet.

Using Valence Electrons to Draw Electron-Dot Formulas

Molecule or Poly atomic Ion	Total Valence Electrons	Form Single Bonds to Attach Atoms (electrons used)	Electrons Remaining	Completed Octets (or H:)
Cl_2	$2(7) = 14$	Cl—Cl ($2e^-$)	$14 - 2 = 12$	:C̤̈l—C̤̈l:
HCl	$1 + 7 = 8$	H—Cl ($2e^-$)	$8 - 2 = 6$	H—C̤̈l:
H_2O	$2(1) + 6 = 8$	H—O—H ($4e^-$)	$8 - 4 = 4$	H—Ö̤—H
PCl_3	$5 + 3(7) = 26$	Cl—P(—Cl)—Cl ($6e^-$)	$26 - 6 = 20$	:C̤̈l—P̤(—C̤̈l:)—C̤̈l:
ClO_3^-	$7 + 3(6) + 1 = 26$	[O—Cl(—O)—O]⁻ ($6e^-$)	$26 - 6 = 20$	[:Ö̤—C̤l(—Ö̤:)—Ö̤:]⁻
NO_2^-	$5 + 2(6) + 1 = 18$	[O—N—O]⁻ ($4e^-$)	$18 - 4 = 14$	[:Ö̤—N̈=Ö:]⁻ ↕ [:Ö=N̈—Ö̤:]⁻

The following are more specific rules for the most common elements in Lewis structures:

- Carbon: four bonds total, zero lone pairs. (e.g., CH_4, CO_2)
- Oxygen:

  O: two bonds total, two lone pairs (e.g., H_2O, O_2)

  O^{1-}: one bond, three lone pairs, a formal charge of –1

  O^{1+}: three bonds, one lone pair, a formal charge of +1

- Nitrogen:

  N: three bonds total, one lone pair (e.g., amines, ammonia NH_3)

  N^+: four bonds, zero lone pairs, a formal charge of +1 (e.g., ammonium NH_4^+)

- Halogens: one bond, three lone pairs (e.g., CCl_4)
- Hydrogen: one bond, zero lone pairs (an exception to the octet rule)
- Carbocation: C^+ has three bonds, no lone pairs
- Carbanion: C^- has three bonds, one lone pair
- Boron: three bonds, zero lone pairs (an exception to the octet rule) (e.g., BH_3)

Formal Charge

The *formal charge* is the individual charge assigned to each atom in a Lewis structure by equal sharing of bonded electron pairs. Formal charges are used in drawing resonance structures (multiple Lewis structures that collectively describe a single molecule) of covalently bonded molecules. This technique describes, compares, and assess resonance structures, like how oxidation numbers balance chemical equations in oxidation-reduction reactions.

The formal charge is calculated by the formula:

Formal charge = valence e^- in a neutral atom – (unshared valence e^- + ½ of the shared e^-)

where e^- represents electrons.

Every atom in a molecule has a formal charge. In a neutral molecule, the sum of atomic charges is zero. That does not mean the atoms are not charged; they cancel to create a neutral molecule.

In charged molecules, the charge is not evenly distributed, and formal charges can help determine where the charge rests.

Some tips for determining formal charge:

Formal charge =

valence e^- in a neutral atom – (unshared valence e^- + ½ of the shared e^-)

For Lewis structures:

Formal charge = [# of valence e^-] – [dots around atom + lines connected to atom]

The number of valence electrons is generally equal to the atom's group number on the periodic table (e.g., N has five valence electrons, O has six valence electrons, and F has seven valence electrons).

The dots around the atom represent electrons that are held entirely by the atom.

The lines connected to the atom represent bonding electron pairs, in which the atom only gets one of the two electrons.

For formal charges other than zero, label the atom with the formal charge.

Some examples of atoms with formal charges:

- Oxygen with only a single bond: –1
- Oxygen with no bonds but with an octet: –2
- Carbon with only three bonds: +1 as a carbocation

    Carbanion: –1

- Nitrogen with four bonds: +1

    Nitrogen with three bonds: –1

- Halogen with no bonds, but has an octet: –1
- Boron with 4 bonds: –1. (e.g., $^-BH_4$)

Carbon:

=C= —C≡ >C= —C— (with bonds above and below) 4 covalent bonds: Formal charge = 0

Nitrogen:

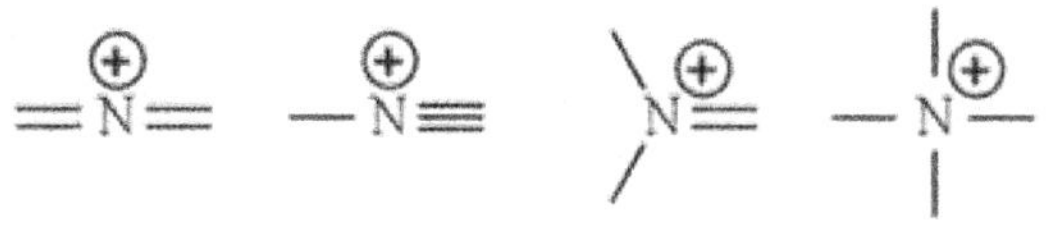

$=\overset{\oplus}{N}=$ $—\overset{\oplus}{N}≡$ $>\overset{\oplus}{N}=$ $—\overset{\oplus}{N}—$ (with bonds above and below) 4 covalent bonds: Formal charge = +1

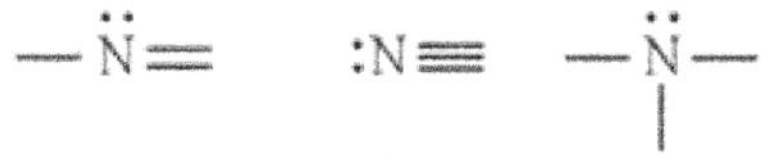

$—\ddot{N}=$:N≡ $—\ddot{N}—$ (with bond below) 3 covalent bonds, 1 lone pair: Formal charge = 0

$—\overset{\ominus}{:\!\ddot{N}}—$ $\overset{\ominus}{:\!N}=$ 2 covalent bonds, 2 lone pairs: Formal charge = −1

Oxygen:

$—\overset{\oplus}{O}=$ $\overset{\oplus}{:\!O}≡$ $—\overset{\oplus}{\ddot{O}}—$ (with bond below) 3 covalent bonds, 1 lone pair: Formal charge = +1

$—\ddot{O}—$ $\ddot{O}=$ 2 covalent bonds, 2 lone pairs: Formal charge = 0

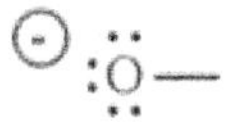

$\ominus\,:\!\ddot{O}—$ 1 covalent bond, 3 lone pairs: Formal charge = −1

The number of valence (outer) shell electrons that an atom needs to gain or lose to achieve a complete octet is "*valence*." In covalent compounds, the number of bonds that are characteristically formed by a given atom is equal to that atom's valence number.

Atom	H	C	N	O	F	Cl	Br	I
Valence	1	4	3	2	1	1	1	1

The valences in the chart above represent the most common form of these elements in organic compounds. Many elements (e.g., chlorine, bromine, and iodine) exist in several valence states in different inorganic compounds.

If the number of covalent bonds to an atom is higher than its typical valence, it carries a positive formal charge.

If the number of covalent bonds to an atom is lower than its typical valence, it carries a negative formal charge.

The following is a step-by-step approach to find the formal charges of each atom in a molecule of nitric acid (HNO_3), using the structure displayed in the image below.

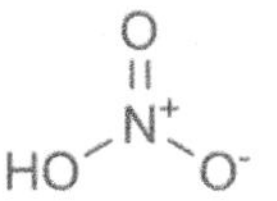

The formal charge for nitric acid (HNO_3) with formal charge shown on the atoms

Formal charge on H

- Hydrogen shares two electrons with oxygen.
- Assign one electron to H and one to O.
- Hydrogen has one valence electron, and $1 - 1 = 0$.
- Therefore, the formal charge of H in nitric acid is zero.

Formal charge on the O bonded to N and H

- Oxygen has four electrons in covalent bonds (2 with N, 2 with H).
- Assign two of these four electrons to O.
- Oxygen has two unshared pairs; assign all four electrons to O.
- Therefore, the total number of electrons assigned to O is $2 + 4 = 6$.
- Oxygen has six valence electrons, and $6 - 6 = 0$.
- The formal charge of the O bonded to H and N in nitric acid is zero.

Formal charge on the double-bonded O

- Oxygen has four electrons in covalent bonds with N.
- Assign two of these four electrons to O.
- Oxygen has two unshared pairs; assign all four electrons to O.
- Therefore, the total number of electrons assigned to O is $2 + 4 = 6$.
- Oxygen has six valence electrons, and $6 - 6 = 0$.

- The formal charge of the double-bonded O in nitric acid is 0.

Formal charge on the single-bonded O

- O has two electrons in a covalent bond.
- Assign one of those electrons to O.
- O has three unshared pairs. Assign all six of these electrons to O.
- Therefore, the total number of electrons assigned to O is 1 + 6 = 7.
- Oxygen has six valence electrons, and 6 – 7 = –1.
- The typical charge of the single-bonded O in nitric acid is –1.

Formal charge of N

- N has eight electrons in covalent bonds.
- Assign four of those electrons to N.
- Nitrogen has five valence electrons, and 5 – 4 = 1.
- Therefore, the formal charge of the N in nitric acid is +1.

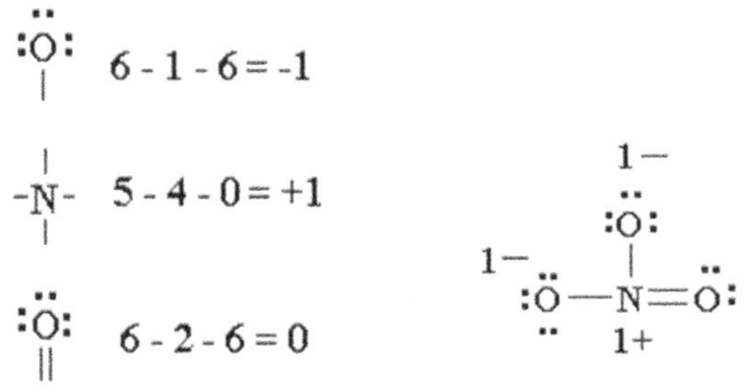

$6 - 1 - 6 = -1$

$5 - 4 - 0 = +1$

$6 - 2 - 6 = 0$

If atoms in a molecule have a formal charge, it does not necessarily mean that the entire molecule has a charge. In the structure of ozone (O_3) below, the central oxygen atom has three bonds and is positively charged. The right-hand oxygen has a single bond and is negatively charged. The overall charge of the ozone molecule is, therefore, zero.

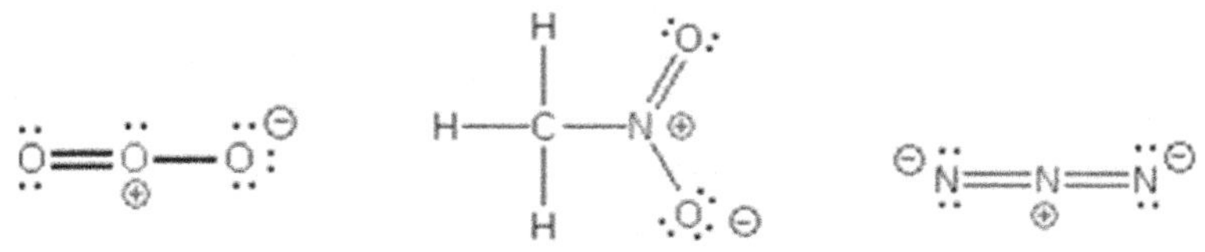

ozone *nitromethane* *azide ion*

Nitromethane (CH_3NO_2) has positively charged nitrogen and negatively charged oxygen; the total molecular charge is again zero.

The azide (N_3^-) anion has two negatively charged nitrogens (2 bonds each) and one positively charged nitrogen (four bonds), and the total charge is minus one.

Resonance structures

When there is more than one possible relatively stable structure for a molecule, it can exist in various forms as *resonance structures*.

Resonance is the averaging of electron distribution over two or more hypothetical contributing structures to produce a hybrid electronic structure. Resonance structures can be used to describe molecules that may contain fractional bonds and charges. The resonance structures must have the same number of paired and unpaired electrons.

No atoms change their positions within the common structural framework, as this would break any existing chemical bond. Only electrons (not atoms, as is the case for isomers) move. Visualize the molecule as quickly "shifting" between each of its possible resonance structures, and intermittently existing in several configurations at any time. However, the molecule spends more time in the most stable resonance form. More accurately, the structure of the molecule is a "combination" (or a weighted average) of its resonance structures, borrowing from the most stable resonance structures.

In a molecule with both a single-and a double-bond resonance structure, the bond length is between a single- and a double-bond length.

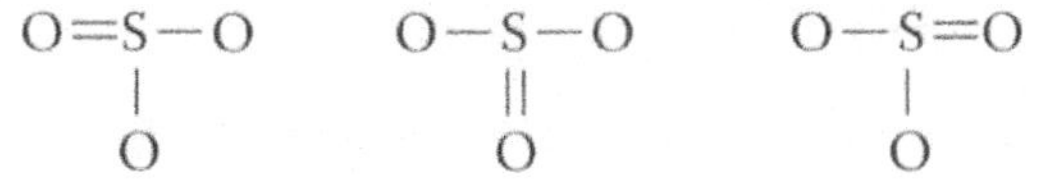

O=S–O (with –O below S)   O–S–O (with =O below S)   O–S=O (with –O below S)

Resonance structures for SO_3^{2-} (single-bonded O with − charge). The images above do not include the formal charge or the double-headed arrows to show the individual contributing resonance structures.

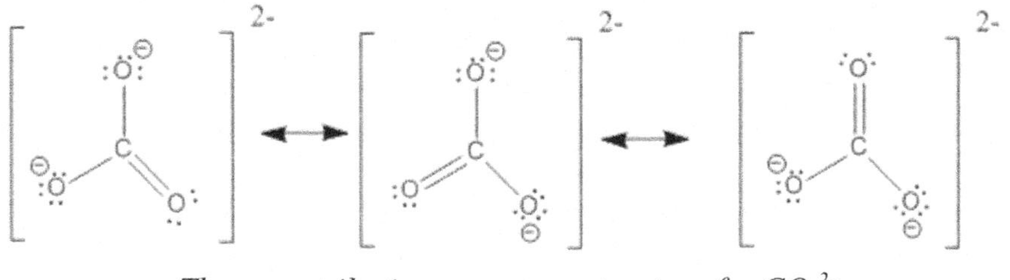

Three contributing resonance structure for CO_3^{2-}

The double-headed arrow seen in the image above for CO_3^{2-} acts as a notation symbol for resonance structures that contribute to the hybrid structure. The principles of resonance are especially useful in rationalizing the chemical behavior of many compounds. The electron delocalization described by resonance significantly enhances the stability (lowers potential energy) of the molecules.

Determine the most stable resonance structure of a molecule when the properties of a stable molecule are known. In stable molecules, the octet group is satisfied in each atom (aside from hydrogen and the other exceptions, like boron).

Neutral molecules are more stable, and if the formal charges on atoms are spread accordingly (i.e., + charge on a positive atom, – charge on electronegative atoms).

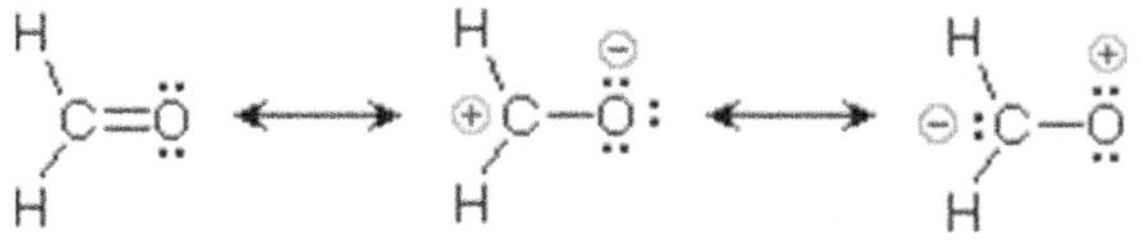

Resonance structure for formaldehyde (CH_2O)

In the example above, the neutral molecule is the most stable. For the charged species, the preferred charge distribution has a positive charge on the less electronegative atom (carbon). A negative charge is held by the more electronegative atom (oxygen). Therefore, the middle formula represents a more reasonable and stable structure than the one on the right. The stable structure(s) indicates that the molecule likely spends most of the time in this possible resonance structure(s) over the others.

If the double bond is broken heterolytically (both electrons going to one of the bonding atoms), formal charge pairs result, as shown in the other two structures. Since the middle, the charge-separated contributor has an electron-deficient carbon atom; this explains the tendency of electron donors (nucleophiles) to bond at this site.

The application of resonance to this example requires a weighted average of these structures. Consequently, if one structure has much greater stability than the others, the hybrid closely resembles it both electronically and energetically. If two or more forms have identical low energy structures, the resonance hybrid is exceptionally stable.

Examples of low energy molecules stabilized by resonance hybrid structures include sulfur dioxide (SO_2) and nitric acid (HNO_3).

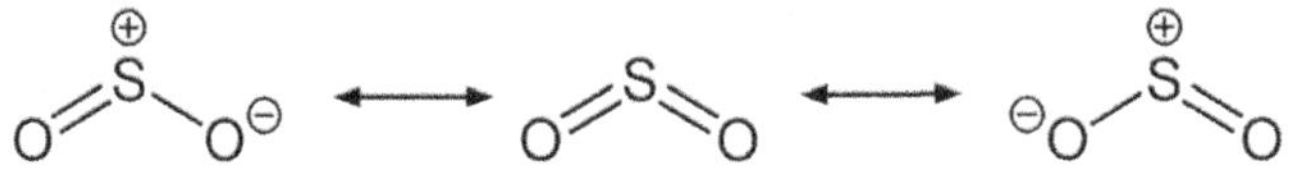

Three resonance forms of sulfur dioxide (SO_2) are shown; the central structure is neutral and therefore the greatest contributor (most molecules exist in this state)

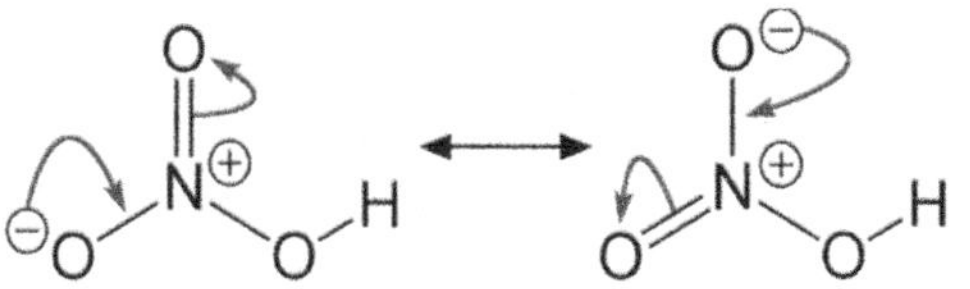

Two resonance forms of nitric acid (HNO_3); both structures are equal in energy and have the nitrogen as positive, and the negative charge is shared between the two oxygens.

Three considerations for drawing resonance structures:

1. The number of covalent bonds in a structure - the greater the bonding, the more essential and stable the contributing structure.

2. Formal charge separation - charge separation decreases the stability and importance of the contributing structure.

3. Evaluate the electronegativity of charge bearing atoms and charge density. The positive charge is accommodated on atoms of low electronegativity. The negative charge is accommodated on highly electronegative atoms.

Lewis acids and bases

Molecules can be categorized based on the number of their free electron pairs:

Lewis acids accept electron pairs.

Lewis bases donate electron pairs.

Acids have vacant orbitals and do not have lone pairs on the central atom (e.g., BF_3), while Lewis bases do have lone pair electrons (e.g., NH_3).

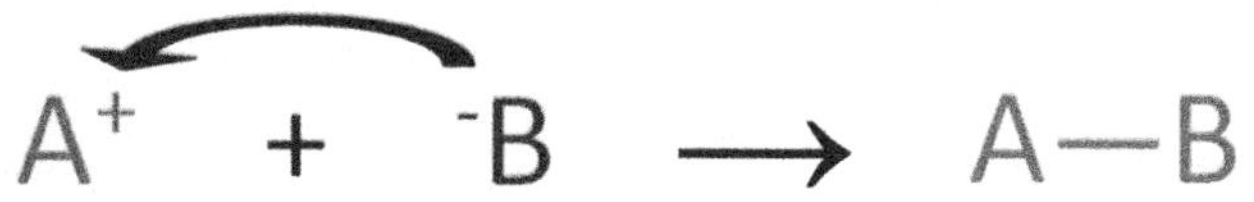

$$A^{+} + {}^{-}B \longrightarrow A{-}B$$

Lewis acid (A): electron acceptor   Lewis base (B): electron donor

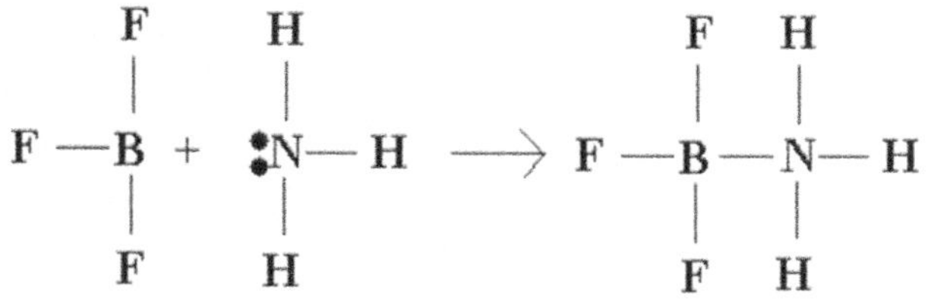

$$F_3B + {:}NH_3 \longrightarrow F_3B{-}NH_3$$

Acid   *Base*

Acids and bases are discussed in the acids and bases chapter.

Notes

Partial Ionic Character

Covalent bonds can be categorized into polar covalent bonds and nonpolar covalent bonds.

Nonpolar covalent bonds occur between atoms of the same element or for atoms with similar electronegativity.

Polar covalent bonds form between atoms with differences in electronegativity. These bonds have *partial ionic character*, due to the distribution of the electrons. One end of the bond is partially negative (δ^-), while the other end is partially positive (δ^+).

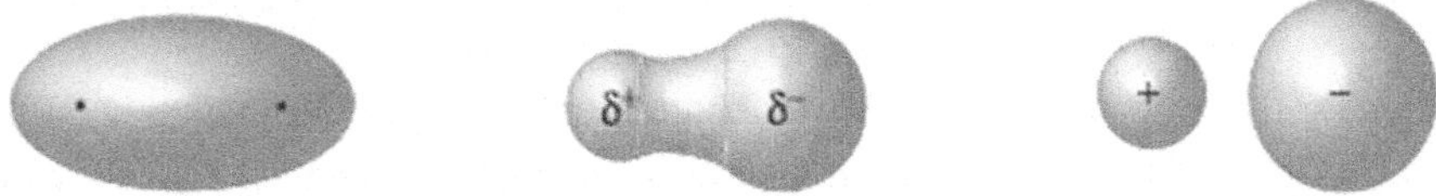

Three types of bonds (covalent, polar covalent, and ionic) are shown above.

The structure above on the left shows a covalent bond between identical atoms.

The center structure is a polar covalent bond between elements of moderately different electronegativities with the notation for the resulting partial positive and partial negative ends of the molecule shown.

The figure on the right shows an ionic bond between elements with significant differences in electronegativities.

Role of electronegativity in determining charge distribution

Electronegativity (EN) is the ability of an element to attract (or hold) onto electrons through bonding.

Due to their differing nuclear charges, and because of shielding by inner electron shells, different atoms of the periodic table have different electronegativities.

The polarity of a molecule is determined by calculating the difference of the EN values of its atoms.

Linus Pauling (Nobel laureate in chemistry, 1954) elucidated the nature of the chemical bond and its application to the structure of complex molecules. Pauling established a quantitative scale of electronegativity values. A larger number (e.g., F = 3.98) signifies a higher affinity for electrons.

H 2.20	**Electronegativity Values for Some Elements**					
Li 0.98	**Be** 1.57	**B** 2.04	**C** 2.55	**N** 3.04	**O** 3.44	**F** 3.98
Na 0.90	**Mg** 1.31	**Al** 1.61	**Si** 1.90	**P** 2.19	**S** 2.58	**Cl** 3.16
K 0.82	**Ca** 1.00	**Ga** 1.81	**Ge** 2.01	**As** 2.18	**Se** 2.55	**Br** 2.96

For the periodic table, EN values increase from left to right, from bottom to top

Higher electronegativity values are located towards the nonmetal side (right side) of the periodic table. Fluorine has the highest electronegativity (3.98). The heavier alkali metals (e.g., potassium, rubidium, cesium) have lower electronegativity values.

Carbon has an electronegativity of 2.55. This value is roughly mid-range for electronegativity and is slightly more electronegative than hydrogen (2.20).

Of all the subatomic particles (protons, neutrons, electrons) of an atom, only the electron has the freedom to move within the atom. When two atoms bond, most of their electrons congregate between the two nuclei.

However, the distribution depends on the electronegativity of the atoms.

Shared electrons in a bond are attracted to the more electronegative atom. This results in a shift of electron density toward the more electronegative atom.

Nonpolar covalent bonds form between identical atoms as *homonuclear diatomic molecules*. For example, in diatomic hydrogen (H_2), both atoms are identical; each has the identical EN value. The electrons, therefore, are evenly distributed between the two hydrogens.

Hydrogen fluoride (HF) has atoms of two different EN values – fluorine (3.98) is much more electronegative than hydrogen (2.20). Therefore, the electrons are more attracted to the fluorine nucleus and are closer to the F than H.

H : F   H F

The electron density for the H–F bond with the greatest density on the electronegative F

If one atom in a molecule is more electronegative than the other, this causes a higher concentration of electrons to locate on one side of the bond, and the molecule is polar (i.e., dipole present).

The *polarity* in a molecule is proportional to the difference in the electronegativity values between the bonded atoms, as shown below.

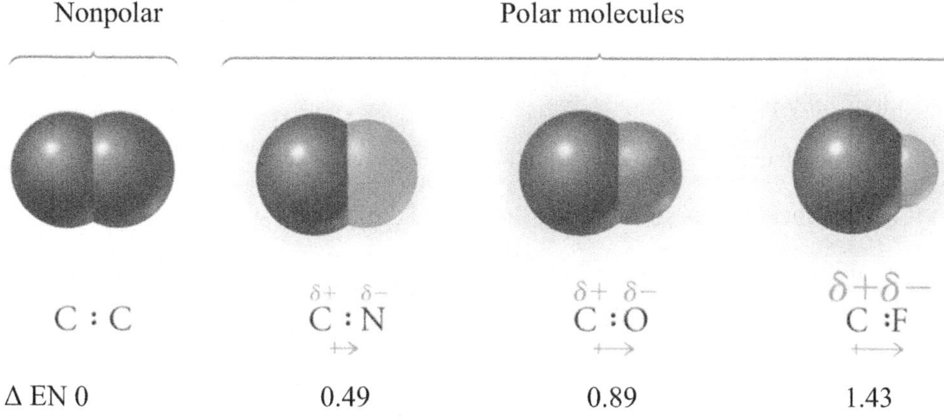

The lowercase Greek letter delta δ denotes partial charges on a polar molecule. A plus or minus sign indicates the partial positive or negative charges.

$$\delta^{+}\ \text{H–Cl}\ \delta^{-}$$

Arrows may also be used to indicate the direction of partial charges. In the diagram below, the arrow is pointing in the direction of electron movement, originating in the positively charged atom and pointing toward the negatively charged atom.

δ- δ+ C—H   δ- δ+ O—H   δ+ δ- C—Cl   δ- δ+ C—Li

The partial positively δ^{+} and negatively δ^{-} charged ends are known as dipoles.

If the difference in electronegativity is high enough in a direction, the molecule acquires a net dipole moment (geometric polarity).

Dipole moment

The *dipole moment* is the measure of polarity in a molecule; it is calculated using the difference in electronegativity (EN) values of each participating or bonded atom.

The geometric shape of a molecule also affects its overall dipole moment.

A molecule might have polar bonds, but if those bonds are arranged in such a way that cancels out the dipole moments, the molecule has a zero net dipole moment and is nonpolar.

A molecule is only nonpolar if:

1) It is symmetrical.

2) It has identical atoms surrounding the central atom.

3) Any non-bonding electron pairs are evenly distributed and cancel (as seen in the illustration below).

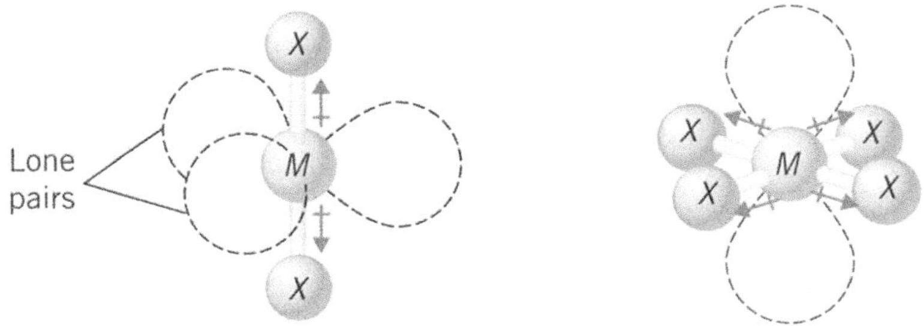

Symmetrical molecules have no net dipole even with individual bond polarity

In the right-hand linear configuration of the molecule below (bond angle 180°), the bond dipoles cancel, and the molecular dipole is zero.

For other bond angles (90° to 120°), the molecular dipole varies in size.

The molecular dipole is the largest at the 90° configuration.

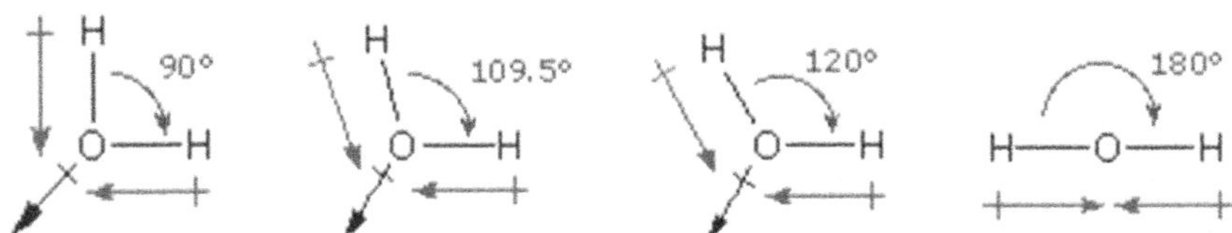

For example, boron trichloride (BCl_3) has three B−Cl bonds. B−Cl is a polar bond, but because the molecule is symmetrical and charges evenly distribute in a single plane, the charges cancel (shown below). BCl_3 does not have any non-bonding electron pairs. As a result, the symmetrical BCl_3 molecule has a zero dipole moment.

Boron trichloride BCl_3 is a symmetric molecule with a zero dipole moment

The configurations of methane (CH_4) and carbon dioxide (CO_2) may be deduced from their zero molecular dipole moments.

Since the bond dipoles have canceled, the configurations of these molecules must be tetrahedral and linear, respectively.

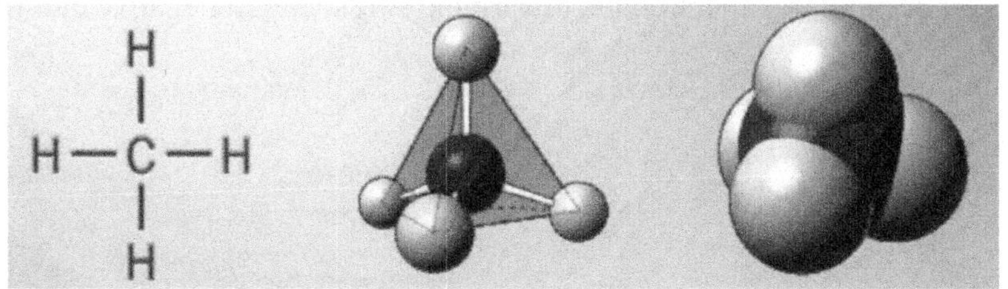

Methane (l to r) as simple drawing, ball-and-stick model, and a space-filling model

Carbon dioxide (CO_2) is a linear molecule and exhibits a net dipole of zero

The methane molecule provides insight into other proposals that confirm its tetrahedral configuration. The substitution of one hydrogen by a chlorine atom gives the methyl chloride CH_3Cl molecule shown below. Since the tetrahedral, square-planar, and square-pyramidal configurations have structurally equivalent hydrogen atoms, they each give a single substitution product. In the trigonal-pyramidal configuration, one chlorine atom at the apex has a different electronegativity than the three hydrogens at the pyramid base.

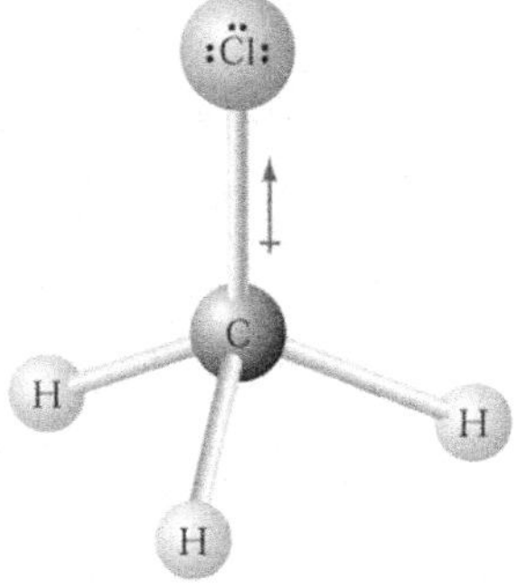

Methyl chloride (CH_3Cl; chloromethane) with dipole pointing towards the single chlorine

Substitution, in this example, should give two CH_3Cl compounds when hydrogen reacts. The tetrahedral configuration of methane leads to a single dichloromethane CH_3Cl product below.

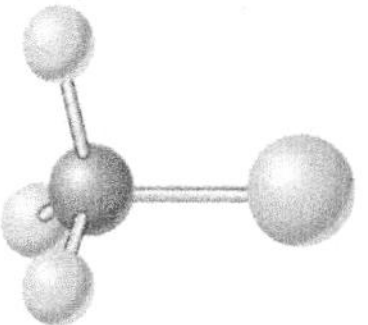

Methyl chloride CH_3Cl is a refrigerant and is also used to manufacture synthetic rubber and silicones. The dipole moment points toward the single electronegative chlorine atom.

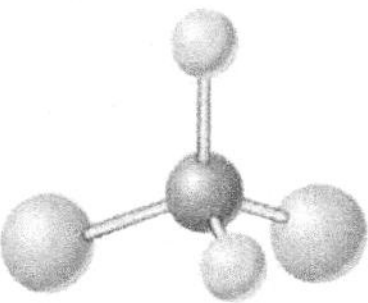

Dichloromethane (methyl chloride; above) CH_2Cl_2 is an industrial solvent used to remove paint and varnish. The dipole moment points toward the two electronegative chlorine atoms. The vector is resolved as between the two chlorines and away from the two hydrogens.

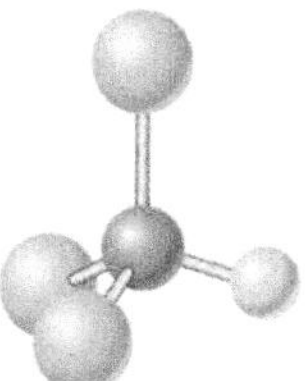

Trichloromethane (chloroform) $CHCl_3$ is a solvent to remove resins and adhesives. The dipole moment points toward the three electronegative chlorine atoms. The vector is resolved as between the three chlorines and away from the hydrogens.

Most covalent compounds show some degree of local charge separation, resulting from differences in electronegativity between the atoms.

If a molecule has polar bonds arranged so that the bonds are polar in a specific direction, then the molecule has a net dipole in that same direction. Some examples of polar molecules are PCl_3 and HCN.

Phosphorous trichloride (PCl_3) has a trigonal pyramidal molecular shape

The phosphorous trichloride molecule has three chlorine atoms positioned on one side. Since chlorine is very electronegative, the electrons are attracted to the chlorine side, creating a dipole moment.

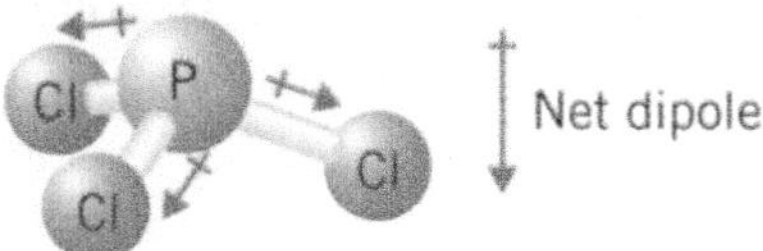

Phosphorous trichloride exhibits a net dipole because of the asymmetric pyramidal shape

Hydrogen cyanide (HCN) is a symmetrical molecule with one atom on each side of the central carbon. Nitrogen has stronger electronegative properties than hydrogen, which means electrons concentrate on the nitrogen side of the molecule.

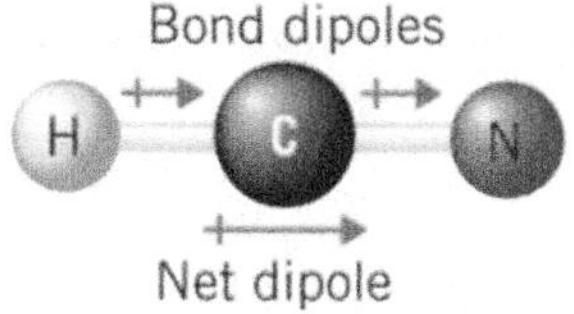

Some physical properties are affected by molecular polarity, including melting and boiling points. Polar molecules tend to have stronger intermolecular forces, which results in higher boiling and melting points (displayed in the table below).

Boiling Points of Polar (high BP) and Nonpolar (low BP) Substances

Substance	**Boiling Point (°C)**
Polar	
Hydrogen fluoride, HF	20
Water, H_2O	100
Ammonia, NH_3	–33
Nonpolar	
Hydrogen, H_2	–253
Oxygen, O_2	–183
Nitrogen, N_2	–196
Boron trifluoride, BF_3	–100
Carbon dioxide, CO_2	–79

A critical example of polarity is the water (H_2O) molecule (shown below). The center oxygen of water has two pairs of non-bonding electrons, two hydrogens (one on each side), and is theoretically symmetrical if the two lone pairs are ignored.

VSEPR theory predicts that water has a bent shape, and the electron pairs will be grouped, as seen in the images below.

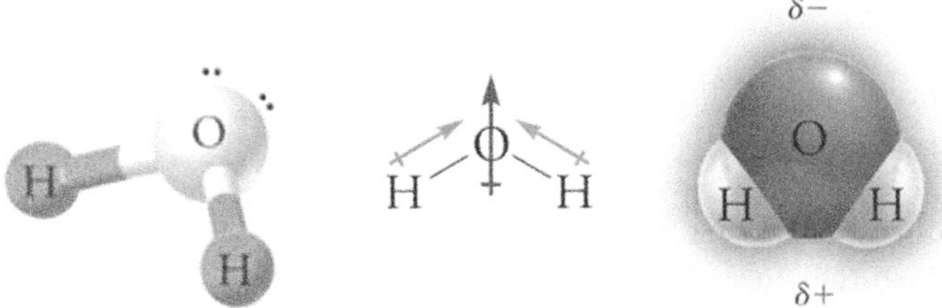

Water (left image) is shown with the two lone pairs. The dipole moment is indicated in the central structure. The space-filling model of water with the partial negative and partial positive ends of the molecule is shown on the right.

For water, the electron pairs are not evenly distributed; they are shifted downwards and concentrated on one side of the oxygen atom. That side has a partial negative charge, while the opposing side has a partial positive charge. Therefore, the water molecule has a net dipole moment and is a polar molecule.

The polar water molecules exert strong intermolecular forces on each other, known as hydrogen bonds.

Hydrogen bonds are a particular class of strong dipole-dipole interactions that occurs when a hydrogen atom is bonded *directly* to one of the three most electronegative atoms (F, Cl, N).

Practice Questions

1. Unhybridized *p* orbitals participate in π bonds as double and triple bonds. How many distinct and degenerate *p* orbitals exist in the second electron shell, where n = 2?

A. 3 **B.** 2 **C.** 1 **D.** 0 **E.** 4

2. Which type of attractive forces occurs in all molecules regardless of the atoms they possess?

A. Dipole–ion interactions
B. Ion–ion interactions
C. Dipole–dipole attractions
D. Hydrogen bonding
E. London dispersion forces

3. Ignoring other factors, it is possible to make a rough approximation of the energy released when an ionic bond forms, simply by considering how much the electrostatic potential energy changes while forming the bond (i.e., bringing the two ions together from infinity). Using this approximation, what is the amount of energy released when 1 mole of NaCl is formed? For two ions with charges q_1 and q_2 that are separated by distance r, the electrostatic potential energy is given by $U_e = k \cdot q_1 q_2 / r$, where the electrostatic constant $k = 9 \times 10^9$ J·m / C^2. (The charge on $Na^+ = +e$, the charge on $Cl^- = -e$, the charge of electron $e = 1.602 \times 10^{-19}$ C, and the separation distance $r = 282$ pm)

A. –493 kJ/mol
B. –288 kJ/mol
C. +91.2 kJ/mol
D. +421 kJ/mol
E. –464 kJ/mol

4. During strenuous exercise, why does perspiration on a person's skin forms droplets?

A. Ability of H_2O to dissipate heat
B. High specific heat of H_2O
C. Adhesive properties of H_2O
D. Cohesive properties of H_2O
E. High NaCl content of perspiration

5. Based on the Lewis structure, how many polar and nonpolar bonds are present in H_2CO?

A. 1 polar bond and 2 nonpolar bonds
B. 2 polar bonds and 1 nonpolar bond
C. 3 polar bonds and 0 nonpolar bonds
D. 0 polar bonds and 3 nonpolar bonds
E. 2 polar bonds and 2 nonpolar bonds

6. Which of the following occur(s) naturally as nonpolar diatomic molecules?

I. sulfur II. chlorine III. argon

A. I only
B. II only
C. I and III only
D. I and II only
E. I, II and III

7. In the nitrogen monoxide molecule, the dipole moment is 0.16 D, and the bond length is 115 pm. What is the sign and magnitude of the charge on the oxygen atom? (Use the conversion factor of 1 D = 3.34×10^{-30} C·m and charge of 1 electron = 1.602×10^{-19} C)

A. $-0.098\ e$ **B.** $-0.71\ e$ **C.** $-1.3\ e$ **D.** $-0.029\ e$ **E.** $+1.3\ e$

8. From the electronegativity below, which single covalent bond is the most polar?

Element:	H	C	N	O
Electronegativity	2.1	2.5	3.0	3.5

A. O–C **B.** O–N **C.** N–C **D.** C–H **E.** C–C

9. Under what conditions is graphite converted to diamond?

A. low temperature, high-pressure
B. high temperature, low-pressure
C. high temperature, high-pressure
D. low temperature, low-pressure
E. none of the above

10. An ion with an atomic number of 34 and 36 electrons has what charge?

A. +2 **B.** –36 **C.** +34 **D.** –2 **E.** neutral

11. Which of the following pairings of ions is NOT consistent with the formula?

A. Co_2S_3 (Co^{3+} and S^{2-})
B. K_2O (K^+ and O^-)
C. Na_3P (Na^+ and P^{3-})
D. BaF_2 (Ba^{2+} and F^-)
E. KCl (K^+ and Cl^-)

12. The ability of an atom in a molecule to attract electrons to itself is:

A. ionization energy
B. paramagnetism
C. electronegativity
D. electron affinity
E. hyperconjugation

13. All of the following are examples of polar molecules, EXCEPT:

A. H_2O **B.** CCl_4 **C.** CH_2Cl_2 **D.** HF **E.** CO

14. When NaCl dissolves in water, what is the force of attraction between Na^+ and H_2O?

A. ion–dipole
B. hydrogen bonding
C. ion–ion
D. dipole–dipole
E. van der Waals

15. Which element likely forms a cation with a +2 charge?

A. Na **B.** S **C.** Si **D.** Mg **E.** Br

Detailed Explanations

1. A is correct.

Three degenerate *p* orbitals exist for an atom with an electron configuration in the second shell or higher. The first shell only has access to *s* orbitals.

The *d* orbitals become available from $n = 3$ (third shell).

2. E is correct.

London dispersion forces result from the momentary flux of valence electrons and are present in all compounds; they are the attractive forces that hold molecules together.

They are the weakest of all the intermolecular forces, and their strength increases with increasing size (i.e., surface area contact) and polarity of the molecules involved.

3. A is correct.

Calculate potential energy for each molecule of NaCl:

$$U_e = k\, q_1 q_2 / r$$

$$U_e = [(9 \times 10^9\ \text{J·m·C}^{-2})\cdot(1.602 \times 10^{-19}\ \text{C})\cdot(-1.602 \times 10^{-19}\ \text{C})] / [(282 \times 10^{-12}\ \text{m})]$$

$$U_e = -8.19 \times 10^{-19}\ \text{J per molecule of NaCl}$$

Calculate the energy released for one mole of NaCl:

$$[(-8.19 \times 10^{-19}\ \text{J}) \times (6.02 \times 10^{23}\ \text{mol}^{-1})] = -493{,}077\ \text{J} \approx -493\ \text{kJ/mol}$$

This approximation disregards a few important effects.

For example, the orientation of the crystal lattice will affect the energy, and there are quantum mechanical effects not considered in this model.

4. D is correct.

Water molecules stick to each other (i.e., cohesion) due to the collective action of hydrogen bonds between individual water molecules (intermolecular bonds).

These hydrogen bonds are continually breaking and reforming, a large portion of the molecules are held together by these bonds.

Water also sticks to surfaces (i.e., adhesion) because of water's polarity.

On an extremely smooth surface (e.g., glass), the water may form a thin film because the molecular forces between glass and water molecules (adhesive forces) are stronger than the cohesive forces between the water molecules.

5. A is correct.

The electronegative oxygen pulls electron density away from the carbon atom and creates a net dipole towards the oxygen, resulting in a polar covalent bond. The electronegativity values of carbon and each hydrogen are similar and result in a covalent bond (i.e., about equal sharing of bonded electrons).

O
||
C
H H

6. B is correct.

The octet rule states that atoms of main-group elements tend to combine in a way that each atom has eight electrons in its valence shell.

7. D is correct.

Formula to calculate dipole moment:

$$\mu = qr$$

where μ is dipole moment (coulomb-meter or C·m), q is charge (coulomb or C), and r is radius (meter or m)

Convert the unit of dipole moment from Debye to C·m;

$$0.16 \times 3.34 \times 10^{-30} = 5.34 \times 10^{-31} \text{ C·m}$$

Convert the unit of radius to meter:

$$115 \text{ pm} \times 1 \times 10^{-12} \text{ m/pm} = 115 \times 10^{-12} \text{ m}$$

Rearrange the dipole moment equation to solve for q:

$$q = \mu / r$$

$$q = (5.34 \times 10^{-31}\ \text{C·m}) / (115 \times 10^{-12}\ \text{m})$$

$$q = 4.65 \times 10^{-21}\ \text{C}$$

Express the charge in terms of electron charge (e).

$$(4.65 \times 10^{-21}\ \text{C}) / (1.602 \times 10^{-19}\ \text{C/e}) = 0.029\ \text{e}$$

In this NO molecule, oxygen is the more electronegative atom.

Therefore, the charge experienced by the oxygen atom is negative: –0.029 e.

8. A is correct.

The greater the difference in electronegativity between two atoms in a compound, the more polar is the bond formed. The atom with the higher electronegativity is the partial (δ or delta) negative end of the dipole.

9. C is correct.

The substantial difference between the two forms of carbon (i.e., graphite and diamonds) is mainly due to their crystal structure, which is hexagonal for graphite and cubic for diamond.

The conditions to convert graphite into diamonds are high pressure and high temperature.

That is why creating synthetic diamonds is time-consuming, energy-intensive, and expensive since carbon is forced to change its bonding structure

10. D is correct.

The valence shell is the outermost shell (i.e., highest principal quantum number, n) of an atom.

Valence electrons are those electrons of the outermost electron shell that can participate in a chemical bond.

11. B is correct.

If the charge of O is −1, the proper formula of its compound with K of +1 is KO.

12. C is correct.

Electronegativity is a chemical property that describes an atom's tendency to attract electrons to itself.

The most common use of electronegativity pertains to polarity along the *sigma* (single) bond between bonding atoms (moieties).

The most electronegative atom is F, while the least electronegative atom is Fr.

The trend for increasing electronegativity within the periodic table is up and toward the right (i.e., fluorine).

13. B is correct.

The greater the difference in electronegativity between two atoms in a compound, the more polar of a bond these atoms form. The atom with the higher electronegativity is the partial (delta) negative end of the dipole.

Each C–Cl bond is very polar. The dipole moments of each of the four bonds in CCl_4 (carbon tetrachloride) cancels because the molecule is a symmetrical tetrahedron.

14. A is correct.

When NaCl dissolves in water, ion–dipole is the force of attraction between Na^+ and H_2O.

15. D is correct.

Group 1 (IA) elements (e.g., Li, Na, and K) tend to lose 1 electron to achieve a complete octet to be cations with a +1 charge.

Group 2 (IIA) elements (e.g., Mg and Ca) tend to lose 2 electrons to achieve a complete octet to be cations with a +2 charge.

Group 7 (VIIA) elements (halogens such as F, Cl, Br, and I) tend to gain 1 electron to achieve a complete octet to be anions with a –1 charge.

Chapter 3

Phases and Phase Equilibria

- **Gas Phase**
- **Intermolecular Forces**
- **Phase Equilibria**

Gas Phase

Gas is a fundamental state of matter, distinguished by the vast separation of the individual gas particles. Gases, unlike solids and liquids, have an indefinite shape and volume because they expand to fill their containers. As a result, gases are subject to changes in pressure, volume, and temperature.

Some common elements and compounds exist in the gaseous state under normal conditions of pressure and temperature, such as oxygen (O_2), nitrogen (N_2), and carbon dioxide (CO_2).

A vapor is the gaseous form of a substance that usually exists as a liquid or solid at ordinary pressures and temperatures (e.g., water vapor).

Four variables are needed to describe the gaseous state:

1. Quantity of gas, n (in moles)
2. The temperature of the gas, T (in Kelvin)
3. The volume of gas, V (in liters)
4. The pressure of gas, P (in kPa)

Absolute temperature, Kelvin scale

Temperature is an objective measure of the kinetic energy of particles, and the volume of a gas (i.e., the space that it occupies) is proportional to its temperature. In the 19th century, experiments revealed that gases expand as temperature rises and compress when temperature drops, which led to the determination of the expansion coefficient of gases per degree Celsius. This posed an interesting question of what happens if the temperature is low enough to reach a point where the calculated volume of gas is zero?

This *absolute zero* temperature was calculated to be –273.15 °C. Gases do not reach zero volume because all gases liquefy or solidify before reaching this temperature. Additionally, gas cannot have a volume of zero because the atoms of the gas molecules occupy a certain amount of space.

In 1848, British physicist Lord William Thomson Kelvin (1824-1907) proposed a temperature scale based on this absolute zero value. The Kelvin (K) scale starts with absolute zero as 0 K, and the increments in this scale are identical to the Celsius scale.

To convert from Kelvin to Celsius, add 273.15; and for the opposite conversion, subtract 273.15. The temperature in Kelvin should be converted to Celsius before converting it into Fahrenheit. Also, by convention, Kelvin temperatures are not written with a degree symbol (273 K vs. 273 °C or 273 °F).

Common temperatures expressed in K, °C, and °F

	K	°C	°F
Absolute zero	0	–273	–460
Freezing point of water / melting point of ice	273	0	32
Room temperature	298	25	77
Body temperature	310	37	99
Boiling point of water / condensation of steam	373	100	212

Pressure, simple mercury barometer

The *pressure* is the average force applied to or experienced by a given area. Gas molecules are always in motion, so if gas is put into a container, the gas molecules continuously move around and collide with the walls of the container. Since the force is applied over a specific area, this force is the gas pressure within a container.

Newton's Second Law, $F = ma$, can be used to determine the units of pressure. The SI units of mass (m) and acceleration (a) are the kilogram (kg) and the meter per second squared (m/s^2), respectively.

The unit of force, derived from the equation $F = ma$, is $kg \cdot m/s^2$. This derived unit is Newton (N). The pressure is the force exerted over a specific area. The SI unit of length is a meter (m), so the SI unit for area is m^2. Therefore, the unit for pressure is N/m^2, also known as the Pascal (Pa).

A *barometer* is a device to measure Earth's atmospheric pressure. Dry air creates more pressure than wet air, so dry air is more massive than the same amount of wet air. Consequently, high barometer readings indicate dry air and fair weather. Low readings indicate an increased chance of rain or showers.

A mercury barometer is made by inverting a column that is filled with mercury and placing it in a mercury dish, without allowing any air to enter the tube. Some of the mercury flows out when the tube is inverted, leaving a space in the tube above the mercury. This space is nearly a vacuum.

Therefore, there is no pressure pushing down on the column. The pressure exerted by the mercury column balances the pressure of the atmosphere. The height of the mercury column is about 760 mm above the surface of the mercury in the dish at sea level.

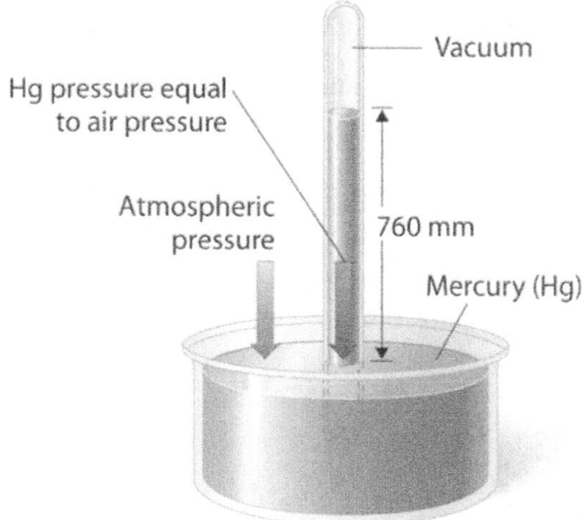

Any liquid may be used to measure atmospheric pressure; the height of the column depends on the density of the liquid. A comparison of mercury (density = 13.6 g/cm^3) and water (density = 1 g/cm^3) indicates that if a barometer were filled with water, the column would be almost 34 feet high.

The pressure exerted by the mercury barometer is recorded in units of millimeters of mercury (mmHg), a unit sometimes called the torr in honor of the Italian physicist Evangelista Torricelli (1608-1647), who invented the mercury barometer in 1643.

The standard atmosphere (atm) is another unit of pressure, and its conversion factors are listed below:

1 atm = 101,325 Pa

1 atm = 760 mm Hg = 760 torr = 76 cm Hg

1 atm = 14.7 lb/in^2 (psi, or pound per square inch)

A bar is a metric unit of pressure equal to 100,000 Pa; it is about equal to atmospheric pressure (101,325 Pa).

A *manometer* is an instrument that also measures the pressure of gases. It is simply a bent piece of tubing. Its principle of operation is like that of a barometer.

There are two major types of manometers.

A closed-tube manometer (shown on the left below) measures pressures below atmospheric pressures, usually the pressure of the gas in a container. Since the end is sealed, it contains a vacuum. The pressure is the difference in the heights of the mercury levels in the two arms.

The open-tube manometer (shown on the right below) measures gas pressures that are near atmospheric pressures. The difference in the heights of the mercury levels in the two arms of the manometer relates the gas pressure to the atmospheric pressure.

The closed tube manometer (left) and open tube manometer (right) are compared.

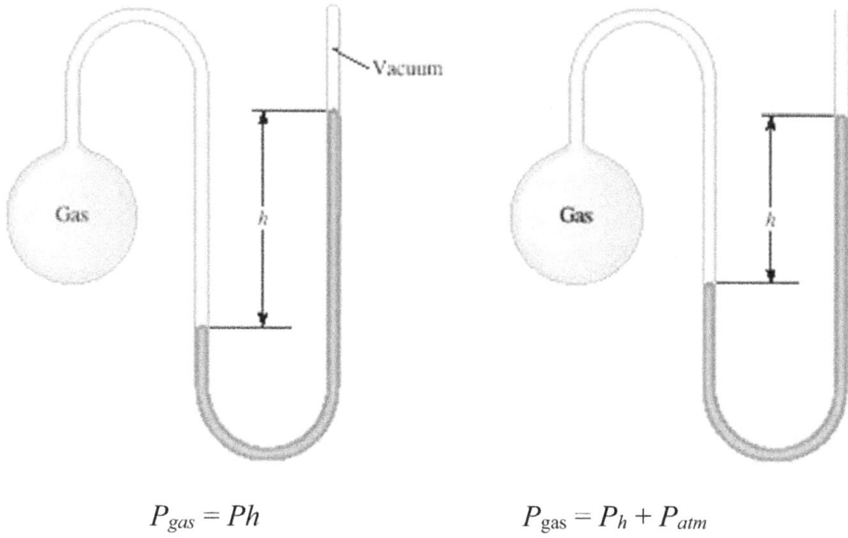

$P_{gas} = Ph$ $\qquad$ $P_{gas} = P_h + P_{atm}$

Two types of manometers used to measure gas pressures.

Left: gas pressure is less than atmospheric pressure (closed tube).

Right: gas pressure is higher than atmospheric pressure (open tube).

Molar volume at 0 °C and 10^5 Pa = 22.7 L/mol

In 1811, Italian scientist Amedeo Avogadro suggested that "equal volumes of gases measured at the same conditions of temperature and pressure contain the same number of molecules."

In chemistry, *molar volume* is usually measured at "standard temperature and pressure" (i.e., STP), which is a temperature of 273.15 K (0 °C, 32 °F) and absolute pressure of exactly 10^5 Pa (100 kPa, 1 bar). At these standard conditions, a mole of gas occupies 22.7 L of space.

Prior to 1982, when the International Union of Pure and Applied Chemistry (IUPAC, the organization that governs chemistry standards) changed their definition of STP, the standard pressure was defined as 101,325 Pa (1 atm). In these conditions, the volume of 1 mol of gas occupies 22.4 L. For general chemistry problems, use the volume of one mole of gas as 22.7 L, unless indicated otherwise.

Kinetic Molecular Theory of Gases

The *kinetic molecular theory* (KMT) uses atomic theory to describe the behavior of gases. This theory was developed throughout over 100 years, eventually culminating in a paper written by German physicist Rudolf Clausius (1822-1888) published in 1857.

The kinetic molecular theory explains the macroscopic properties of a gas (e.g., pressure and temperature) in terms of its microscopic components (e.g., molecules or atoms).

The kinetic molecular theory of gases originated in the ancient idea that matter consists of tiny invisible atoms in rapid motion. Between the 17th and 19th century, the idea was revived and developed to explain the properties of gases.

The kinetic theory for the behaviors of gases requires five critical assumptions.

1. Gas molecules move in random molecular motion. They continue in a straight line until they collide with something.

2. The gas molecule is a "point mass," which is a particle so small that its mass is nearly zero. An ideal gas particle has a negligible volume.

3. No attractive intermolecular molecular forces for gases participate.

4. Collisions between gas molecules are "perfectly elastic." Therefore, the total kinetic energy (*KE*) of the gas molecules remains constant before and after the collision, since intermolecular attractive and repulsive forces are nonexistent.

5. The average kinetic energy (*KE*) for all gas particles is proportional only to the absolute temperature, regardless of the chemical identity or atomic mass. At 0 K (absolute zero or –273 °C), the molecules (and even the orbiting electrons) are not moving and have no volume.

The final assumption (#5) for KE is expressed by the equation:

$$\text{KE} = \tfrac{1}{2} mv^2 = (3/2)\, k_B T$$

where *KE* is kinetic energy, *m* is mass, *v* is velocity, *T* is the temperature (Kelvin) and k_B is *Boltzmann's constant* ($k_B = 1.38 \times 10^{-23}$ J/K). Boltzmann's constant relates the macroscopic and microscopic behavior of gases to their particles.

The above equation is significant because it states that the average kinetic energy of a gas particle is proportional to the absolute temperature of the gas. Increasing the temperature of the gas particles increases their total speed and overall energy.

Since gas atoms are infinitesimally small, it is nearly impossible to measure the speed of any given particle accurately. The speed of gas particles is defined by the root-mean-square speed (u_{rms}).

The following equation gives the root-mean-square speed:

$$u_{rms} = \sqrt{\frac{3RT}{MM}}$$

where R is the ideal gas constant, T is the absolute temperature (K), and MM is the molar mass of the gas sample.

Root-mean-square speed can be written in terms of the Boltzmann constant, k_B:

$$u_{rms} = \sqrt{\frac{3kT}{m}}$$

where k is the Boltzmann constant, T is the absolute temperature (K), and m is the mass of one molecule of gas.

At a given temperature, a plot of the distribution of speeds of each gas particle shows a slightly asymmetric curve. This speed-distribution curve is known as the *Maxwell-Boltzmann distribution curve.*

The Boltzmann distribution plot below shows that the peak of the curve corresponds to the most probable speed.

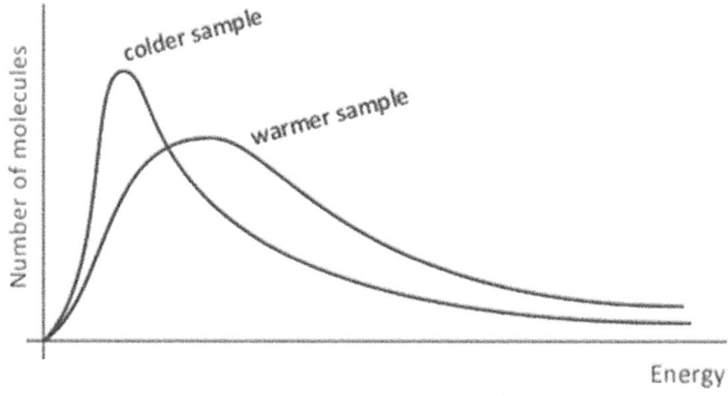

Maxwell-Boltzmann Distribution curves of molecular speeds at two temperatures

If the temperature (K) remains constant, the total kinetic energy remains unchanged. However, the energy can be distributed in many ways, and the gas particles can be traveling at many different speeds at any given point in time. This distribution changes continually as the gas atoms collide with each other and with the container walls.

The curve flattens (i.e., net area under the curve is greater) and shifts to the right at higher temperatures, indicating that a greater number of gas molecules are moving at higher speeds, and therefore possess more kinetic energy.

Diffusion, the process whereby a substance (solute or particle) spreads from a region of high concentration to one of lower concentration, can be explained using the kinetic theory of motion.

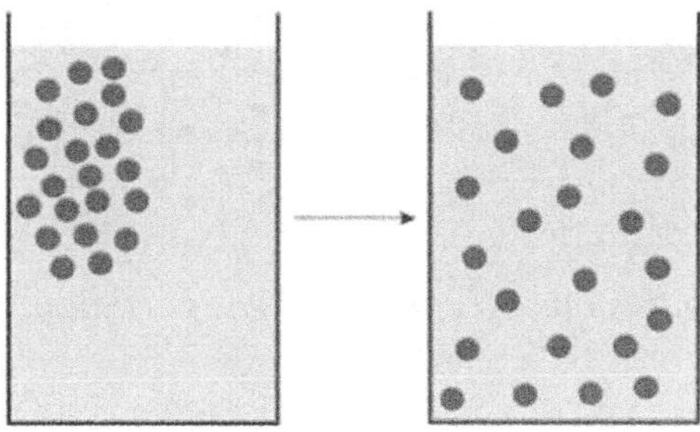

Diffusion of solutes in a solvent

According to the kinetic theory, heavier gases diffuse more slowly than lighter gases because of the difference in the speed at which they travel.

Scottish chemist Thomas Graham (1805-1869) developed *Graham's Law*, which states that *at constant temperature and pressure conditions, the rates at which two gases diffuse are inversely proportional to the square root of their molar masses.*

Graham's Law is expressed by the equation:

$$\frac{r_1}{r_2} = \sqrt{\frac{MM_2}{MM_1}}$$

where r_1 and r_2 are the diffusion rates of gas 1 and gas 2, respectively, and MM_1 and MM_2 are the molar masses of the gases, respectively.

Effusion is the flow of gas particles under pressure from one compartment to another through a small opening, as shown in the figure below.

Graham's Law also applies to the effusion of gas particles, and the equation for effusion is the same as the equation for diffusion.

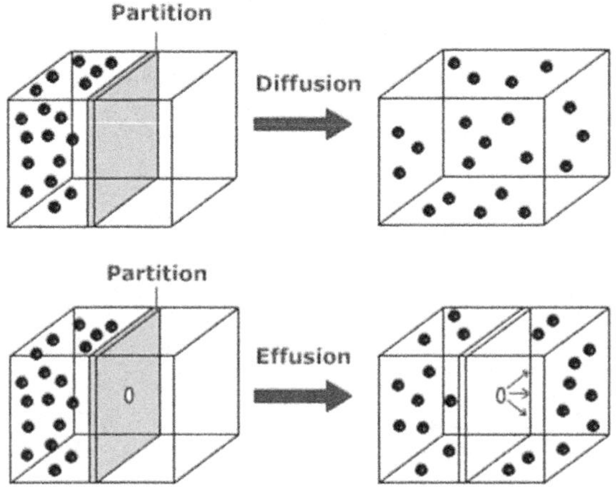

Diffusion and effusion of gas particles

Ideal gas - Definition

The kinetic molecular theory (KMT) described previously is based on a theoretical ideal gas.

An *ideal gas* follows the five assumptions of KMT: gas molecules move in random motion, are point masses with no volume, there are no intermolecular forces, all collisions are elastic, and the kinetic energy of molecules is proportional to temperature.

In comparison to an ideal gas, the behavior of a real gas is complicated. By conceptualizing ideal gas behavior, real gas behavior is easier to comprehend.

In the 17th and 18th centuries, scientists began to realize that there are several relationships between the properties of gases; these are the gas laws.

Boyle's Law, Charles' Law, and Avogadro's Law all relate the properties of pressure, volume, and temperature. These laws were developed empirically and are individual cases of the ideal gas equation, discussed later.

Boyle's Law

In 1661, Robert Boyle (1627-1691) studied the compressibility of gases. He observed that *the volume of a fixed amount of gas at a given temperature is inversely proportional to the pressure exerted on the gas.*

Boyle's Law is expressed by the equation:

$$P_1V_1 = P_2V_2$$

where the subscripts 1 and 2 refer to the same sample of gas under two different sets of pressure and volume conditions.

A plot of volume vs. pressure for a gas illustrates that Boyle's Law is indeed a particular case of the ideal gas law, where n (moles of gas) and T (temperature) are constant, as shown in the following figure.

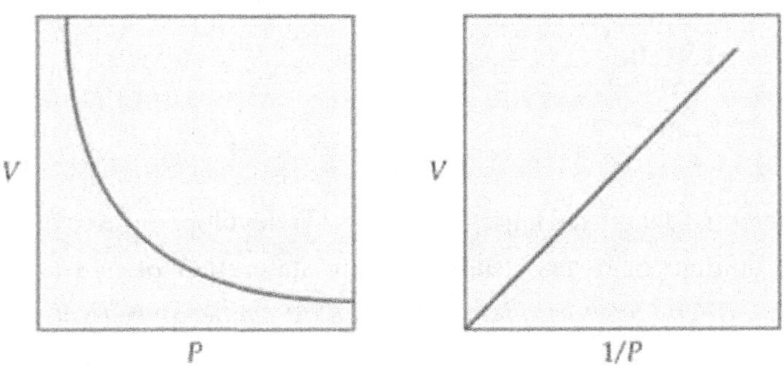

Boyle's Law: as pressure increases, volume decreases

What is the new pressure in a container if gas in a 15.0 L container at a pressure of 5.00 atm has the volume decreased to 0.500 L.?

Solution: Substitute the known quantities (P_1, V_1, and V_2) into the equation for Boyle's Law to solve for P_2. Note: the volume units need to be consistent. In this example, both volumes are expressed in units of liters.

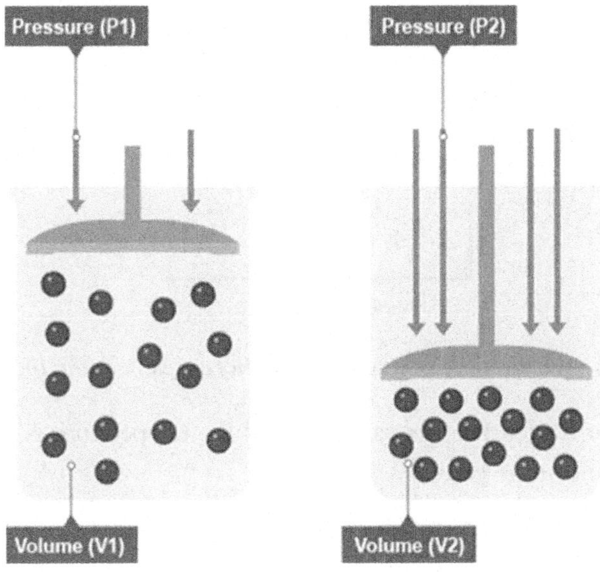

Boyle's Law:

$$P_1V_1 = P_2V_2$$

Solve for P_2:

$P_1V_1 / V_2 = P_2$

$[(5.00 \text{ atm})\cdot(15.0 \text{ L})] / 0.500 \text{ L} = P_2$

$P_2 = 150 \text{ atm}$

Charles' Law

French scientist Jacques Charles (1746-1823) developed the relationship between temperature and volume of a gas. *Charles' Law* states that *at constant pressure, the volume of a gas is directly proportional to its absolute temperature (Kelvin).*

Charles' Law is expressed by the equation:

$$\frac{V_1}{T_1} = \frac{V_2}{T_2}$$

where the subscripts 1 and 2 refer to the same sample of gas but under different conditions for temperature and volume.

A plot of *temperature vs. volume* for a gas illustrates that Charles' Law is another application of the ideal gas law, where n and P are constant, as shown below.

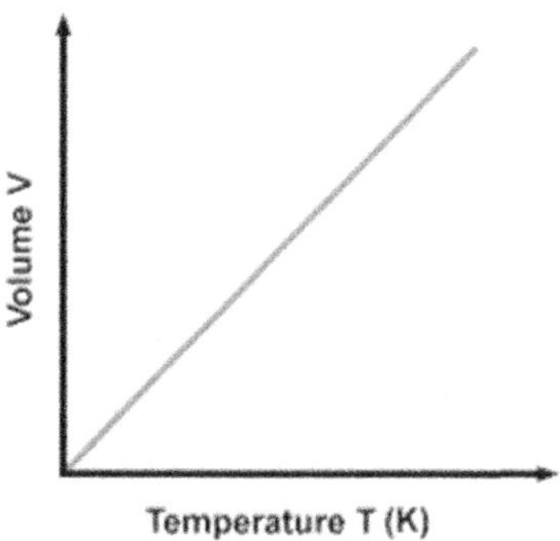

Charles' Law states that as temperature increases, volume increases

Suppose there is 25.0 L of gas at 0 °C, and the temperature is raised to 100 °C. What is the new volume of the gas?

First, the temperature needs to be converted to Kelvin.

$T_1 = 0 \text{ °C} + 273 = 273 \text{ K}$

$T_2 = 100 \text{ °C} + 273 = 373 \text{ K}$

Substitute the known quantities into the equation for Charles' Law.

$V_1 / T_1 = V_2 / T_2$

Solve for V_2:

$V_1\ T_2\ /\ T_1 = V_2$

$(25.0\ L)\cdot(373\ K)\ /\ 273\ K = V_2$

$V_2 = 34.2\ L$

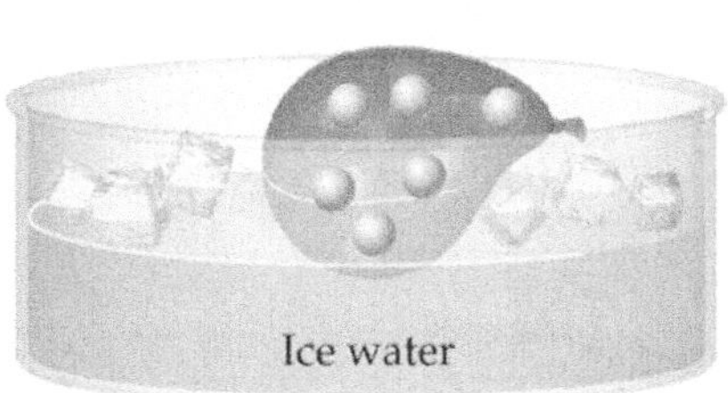

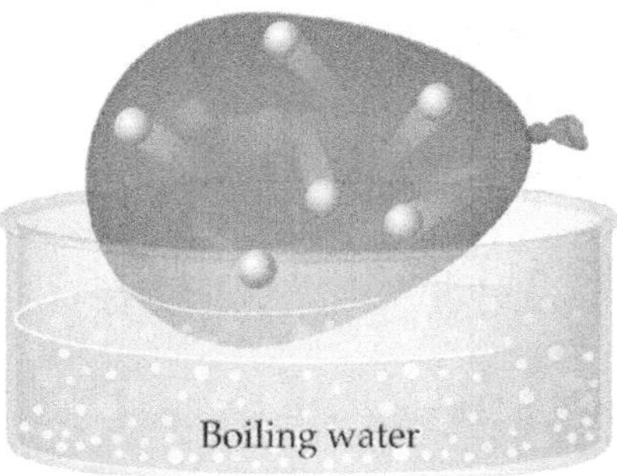

Charles' law states that the volume of a gas is proportional to temperature

Gay-Lussac's Law

In 1802, French chemist Louis Joseph Gay-Lussac (1778-1850) proposed the *Gay-Lussac's Law* that states that *at a constant volume, the pressure of a given sample of gas is directly proportional to its absolute temperature (Kelvin).*

Gay-Lussac's Law is expressed by the equation:

$$\frac{P_1}{T_1} = \frac{P_2}{T_2}$$

where the subscripts 1 and 2 refer to the same sample of gas but under two different sets of temperature and pressure conditions.

A plot of *temperature vs. pressure* for a gas illustrates that Gay-Lussac's Law is another particular case of the ideal gas law, where n and V are constant.

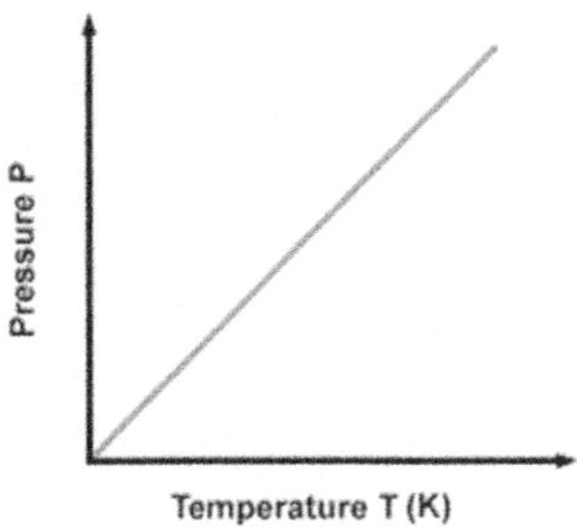

Gay-Lussac's Law states that as temperature increases, pressure increases

Suppose a gas is at 30.0 atm pressure and 100 °C, and the temperature changes to 400 °C. What is the new pressure of the gas?

Convert the temperature to units of Kelvin.

$T_1 = 100\ °C + 273 = 373\ K$

$T_2 = 400\ °C + 273 = 673\ K$

Gay-Lussac's Law solves for P_2 by substituting the given quantities into the equation.

$$\frac{P_1}{T_1} = \frac{P_2}{T_2}$$

$P_2 = [(30.0\ atm)\cdot(673\ K)]\ /\ 373\ K$

$P_2 = 54.1\ atm$

Avogadro's Law

The volume of a gas in a container is affected by pressure, temperature, and amount of gas. The relationship between the quantity of a gas and its volume was derived by the Italian scientist Amadeo Avogadro (1776-1856).

Avogadro's Law states that *all gases at a given temperature and pressure occupy a volume that is directly proportional to the number of moles of gas present.*

Avogadro's Law is expressed by the equation:

$$\frac{n_1}{V_1} = \frac{n_2}{V_2}$$

where n_1 and n_2 are the numbers of moles of gas 1 and gas 2, and V_1 and V_2 are the volumes of the gas 1 and gas 2.

A plot of *volume vs. moles* of gas (shown below) illustrates that Avogadro's Law is a special case of the ideal gas law where T and P are held constant.

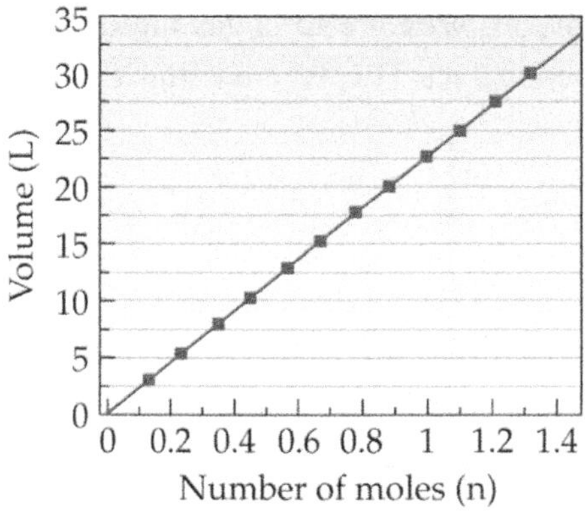

Avogadro's Law states that as moles of a gas increase, volume increases

Suppose that 8.00 moles of a gas occupies a volume of 4.00 L at constant pressure and temperature. What volume of gas would 16.0 moles of this gas occupy at the same temperature and pressure?

Use *Avogadro's Law*

$n_1 / V_1 = n_2 / V_2$

Solve for the V_2

$V_2 = (V_1 \times n_2) / n_1$

$V_2 = [(4.00 \text{ L})\cdot(16.0 \text{ mol})] / 8.00 \text{ mol}$

$V_2 = 8 \text{ L}$

Combined Gas Law

The *combined gas law* is essentially an amalgam of the other gas laws. The combined gas law is *not* the ideal gas law. The combined gas law is expressed as an equation to relate changes in temperature, volume, and pressure of the gas.

The *combined gas law* is expressed by the equation:

$$\frac{P_1V_1}{T_1} = \frac{P_2V_2}{T_2}$$

where the subscripts 1 and 2 refer to the same sample of gas but under different temperature, pressure, and volume.

Suppose a gas is at 15.0 atm pressure, with a volume of 25.0 L and a temperature of 300.0 K. What would the volume of the gas be at standard temperature and pressure? (Use standard pressure = 1.00 atm and standard temperature = 273.0 K)

Use the *combined gas law* at standard temperature (T_2) and pressure (P_2):

$(P_1V_1) / T_1 = (P_2V_2) / T_2$

Solve for V_2

$(P_1 \times V_1 \times T_2) / (T_1 \times P_2) = V_2$

$V_2 = [(15.0 \text{ atm})\cdot(25.0 \text{ L})\cdot(273.0 \text{ k})] / [(300.0 \text{ K})\cdot(1.00 \text{ atm})]$

$V_2 = 341 \text{ L}$

Ideal Gas Law

The *ideal gas law* was proposed in 1834 by French physicist Émile Clapeyron (1799-1864). It is a combination of the gas laws discussed above, and it is the equation regarding the state of a hypothetical gas.

The *ideal gas law* is expressed by the equation:

$$PV = nRT$$

where P is the pressure, V is the volume, T is the temperature, n is the number of moles, and R is the gas constant (use $R = 0.082$ L·atm/K·mol).

Suppose 2.5 moles of gas are at standard temperature and pressure. What is the volume occupied by the gas? (Use standard pressure = 1.00 atm and standard temperature = 273.0 K)

Use the ideal gas law

$$PV = nRT$$

Solve for V

$$V = (nRT) / P$$

$$V = [(2.5 \text{ mol})\cdot(0.082 \text{ L}\cdot\text{atm/K}\cdot\text{mol})\cdot(273.0 \text{ K})] / (1.00 \text{ atm})$$

$$V = 56 \text{ L}$$

From the ideal gas law ($PV = nRT$), several useful expressions are derived that relate the molar mass and density of gases to the pressure and temperature. This involves substituting a different known expression for one of the variables in the ideal gas law.

The moles of a gas n can be expressed as the mass of the gas in grams over the molar mass of that gas:

$$n = m / MM$$

where MM = molar mass, m = mass of the gas in grams and n = moles of gas

Substitute this expression (m/MM) into the ideal gas law for moles (n):

$$PV = nRT$$

$$PV = mRT / MM$$

Multiplying both sides by the molar mass (MM):

$$(MM)PV = mRT$$

This derived equation determines the molar mass of gas from experimental data, where the mass, pressure, volume, and temperature of the gas is measured.

Divide both sides of the above expression by the volume (V):

$$(MM)P = gRT / V$$

Density:

$$\rho = g/V$$

Substitute density for g/V in the derived equation:

$$(MM)P = \rho RT$$

This other derived equation is useful for relating the pressure, density, and temperature of a gas, like the other empirical gas laws.

The density (ρ) of a gas is measured at 1.855 g/L at 0.950 atm and 297.0 K. What is its molar mass of the gas? (Use R = 0.0820 L·atm/K·mol)

Substitute the given quantities into the derived equation above.

$$(MM)P = \rho RT$$

$$(MM)\cdot(0.950 \text{ atm}) = (1.855 \text{ g/L})\cdot(0.0820 \text{ L}\cdot\text{atm/K}\cdot\text{mol})\cdot(297.0 \text{ K})$$

Solve for the molar mass (MM) by dividing both sides of the equation by P (0.95 atm).

$$MM = [(1.855 \text{ g/L})\cdot(0.0820 \text{ L}\cdot\text{atm/K}\cdot\text{mol})\cdot(297.0 \text{ K})] / (0.950 \text{ atm})$$

$$MM = 47.6 \text{ g/mol}$$

Deviation of real-gas behavior from the ideal gas law: a) qualitative and b) quantitative (van der Waals equation)

The kinetic molecular theory and the ideal gas law are approximations of gas behavior. These theories are based on the theoretical ideal gas, while real gases have deviations from this behavior (five critical assumptions of an ideal gas).

When molecules are far apart (under conditions of low pressure and high temperature), a real gas behaves more like an ideal gas.

When molecules are brought close together (under conditions of high pressure and low temperature), the gas molecules experience an intermolecular attraction. Therefore, the gas may deviate significantly from the behavior of ideal gas.

The most substantial deviations from the ideal occur at high pressure and low temperature. At these conditions, the gas molecules are "squished" together, and thus experience intermolecular interactions. The molecular volume becomes significant when the total volume is *squished*. The intermolecular attractions cause collisions to be sticky and inelastic. Furthermore, at extremely high pressures and low temperatures, gases condense into liquids.

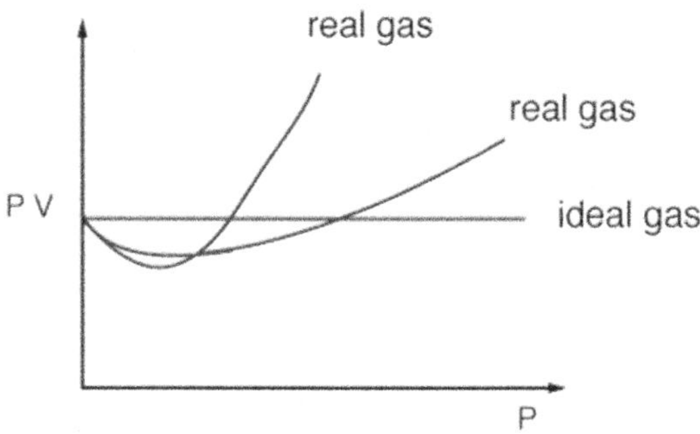

An ideal gas is a point mass. A point mass is a particle so small that the volume of that particle is negligible. A real gas particle has a volume because real gases liquefy as they cool and cannot compress to zero volume; the molecules of a gas are not dimensionless points. For an ideal gas, the collisions between gas particles are "elastic" with no attractive or repulsive forces. Thus, no energy is exchanged during the collisions. For a real gas, collisions are inelastic.

In 1873, Dutch theoretical physicist Johannes Diderik van der Waals (1873-1923) developed an equation that describes the behavior of real gases and their condensation to the liquid phase.

The *ideal gas equation* for comparison:

$$PV = nRT$$

"Corrections" are made to the pressure (P) and volume (V) terms where P becomes $P + (n^2a) / V^2$ and V becomes $V - nb$.

The *van der Waals equation* is expressed by:

$$[P + (n^2a) / V^2] \times (V - nb) = nRT$$

Since the collisions of real gases are inelastic, the term n^2a / V^2 corrects for the interactions between gas particles.

Since real gas particles occupy space (volume), the nb term corrects for the volume neglected by the ideal gas law.

The values of a and b are constant and are determined experimentally for each gas. The constants a and b would be given in a problem, and only pressure or temperature can be solved easily. Solving for the volume is nontrivial and involves solving a cubic polynomial equation.

For pressure, the van der Waals equation is rearranged by dividing both sides of the equation by the volume term ($V - nb$):

$$P + (n^2a) / (V^2) = (nRT) / (V - nb)$$

Subtract the intermolecular interaction term (n^2a / V^2) from both sides of the equation:

$$P = [(nRT) / (V - nb)] - [(n^2a) / (V^2)]$$

Repulsion   Attraction

For example, use the real gas law to find the pressure of 2.00 moles of carbon dioxide (CO_2) gas at 298 K in a 5.00 L container. (Use the van der Waals constants for CO_2 whereby $a = 3.592$ L^2·atm/mol^2 and $b = 0.04267$ L/mol).

Substitute the variables into the equation and solve for the pressure (P):

$$P = [(nRT) / (V - nb)] - [(n^2a) / (V^2)$$

$$P = [(2 \text{ mol})\cdot(0.0821 \text{ L}\cdot\text{atm})\cdot(298 \text{ K})] / [(5 \text{ L}) - (2.00 \text{ mol})\text{mol}\cdot\text{K}(0.04267 \text{ L})]$$

$$- [(2 \text{ mol})^2\cdot(3.592 \text{ L}^2\cdot\text{atm}) / (5.00 \text{ L})^2\cdot\text{mol}^2]$$

$$P = 9.38 \text{ atm}$$

Compare this value (9.38 atm) to the calculated pressure using the ideal gas law:

$$PV = nRT$$

$$P = nRT / V$$

$$P = [(2.00 \text{ mol})\cdot(0.0821 \text{ L}\cdot\text{atm})\cdot(298 \text{ K})] / (5.00 \text{ L})\text{mol}\cdot\text{K}$$

$$P = 9.77 \text{ atm}$$

Although the results from the two equations are similar, they are not identical, illustrating the difference in the behavior of a real gas (not obeying the five stated assumptions) from an ideal gas.

Partial pressure, mole fraction

There is so much distance between the molecules of gas that the molecules of another gas share the space. Different gases combine to form homogeneous mixtures.

The *partial pressure* of a gas is the pressure that the gas would exert if it were the only gas in the container. Therefore, each gas behaves independently of the other and makes its unique contribution to the total pressure.

The *mole fraction* indicates the ratio between moles of a specific gas component and the total number of moles in the homogeneous mixture.

The symbol X_A expresses the mole fraction of the component gas in the mixture.

$$X_A = (\text{moles of gas A}) / (\text{total moles of the gas mixture})$$

Dalton's Law relating partial pressure to the composition

When two or more nonreactive gases are present in the same container, they behave relatively independently.

Dalton's Law of partial pressures states that the total pressure of a given mixture of gases is equal to the sum of the partial pressures of the individual gas components.

Dalton's Law of partial pressures is expressed by:

$$P_T = P_A + P_B + P_C \ldots$$

where P_T is the total pressure in the container, and P_A, P_B and P_C equal the partial pressures of gases *A*, *B,* and *C*, respectively.

The partial pressure of a gas is related to its mole fraction by:

$$P_A = X_A P_T$$

where X_A = (moles of gas A) / (total moles of gas)

Suppose there is 1.0 L oxygen at 1.0 atm pressure in a container, 1.0 L nitrogen at 0.5 atm pressure in another container, and 1.0 L hydrogen at 3.0 atm pressure in a third container. What is the total pressure if the gases combine in a single 1.0 L container?

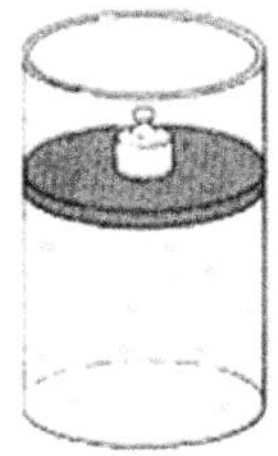

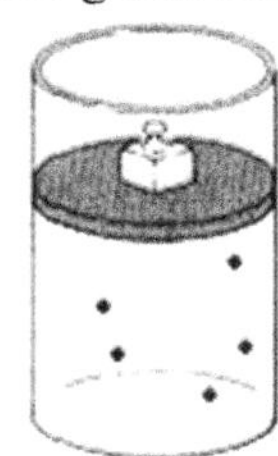

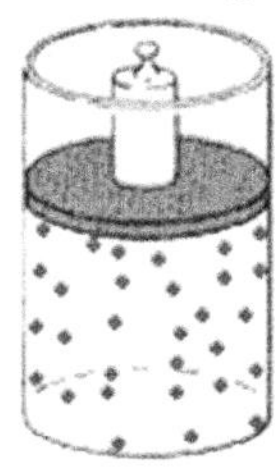

P_{O_2} = 1.0 atm P_{N_2} = 0.5 atm P_{H_2} = 3.0 atm

For a single 1.0 L container, the total pressure is the sum of the partial pressures (X_X) of each individual gas component:

$$X_X = 1.0 \text{ atm} + 0.5 \text{ atm} + 3.0 \text{ atm} = 4.5 \text{ atm}$$

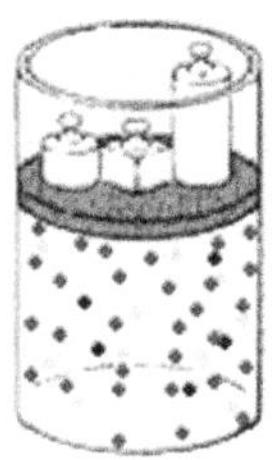

$$P_{Total} = P_{O_2} + P_{N_2} + P_{H_2}$$

$$P_{Total} = 4.5 \text{ atm}$$

Dalton's Law of partial pressures states that *the total pressure of a mixture of gases equals the sum of the individual gas pressures.*

A critical application of the law of Dalton's Law of partial pressures is to determine *vapor pressure* (the pressure exerted by a vapor in thermodynamic equilibrium with its liquid or solid phase). The vapor pressure determines the amount of a water-insoluble gaseous reaction product or a slightly soluble gas such as hydrogen or oxygen.

A typical experimental method (illustrated below) to determine the amount of gas present is by collecting it over water and measuring the height of displaced water; this is accomplished by placing a tube into an inverted bottle, the opening of which is immersed in a larger container of water. As the gas bubbles into the test tube, it displaces the water until the test tube is full. The collected gas is not the only gas in the test tube since liquid water is always in equilibrium with its vapor, so the collected gas in the test tube is a mixture of two gases: the gas collected and the water vapor.

The partial pressure of water is the *vapor pressure of water* and is dependent on temperature. The volume of gas collected consists of a mixture of the gas collected and water vapor.

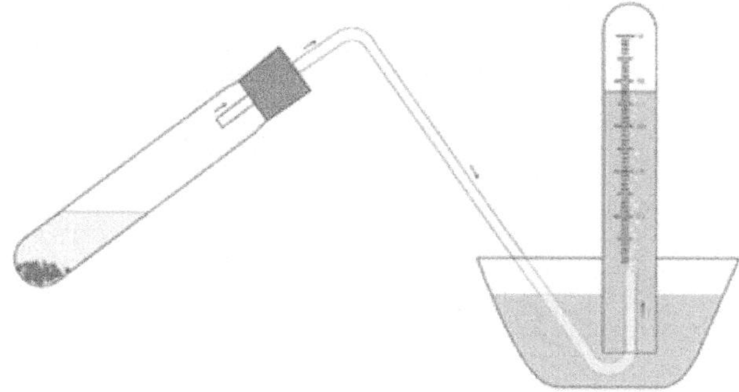

An experimental method to determine the amount of gas present

The total pressure is the sum of the two contributing partial pressures.

From a table of water vapor pressure values at various temperatures, the value of the pressure of dry gas collected can be calculated. The partial pressure of the dry gas collected is calculated by subtracting the reference vapor pressure of H_2O.

The equalizing of the atmospheric pressure yields the partial pressure of the gaseous product collected. With known volume and temperature, the amount of gas product can be determined.

Vapor Pressure of Water at Specific Temperatures

Temperature (°C)	Vapor Pressure (kPa)	Temperature (°C)	Vapor Pressure (kPa)
0	0.61	26	3.36
5	0.87	27	3.57
10	1.23	28	3.78
15	1.71	29	4.00
16	1.82	30	4.24
17	1.94	35	5.62
18	2.06	40	7.38
19	2.20	45	9.58
20	2.34	50	12.33
21	2.49	60	19.92
22	2.64	70	31.16
23	2.81	80	47.34
24	2.98	90	70.10
25	3.17	100	101.3

For example, a small piece of zinc (Zn) reacts with dilute hydrochloric acid (HCl) to form hydrogen (H_2) gas, which is collected over water at 16.0 °C. The total pressure is adjusted to a barometric pressure of 100.24 kPa, and the volume of hydrogen gas is measured as 1,495 cm^3. What is the partial pressure of the hydrogen gas and the mass of the hydrogen gas?

The partial pressure of hydrogen gas:

$$P_{HCl} = 100.24 \text{ kPa} - 1.82 \text{ kPa}$$

$$P_{HCl} = 98.42 \text{ kPa}$$

To calculate the mass of the hydrogen gas, use the ideal gas equation (PV = nRT) and solve for the number of moles. For hydrogen (H_2) gas, multiply the mass of one hydrogen atom (1.008 g/mol) by two to obtain the molecular mass of 2.016 g/mol.

Temperature and volume were given. The partial pressure of hydrogen was calculated in the first part of the problem; the ideal gas equation can now be used.

Determine the number of moles of H_2 gas:

$$PV = nRT$$

$$n = PV / RT$$

$$n = [(98.42 \text{ kPa})\cdot(1.495 \text{ L})] / [(8.314 \text{ L}\cdot\text{kPa}\cdot\text{K}^{-1}\cdot\text{mol}^{-1})\cdot(289 \text{ K})]$$

$$n = 0.061 \text{ mol of } H_2 \text{ gas}$$

Calculate the mass of the H_2 gas formed:

$$\text{mass} = (\text{moles of } H_2 \text{ gas}) \times (\text{molecular mass of } H_2 \text{ gas})$$

$$\text{mass} = 0.061 \text{ mol } H_2 \text{ gas} \times 2.016 \text{ g/mol } H_2 \text{ gas}$$

$$\text{mass} = 0.123 \text{ g of } H_2 \text{ gas}$$

Notes

Intermolecular Forces

Intermolecular forces are attractive or repulsive forces that occur between molecules (e.g., covalent, ionic bonding). However, some other intermolecular interactions are not as strong as covalent or ionic bonds, but still, have a significant role in determining the properties of substances.

Dipole interactions

Molecules with dipole moments (resulting from differences in electronegativity of bonded atoms) are attracted to each other. The more polar the molecule is, the stronger the dipole attraction.

There are four types of dipole attractions:

- *Ion-dipole attractions* form between an ion and a polar molecule. Since a polar molecule has a slight charge (dipole) and an ion is a charged atom, these particles are attracted to each other.

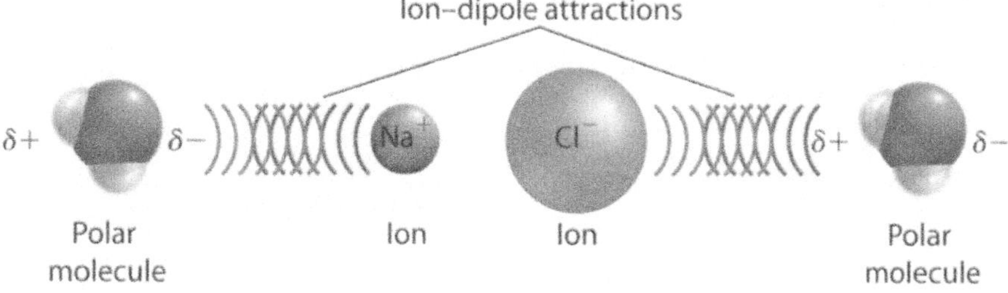

- *Dipole-dipole attraction* form between two polar molecules. The partially positive side of one polar molecule attracts the partially negative side of another polar molecule.

Dipole-dipole

- *Dipole-induced dipole attraction* form between a partially polar molecule and a nonpolar molecule. For example, the polar H_2O molecules can induce the nonpolar O_2 molecules, shifting electron density and inducing a temporary partial positive and negative regions in the nonpolar O_2 molecules.

positive pole ⟶ negative pole

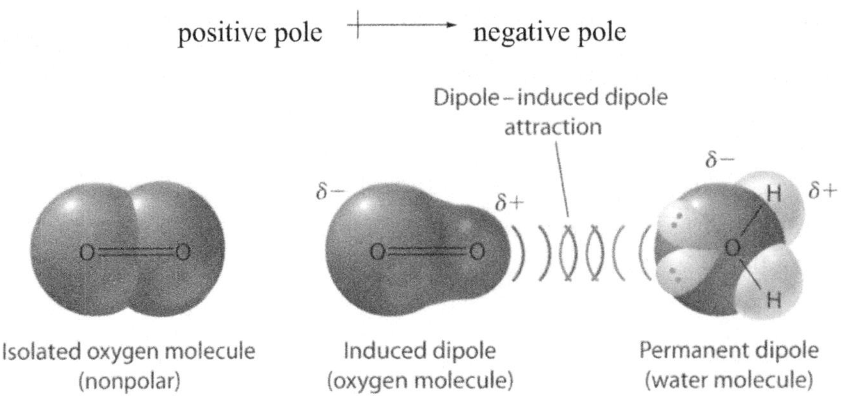

Induced dipole-induced dipole attraction and *London dispersion force* form between two (or more) nonpolar molecules. These nonpolar molecules become temporarily polar (i.e., momentary flux) due to random, instantaneous induction by another polar molecule that results in the probability of finding the negative electron in a specific region within its orbital.

Longer molecules have more surface contact and tend to experience higher London dispersion forces.

For example, octane (C_8H_{18}) has a stronger dispersion force than methane (CH_4).

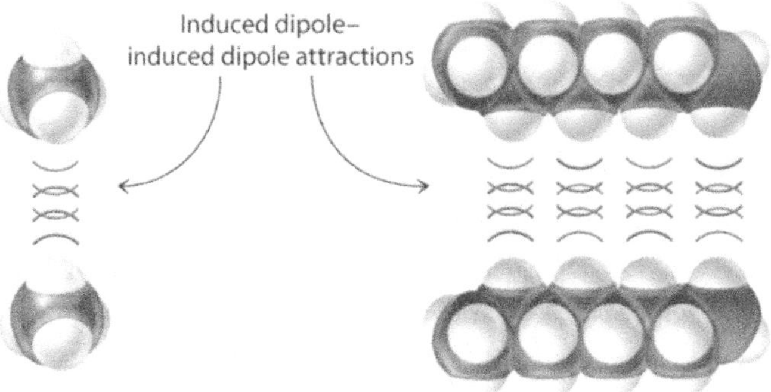

Molecular Attractions Involving Dipoles

Attraction	Relative Strength
Ion–dipole	Strongest
Dipole–dipole	↑
Dipole–induced dipole	
Induced dipole–induced dipole	Weakest

Hydrogen bonding

Hydrogen bonding is a specific type of bond that, based on its strength, is categorized as a dipole-dipole attraction.

Hydrogen (hydrogen bond donor with one valence electron) bonding occurs when hydrogen is covalently bonded directly to the electronegative atom of N, O, or F (hydrogen bond donors). The bonding electrons are closer to the more electronegative N, O, or F atom.

Due to these factors, the hydrogen nucleus is very exposed. The highly positive hydrogen nucleus attracts the electronegative atom from a neighboring molecule. This creates a relatively powerful intermolecular force. The neighboring molecule must possess a lone electron pair to form a hydrogen bond and this molecule that bonds (by the lone pair electrons) to the hydrogen (bonded to F, O, N) is the hydrogen bond acceptor.

The more polar a bond is, the stronger the hydrogen bond.

The H–F bond is the most polar, then the H–O bond, and then the H–N bond.

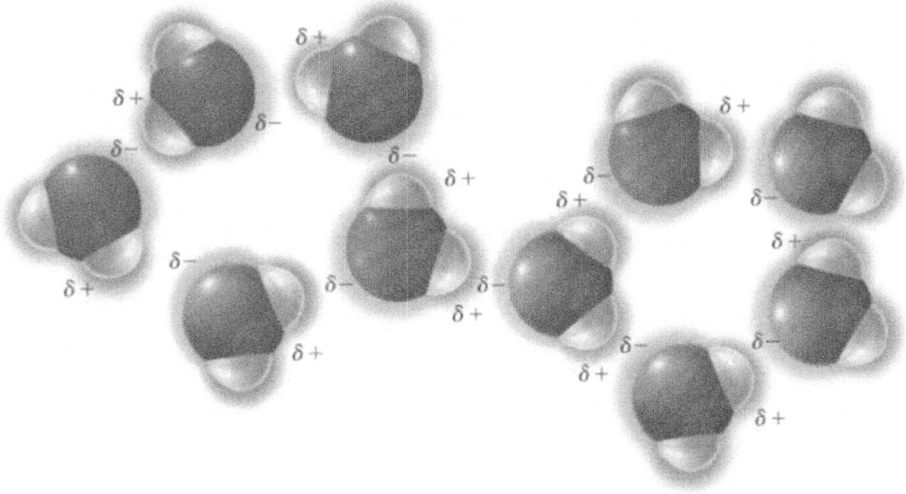

A consequence of hydrogen bonding is differences in physical properties

For example, hydrogen bonding significantly increases the boiling point of compounds. Methanol (CH_3OH) and methane (CH_4) have similar molecular structures.

However, methanol is liquid at room temperature, while methane is gaseous. Methanol is liquid at room temperature because of hydrogen bonding between methanol molecules, which is much stronger than the London dispersion (weak intermolecular) forces between the nonpolar methane molecules. Therefore, methane is a gas at room temperature.

H-bond Donors (have hydrogen) **H-bond Acceptors (have lone pairs)**

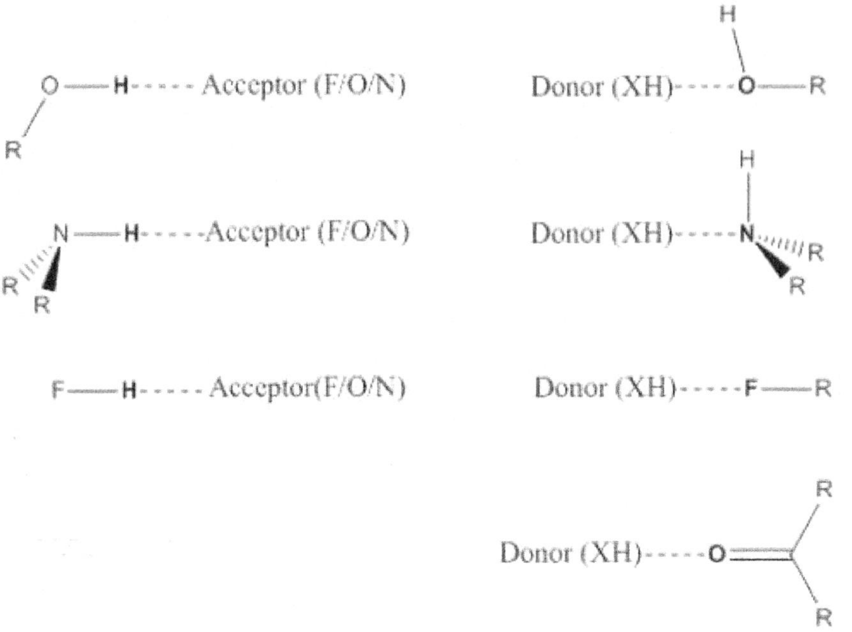

The hydrogen bond donor has a δ^+ hydrogen that is bonded to the electronegative F, O, N.

The hydrogen bond acceptor has a lone pair of electrons used to bond to the δ^+ hydrogen.

H-bond examples

Van der Waals' Forces (London dispersion forces)

Atoms and molecules have a weak intermolecular attraction for each other, known as the *van der Waals forces*. This results from the weak attraction of a molecule's nuclei to another molecule's valence electrons. The van der Waals equation (described previously) accounts for these attractive intermolecular forces. The van der Waals force affects the boiling point of a substance because the boiling point reflects the kinetic energy needed to release a molecule from the cooperative attractions of the liquid.

Nonpolar molecules have fluctuating dipoles that tend to align with those of other molecules from one instant to the next. Van der Waals' forces can result from these temporarily fluctuating molecules ("induced dipole or induced dipole attraction"), and this category of van der Waal's forces is the *London dispersion force*.

London dispersion forces exist for all molecules but are only significant for nonpolar molecules that do not have dipole-dipole, dipole-induced dipole, or hydrogen bonding. London dispersion forces are weak compared to the forces generated between polar molecules. For polar molecules, dipole forces are strong, more influential, and predominant.

In general, larger nonpolar molecules have higher boiling points compared to smaller nonpolar molecules due to the increase in London dispersion forces. The momentary flux of electron density is attracted to the positive charge of nearby nuclei. It decreases dramatically as the distance between the molecules increases. Branched molecules (i.e., hydrocarbons) have weaker London dispersion forces. Also, a larger molecular mass reduces the kinetic energy ($KE = \frac{1}{2}mv^2$) of the larger molecule at the same temperature.

Notes

Phase Equilibria

Phase changes

Most substances can exist in solid, liquid, or gas form.

A *solid* has a definite shape and volume (e.g., ice). The molecules in a solid vibrate about a fixed position, and a solid cannot be compressed.

A *liquid* has a definite volume but takes the shape of its container (e.g., water). The molecules can move about but are close together and bound by intermolecular forces.

A *gas* has neither a definite volume nor a definite shape; it takes both the volume and shape of its container (e.g., steam or water vapor). The molecules fly apart from each other and are not held together by intermolecular forces. Gas is easily compressible.

Freezing point, melting point, boiling point, condensation point

The *freezing point* is the temperature at which a liquid changes state to become a solid (i.e., solidification).

The *melting point* is the temperature when a solid changes state to become a liquid.

Theoretically, the melting point and freezing point of a substance should be identical. Experimentally, a difference in the quantities of the respective processes may be observed.

Vaporization is the phase transition at which a liquid becomes a gas. There are two types of vaporization: boiling and evaporation.

The *boiling point* is when the liquid is heated, and it reaches a temperature (kinetic energy of the molecules) where the vapor pressure is high enough for bubbles to form inside the body of the liquid.

As a liquid is boiling, the temperature remains constant until all the liquid has converted into gas. Since the boiling point is based on pressure, the precise value depends on the environment (e.g., elevation relative to sea level).

For example, at high-altitude, low-pressure locations, the boiling point of a substance is lower. A lower boiling point requires less heat energy to increase the kinetic energy of the vaporizing molecule to break its intermolecular attraction to the adjacent molecule(s).

Evaporation occurs at temperatures below the boiling point. It occurs from the surface of the liquid into a gaseous phase that is not saturated with the substance. This is contrasted with boiling, which occurs from the bulk of the liquid, not just the surface. Evaporation tends to occur more quickly in liquids with higher vapor pressure.

Condensation, the reverse of evaporation, is the change from the gas phase to the liquid phase. Condensation commonly occurs when a vapor is cooled or compressed to its saturation limit, and the temperature at which *condensation* occurs is the condensation point. For example, condensation is the formation of water droplets in atmospheric clouds.

Sublimation, the reverse of deposition, is the phase transition from a solid to a gas, which is essentially evaporation that occurs directly from the solid phase below the melting point.

Deposition, the reverse of sublimation, is the phase transition from a solid directly into a gas (i.e., the matter is not present as an intermediate liquid). This process can be observed in the winter months when the temperature does not rise above freezing. The snow does not thaw to liquid, and yet the snow melts.

Molality

Molality (molal concentration) measures the moles (concentration of solutes) in a solvent in terms of the mass (on kg) of the solvent. The SI unit for molality is mol/kg. Molality is expressed with the symbol *m, which is* different from the uppercase *M* for molarity (i.e., *M* = moles/liter).

Molality = (moles of solute) / (mass in kg of solvent)

For example, if 3.00 moles of sucrose are mixed into 1.00 L water until the sucrose is entirely dissolved (i.e., saturated solution). What is the molality of the solution? (Use the density of water = 1.00 g/mL or 1.00 kg/L).

The mass of the liter of water is 1.00 kg.

Use the solvent mass of 1. Kg/L and solve for molality.

m = 3.00 mol / 1.00 kg

m = 3.00 mol/kg of sucrose

m = 3.00

The molality of the solution is 3.00 m.

More commonly, instead of expressing the number of moles, the mass of the solute is given in grams.

For example, suppose someone had 116.88 grams of sodium chloride (NaCl) and dissolved it into 4.00 kg of water. What is the molality of the solution?

Convert the grams of NaCl into moles (Use the molar mass of NaCl = 58.44 g/mol).

moles = (mass) / (molar mass)

moles = (116.88 g) / (58.44 g/mol)

moles = 2.00 mol NaCl

Solve for molality:

m = (moles of solute) / (mass in kg of solvent)

m = (2.00 mol) / (4.00 kg of solvent)

m = 0.500 mol/kg of the solution

m = 0.500

The molality of the solution is 0.500 mol/kg.

When calculating molality, the mass of the solvent is only of the pure solvent, excluding the mass of the solute. Molarity is the volume of solution, which includes both the mass of the solute and mass of the solvent. The molality is (mol solute/kg solvent), while molarity is (mol solute/L solution).

Colligative properties

Colligative properties are physical properties that depend on the ratio of the solute particles to the solvent, but not on the type of chemical species present.

Colligative properties include:

1) the relative lowering of vapor pressure,

2) the elevation of boiling point,

3) the depression of freezing point, and

4) osmotic pressure.

The colligative properties calculations incorporate the van't Hoff factor (i), which is a measure of the effect a solute on the colligative properties of the solution.

The van't Hoff factor is the ratio of the concentration of particles produced when a substance dissolves to the concentration of a substance calculated from its mass.

Concentration is converted to reflect the total number of particles in the solution. (e.g., 5 g of NaCl dissociates into Na^+ and Cl^-).

For example, glucose ($C_6H_{12}O_6$) has an *i* value of 1 because it does not dissociate in solution. Nonelectrolytes do not dissociate in solution (e.g., CH_4 and O_2). The salt NaCl has an *i* value of 2 because it dissociates into 1 Na^+ and 1 Cl^- (2 resulting ions per one molecule of reactant).

Vapor pressure lowering (Raoult's Law)

In 1888, French chemist Francois-Marie Raoult (1830-1901) proved that the vapor pressure of a solution is equal to the mole fraction of the solvent times the vapor pressure of the pure solvent.

Raoult's Law is expressed by the equation:

$$P = \chi_{solvent} \cdot P°_{solvent}$$

where P is the vapor pressure, $\chi_{solvent}$ is the mole fraction of solvent (moles of solvent / total moles of both solute and solvent), and $P°_{solvent}$ is the vapor pressure of the pure solvent.

Calculate the χ_{solute} using the van't Hoff factor (i.e., 1 mol of NaCl results in 2 mol of ion particles in the solution).

Suppose 2.00 moles of sucrose is added to a pitcher containing 2.00 liters of water. What is the vapor pressure of the sucrose solution? (Use the value 1.00 L water = 1.00 kg, the vapor pressure of pure water = 23.8 mmHg and the molecular mass of H_2O = 18.02 g/mol)

Convert the liters of water into mass:

2.00 L water = 2 Kg

2 Kg = 2,000 g

Convert the mass of the water into moles:

molecular mass = mass / moles

moles = mass / molecular mass

moles = (2,000 g) / (18.02 g/mol)

moles = 110.9 moles H_2O

Solve for the mole fraction, $\chi_{solvent}$

$\chi_{solvent}$ = (moles of solvent) / (total moles of the solute and solvent)

$\chi_{solvent}$ = (110.9 moles H_2O) / (total moles of the solute and solvent)

$\chi_{solvent}$ = 110.9 moles H_2O / (110.9 moles solvent + 2 moles solute)

$\chi_{solvent}$ = 0.98

Use Raoult's Law to determine the pressure.

$P = \chi_{solvent} \cdot P^{\circ}_{solvent}$

$P = (0.98)\cdot(23.8 \text{ mmHg})$

$P = 23.4 \text{ mmHg}$

The addition of a solute caused the vapor pressure to decrease.

Boiling point elevation (ΔTb = Kbm)

Adding a solute to a solvent stabilizes the solvent in the liquid phase, thus lowering the tendency of the solvent molecules to move to the gas or solid phases. Therefore, the boiling point increases.

The *boiling point elevation* is proportional to the decrease of the vapor pressure in a dilute solution:

$$\Delta T_b = k_b \cdot m \cdot i$$

where ΔT_b is the increase in boiling point, k_b is the molal boiling point constant (a value given in the problem because it is determined experimentally), m is the molality (mol solute/kg solvent), and i is the van't Hoff factor.

What is the boiling point elevation if 6.4 g of ammonia is dissolved in 0.3 kg of water? (Use k_b for water = 0.52 °C/m and the molar mass of ammonia = 17.031 g/mol)

Convert the mass of ammonia (NH_3) into moles:

molecular mass = mass / moles

moles = mass / molecular mass

moles = 6.4 g / 17.031 g/mol

moles = 0.38 mol NH_3

Calculate the molality of the solution:

molality = (moles of solute) / (mass (in kg) of solvent)

m = (0.38 mol NH_3) / (0.3 kg of water)

m = 1.27 m

Calculate the boiling point elevation:

$\Delta T_b = k_b \cdot m \cdot i$

$\Delta T_b = (0.52\ °C/m) \cdot (1.27\ m) \cdot (1)$

$\Delta T_b = 0.66\ °C$

The boiling point increased by 0.66 °C due to the addition of 6.4 g of NH_3.

Freezing point depression (ΔTf = Kfm)

Solute particles in mixtures increase the strength of intermolecular (e.g., dipole-dipole) bonds. It is more difficult to boil the solution (boiling point elevation) or freeze the solution (freezing point depression).

The lowering of the freezing point, *freezing point depression*, is calculated by the following equation (the negative sign indicates that the change is a decrease).

$\Delta T_f = -k_f \cdot m \cdot i$

where ΔT_f is the decrease in freezing point, k_f is the molal freezing point constant (a given value that must be determined experimentally), *m* is the molality (mol solute/kg solvent), and *i* is the van't Hoff factor (given since it is determined experimentally).

For example, a 48.0 g sample of a nonelectrolyte (does not dissociate in solution) is dissolved in 500.0 g of water to produce a solution with a freezing point of –3.5 °C. What is the molar mass of the compound? (Use the k_f of water = 1.86 °C/m, the freezing point of pure water = 0 °C)

The freezing point of pure water = 0 °C and the freezing point depression, $\Delta T_f = 3.5\ °C$.

Calculate the moles using the freezing point depression expression:

$\Delta T_f = k_f \cdot m \cdot i$

$3.5\ °C = (1.86\ °C/m) \cdot (x\ /\ 0.5\ kg) \cdot (1)$

$3.5\ °C = (3.72\ °C/m) \cdot (x)$

$3.5\ °C\ /\ (3.72\ °C/m) = x$

$x = 0.94$ mol nonelectrolyte

Calculate the molar mass by dividing the mass of the sample by the moles in the sample.

Molar mass = mass / moles

Molar mass = 48.0 g / 0.94 mol

Molar mass = 51.1 g/mol

The molar mass of the unknown nonelectrolyte is 51.1 g/mol.

Osmotic pressure

Osmotic pressure is the *minimum pressure that needs to be applied to a solution to prevent the inward flow of water across a semipermeable membrane*. The semipermeable membrane allows solvent particles, but not solute particles, to pass.

The *osmotic pressure* is expressed as:

$$\Pi = MRT \cdot i$$

where Π is the osmotic pressure, *M* is the molarity (mol/L), *R* is the ideal gas constant (0.08206 L·atm/mol·K), and *T* is the temperature (K).

Osmotic pressure determines whether and in what direction osmosis occurs. *Osmosis* is the movement of a solvent across a semi-permeable membrane from an area of low solute concentration (high solvent concentration) to an area of high solute concentration (low solvent concentration).

The solvent moves from an area with a low Π value to an area across the semipermeable membrane with a high Π value.

For example, what is the osmotic pressure of a solution prepared by adding 10.5 g of sucrose ($C_{12}H_{22}O_{11}$) to enough water to make 300.0 mL of solution at 25 °C? (Use the molecular mass of sucrose = 342.0 g/mol and the conversion of 300.0 mL = 0.30 L)

Calculate the number of moles of $C_{12}H_{22}O_{11}$ by dividing the mass of the sample by the molar mass of the sample.

moles = (mass) / (molar mass)

moles = (10.5 g $C_{12}H_{22}O_{11}$) / (342.0 g/mol $C_{12}H_{22}O_{11}$)

moles = 0.03 mol $C_{12}H_{22}O_{11}$

Calculate the molarity, *M*:

Molarity = (moles solute) / (liters of solution)

Molarity = (0.03 mol $C_{12}H_{22}O_{11}$) / (0.30 L solution)

Molarity = 0.10 mol/L

Convert temperature to Kelvin:

T = °C + 273

T = 25 + 273

T = 298 K

Calculate osmotic pressure (Π) using $i = 1$ since sucrose does not dissociate.

$$\Pi = MRT{\cdot}i$$

$$\Pi = (0.10 \text{ mol/L}){\cdot}(0.08206 \text{ L}{\cdot}\text{atm/mol}{\cdot}\text{K}){\cdot}(298 \text{ K}){\cdot}(1)$$

$$\Pi = 2.45 \text{ atm}$$

The osmotic pressure for the sucrose solution is 2.45 atm.

Colloids

A *solution* was described as a homogeneous mixture that consists of only one phase (i.e., the solution stays mixed).

A *colloid* is a solution with microscopic insoluble particles dispersed throughout the solution. A colloid has a dispersed phase (i.e., suspended particles) and a continuous phase (i.e., a medium of suspension that, essentially, holds the particles). A colloid stays mixed (i.e., the particles will not settle) unless centrifuged under the influence of gravity.

The typical colloid example is homogenized milk, which consists of butterfat globules dispersed within a water-based solution. Also, when water and oil are vigorously shaken together, an emulsion forms, which is a colloid.

Henry's Law

In 1803, English chemist William Henry (1774-1836) showed that at a constant temperature, the solubility of a gas in a liquid is directly proportional to the partial pressure of the gas above the liquid (when the gas is in equilibrium with the liquid).

Henry's Law is described by the following formula:

$$P_{\text{solute}} = k_H{\cdot}c$$

where P_{solute} is the partial pressure of the solute at the solution's surface, k_H is the Henry's Law constant (unique for each solute-solvent pair), and c is the concentration of the dissolved gas.

How many grams of carbon dioxide (CO_2) gas dissolved in a 0.5 L bottle of carbonated water if the manufacturer uses a pressure of 2.5 atm in the bottling process? (Use the K_H of CO_2 in water = 29.76 atm/(mol/L) and molar mass of CO_2 = 44 g/mol)).

Determine the concentration of CO_2 using Henry's Law:

$P_{solute} = k_H \cdot c$

2.7 atm = [29.76 atm/(mol/L)]$\cdot c$

c = 0.09 mol/L

Calculate the moles of CO_2 in 0.5 L by dividing the mol/L by 2.

moles / grams = molar mass

0.09 / 2 = 0.045 mol

Convert moles to grams by using the molar mass of CO_2, which is 44 g/mol (can be determined from the periodic table).

0.045 mol × (44 g/mol) = 1.98 g

There are 1.98 grams of CO_2 in a 0.5 L bottle of carbonated water.

Notes

Practice Questions

1. Consider the phase diagram for H_2O. The termination of the gas-liquid transition at which distinct or liquid phases do NOT exist is the:

A. critical point
B. endpoint
C. triple point
D. condensation point
E. inflection point

2. When liquids and gases are compared, liquids have [ ] compressibility compared to gases and a [ ] density.

A. lower… lower
B. higher … higher
C. higher… lower
D. lower … higher
E. same…same

3. How does a real gas deviate from an ideal gas?

I. Molecules occupy a significant amount of space
II. Intermolecular forces may exist
III. Pressure is created from molecular collisions with the walls of the container

A. I only
B. II only
C. I and II only
D. II and III only
E. I, II and III

4. Under which conditions does a real gas behave most nearly like an ideal gas?

A. High temperature and high pressure
B. High temperature and low pressure
C. Low temperature and low pressure
D. Low temperature and high pressure
E. If it remains in the gaseous state regardless of temperature or pressure

5. Which of the following compounds has the highest boiling point?

A. CH_3OH
B. $CH_3CH_2CH_2CH_2CH_2OH$
C. $CH_3OCH_2CH_2CH_2CH_3$
D. $CH_3CH_2OCH_2CH_2CH_3$
E. $CH_3CH_2CH_2C(OH)HOH$

6. 15.0 liters of O_2 gas is at a temperature of 23 °C. If the temperature of the gas is raised to 45 °C at constant pressure, the new volume is:

A. 16.1 liters
B. 11.4 liters
C. 8.20 liters
D. 22.8 liters
E. 6.80 liters

7. The boiling point of a liquid is the temperature:

A. where sublimation occurs
B. where vapor pressure of the liquid is less than the atmospheric pressure over the liquid
C. equal to or greater than 100 °C
D. where the rate of sublimation equals evaporation
E. where vapor pressure of the liquid equals the atmospheric pressure over the liquid

8. Which of the following statements about gases is correct?

A. Formation of homogeneous mixtures, regardless of the nature of non-reacting gas components
B. Relatively long distances between molecules
C. High compressibility
D. No attractive forces between gas molecules
E. All of the above

9. What is the proportionality relationship between the pressure of a gas and its volume?

A. directly
B. inversely
C. pressure is raised to the 2nd power
D. pressure raised to the √2 power
E. none of the above

10. How does the volume of a fixed sample of gas change if the pressure is doubled?

A. Decreases by a factor of 2
B. Increases by a factor of 4
C. Doubles
D. Remains the same
E. Requires more information

11. What is the ratio of the diffusion rate of O_2 molecules to the diffusion rate of H_2 molecules, if six moles of O_2 gas and six moles of H_2 gas are placed in a large vessel, and the gases and vessel are at the same temperature?

A. 4:1
B. 1:4
C. 12:1
D. 1:1
E. 2:1

12. Under ideal conditions, which of the following gases is least likely to behave as an ideal gas?

A. CF_4
B. CH_3OH
C. N_2
D. O_3
E. CO_2

Detailed Explanations

1. A is correct.

At a pressure and temperature corresponding to the triple point (point D on the graph) of a substance, all three states (gas, liquid and solid) exist in equilibrium.

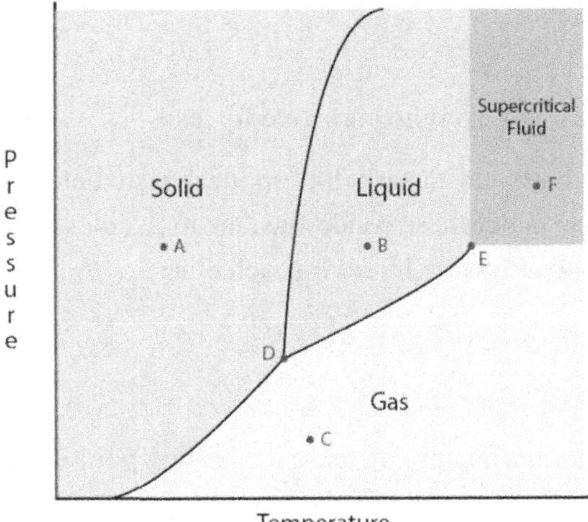

Phase diagram of pressure vs. temperature

The critical point (point E on the graph) is the endpoint of the phase equilibrium curve where the liquid and its vapor become indistinguishable.

2. D is correct.

Density = mass / volume

Gas molecules have a large amount of space between them. Therefore they can be pushed together, and thus gases are very compressible.

There is such a large amount of space between each molecule in a gas. The extent to which the gas molecules can be pushed together is much higher than the extent to which liquid molecules can be pushed together.

Therefore, gases have higher compressibility than liquids.

Gas molecules are further apart than in liquid molecules, which is why gases have a smaller density.

3. C is correct.

The molecules of an ideal gas do not occupy a significant amount of space and exert no intermolecular forces.

In contrast, the molecules of a real gas do occupy space and do exert (weak attractive) intermolecular forces.

Both an ideal gas and a real gas have pressure. The pressure is created from the molecular collisions of the molecules with the walls of the container.

4. B is correct.

The molecules of an ideal gas exert no attractive forces.

Therefore, a real gas behaves most nearly like an ideal gas when it is at high temperature and low pressure because under these conditions, the molecules are far apart from each other and exert little or no attractive forces on each other.

5. E is correct.

Hydroxyl (~OH) groups significantly increase the boiling point because they form hydrogen bonds with ~OH groups of neighboring molecules.

Hydrocarbons are nonpolar molecules, which means that the dominant intermolecular force is London dispersion. This force gets stronger as the number of atoms in each molecule increases. The stronger force increases the boiling point.

The branching of hydrocarbons also affects the boiling point.

Straight molecules have slightly higher boiling points than branched molecules with the same number of atoms. The reason is that straight molecules can align parallel against each other, and all atoms in the molecules are involved in the London dispersion forces.

Another factor is the presence of other heteroatoms (i.e., atoms other than carbon and hydrogen). For example, the electronegative oxygen atom between carbon groups or in an ether (C–O–C) slightly increases the boiling point.

6. A is correct.

Charles' Law (law of volumes) explains how, at constant pressure, gases behave when heated.

$$V \alpha T$$

or $$V / T = \text{constant}$$

Convert the initial temperature from Celsius to Kelvin:

$$23\ °C = 273 + 23 = 296\ K$$

Convert the initial temperature from Celsius to Kelvin:

$45\ °C = 273 + 45 = 318\ K$

Charles' Law:

$(V_1 / T_1) = (V_2 / T_2)$

Solve for the final volume:

$V_2 = (V_1 / T_1) / T_2$

$V_2 = [(15.0\ L) / (296\ K)] \times (318\ K)$

$V_2 = 16.1$ L of O_2

7. E is correct.

Boiling occurs when the vapor pressure of a liquid equals atmospheric pressure.

Vapor pressure is the pressure exerted by a vapor in equilibrium with its condensed phases (i.e., solid or liquid) in a closed system at a given temperature.

Atmospheric pressure is the pressure exerted by the weight of air in the atmosphere.

Vapor pressure is inversely correlated with the strength of the intermolecular force.

With stronger intermolecular forces, the molecules are more likely to stick together in the liquid form, and fewer of them participate in the liquid-vapor equilibrium; therefore, the molecule would boil at a higher temperature.

8. E is correct.

Gases form homogeneous mixtures, regardless of the identities or relative proportions of the component gases.

There is a relatively large distance between gas molecules (as opposed to solids or liquids where the molecules are much closer together).

When pressure is applied to gas, its volume readily decreases, and thus, gases are highly compressible.

There are no attractive forces between gas molecules, which is why molecules of gas can move about freely.

9. B is correct.

Boyle's Law (i.e., pressure-volume law) states that pressure and volume are inversely proportional:

$$(P_1V_1) = (P_2V_2)$$

or

$$P \times V = \text{constant}$$

If the volume of a gas increases, its pressure decreases proportionally.

10. A is correct.

Boyle's Law (i.e., pressure-volume law) states that pressure and volume are inversely proportional:

$$(P_1V_1) = (P_2V_2)$$

or

$$P \times V = \text{constant}$$

If the pressure of a gas increases, its volume decreases proportionally.

Doubling the pressure reduces the volume by half.

11. B is correct.

Graham's Law of Effusion states that the rate of effusion (i.e., escaping through a small hole) of a gas is inversely proportional to the square root of the molar mass of its particles.

Rate 1 / Rate 2 = √(molar mass gas 1 / molar mass gas 2)

The diffusion rate is the inverse root of the molecular weights of the gases.

Therefore, the rate of effusion is:

O_2 / H_2 = √(2 / 32)

rate of diffusion = 1: 4

12. B is correct.

Methanol (CH_3OH) is an alcohol that participates in hydrogen bonding.

Therefore, this gas experiences the strongest intermolecular forces.

Chapter 4

Stoichiometry

- **Molecular Weight**
- **Empirical Formula vs. Molecular Formula**
- **Metric Units Commonly Used in the Context of Chemistry**
- **Description of Composition by Percent Mass**
- **Mole Concept, Avogadro's Number *N***
- **Definition of Density (ρ)**
- **Oxidation Number**
- **Description of Reactions by Chemical Equations**

Molecular Weight

A *molecular compound* is an electrically neutral particle that consists of two or more nonmetals that are covalently bonded together. As described for chemical bonding, a covalent bond arises from sharing electrons between two atoms.

Groups of atoms within a molecule behave like single particles or discrete units. Each molecule, based on the composition and type of atoms, has a specific molecular weight (MW). For example, a water molecule consists of two hydrogen atoms bonded to one oxygen atom.

Molecular weight (MW) is the weight of 1 mole of molecules, where 1 mole equals 6.02×10^{23} particles. This value is Avogadro's number (N_A).

In general, the molecular weight is expressed in grams per mole (g/mol). Molecular weight can be expressed in *atomic mass units* (amu or u), where 1 u = 1 g/mol.

The value of 1 amu = 1 Dalton (Da).

For example, ^{12}C weighs 12 amu = 12 g/mol.

The molecular weight of a substance is obtained by adding the atomic masses of its constituent atoms.

The *atomic mass* of an element is found on the periodic table; it is the number directly below the symbol for the element, as indicated by the arrow for lithium (Li).

For example, determine the MW of H_2O.

Look at the periodic table and observe that the mass of the hydrogen (H) atom is 1.008 amu, and the mass of oxygen (O) atom is 15.999 amu.

Water, from its chemical formula, contains two hydrogens and one oxygen.

The equation to determine the MW of water (H_2O):

The chemical formula for water:

2 (H) + 1 (O)

Molecular weight for water:

MW= 2 H (1.008 g/mol) + 1 O (15.999 g/mol)

MW = 18.015 g/mol

Another example of atomic mass is seen in the graphic below.

A single carbon atom is approximately 12 amu (12 g/mol).

Oxygen exists as a diatomic molecule, O_2 (2 × 15.999 g/mol = 32 g/mol).

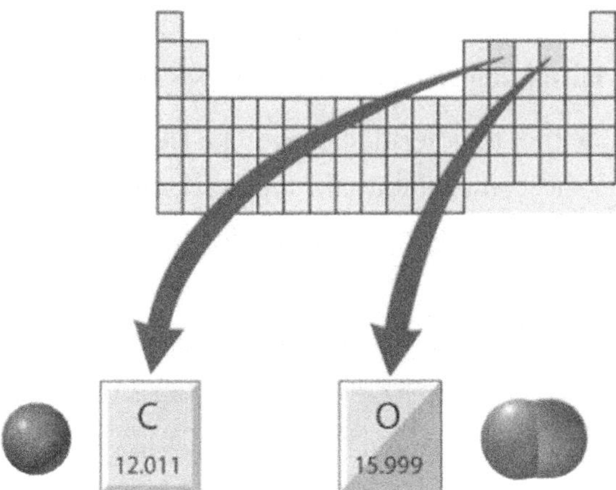

The mass of one carbon atom is approximately 12 amu.

The mass of one O_2 molecule is approximately (16 + 16) 32 amu.

A carbon (C) atom is (12 g/mol) / (32 g/mol), or 3/8 as massive as an O_2 molecule.

Empirical Versus Molecular Formula

There are two ways to express the chemical formula of a molecular compound.

The *molecular formula* describes the atomic composition of a molecule in its naturally occurring form. The molecular formula specifies the number of each type of atom present.

The *empirical formula* describes the *simplest integer ratio* of atoms of a molecule.

The *molecular formula* of hydrogen peroxide is H_2O_2.

The *empirical formula* of hydrogen peroxide is HO.

This chart illustrates the structure, molecular, and empirical formula of glucose.

Molecular Structure	**Molecular Formula**	**Empirical Formula**
CHO H—C—OH H—C—OH H—C—OH H—C—OH CH_2OH	$C_6H_{12}O_6$	CH_2O

Binary molecular compounds are covalently bonded compounds composed of two elements, generally nonmetals. Writing the molecular formula of a binary molecular compound is straightforward using three rules:

1. The first element in the formula is given first, using the full name of the element.

2. The second element is named as if it were an anion.

3. The number of each element in the compound is indicated by a prefix (chart below).

Numerical Prefixes for Naming Compounds

Prefix	Meaning
mono-	1
di-	2
tri-	3
tetra-	4
penta-	5
hexa-	6
hepta-	7
octa-	8

For example, the formula for carbon dioxide is CO_2.

The name indicates a single carbon (no prefix because the *mono-* prefix is omitted when the first element exists as a single atom).

The name indicates two oxygens because of the prefix *di-* (means 2).

Disulfur tetrafluoride has the formula S_2F_4 :

The *di-* prefix indicates two sulfur atoms.

The *tetra-* prefix indicates four fluorine atoms.

Ionic compounds are named differently.

An *ionic compound* consists of a metal atom and nonmetal atom that have transferred valence electron(s) and are bonded by electrostatic forces between the oppositely charged particles.

When writing a chemical formula from the name of an ionic compound, consider the ions (e.g., cations and monatomic and polyatomic anions) in the compound. The inability to recognize these ions is the leading cause of difficulty in writing chemical formulae of inorganic compounds; memorization of the ions is essential.

In the name of an ionic compound (or chemical formula), the cation is listed first, and the anion is second.

For example, sodium chloride is NaCl, which is formed from a sodium cation (Na^+) and a chlorine anion (Cl^-).

Some cations only form one ion (Group 1 (IA), 2 (IIA), and 3 (IIIA) elements, except Thallium). Cations that can form more than one ion have a stock number, shown as a Roman numeral in parentheses (e.g., Cu (I), Fe (II) or Fe (III)).

For monatomic anions, the charge is equivalent to the Group number minus 8.

For example, oxygen has the monatomic anion O^{2-}.

The group number for O is 6, so the charge is 6 – 8 = –2.

The polyatomic anions in the table below should be memorized.

Common Polyatomic Ions

$C_2H_3O_2^-$	acetate	OH^-	hydroxide
NH_4+	ammonium	ClO^-	hypochlorite
CO_3^{2-}	carbonate	NO_3^-	nitrate
ClO_3^-	chlorate	NO_2^-	nitrite
ClO_2^-	chlorite	$C_2O_4^{2-}$	oxalate
CrO_4^{2-}	chromate	ClO_4^-	perchlorate
CN^-	cyanide	MnO_4^-	permanganate
$Cr_2O_7^{2-}$	dichromate	PO_4^{3-}	phosphate
HCO_3^-	bicarbonate	SO_4^{2-}	sulfate
HSO_4^-	bisulfate	SO_3^{2-}	sulfite
HSO_3^-	bisulfite		

Notes

Metric Units Commonly Used in Chemistry

The metric system is a decimal system of measurement referred to as the *International System of Units* (commonly abbreviated as SI).

The SI has seven base units:

> length (meters), mass (kilograms), time (seconds), electric current (amperes), temperature (Kelvin), amount of substance (moles), and luminous intensity (candelas).

Other SI units of measurements, such as volume (liters), density (kg/m^3), and pressure (atmospheres), are derived from the base units.

All the SI units are based on 10 (or multiples of 10); conversion between the units is uniform and straightforward.

The SI system uses prefixes to indicate the magnitude of a measured quantity; the prefix itself gives the conversion factor. Learn some of the common prefixes shown below, as they are used regularly.

Prefix	Symbol	Power	Prefix	Symbol	Power
mega-	M	10^6	centi-	c	10^{-2}
kilo-	k	10^3	milli-	m	10^{-3}
hecto-	h	10^2	micro-	μ	10^{-6}
deca-	D	10^1	nano-	n	10^{-9}
deci-	d	10^{-1}	pico-	p	10^{-12}

The SI units do not allow for double prefixes (e.g., 1,000 g ≠ 1 hectodecagram); instead, only one prefix is used for any given quantity of base units (e.g., 1,000 g = 1 kilogram). Conversions between metric units involve adding or subtracting zeros.

Suppose the mass of a 250 mg aspirin tablet needs to be converted to grams. Start by using the units to set up the problem. If a unit is to be converted (in this case, mg), it is placed in the numerator. That unit must then be in the denominator of the conversion factor for it to cancel:

$$(250\ \text{mg}/1)\cdot(1\times10^{-3}\ \text{g}/1\ \text{mg})$$

$$(250\ \cancel{\text{mg}}/1)\cdot(1\times10^{-3}\ \text{g}/1\ \cancel{\text{mg}}) = 0.250\ \text{g}$$

The units cancel to give grams. The conversion factor is shown as a numerator of 1×10^{-3} because on most calculators, it must be entered this way, not as just 10^{-3}. The mg is assigned the value of 1, and the prefix "*milli-*" applies to the gram unit. 1 mg means 1×10^{-3} g.

Conversions between English/imperial and metric units work similarly. The difference is the conversion factors; they are not powers of ten, and they are different for each unit, which makes conversions between English/imperial and metric units more complex. Therefore, the imperial system is not ideal for scientific calculations.

Memorize a few typical metric-imperial conversions:

Length: 2.54 cm = 1 inch

Mass: 454 g = 1 pound

Volume: 0.946 L = 1 quart

Temperature: °C = (°F − 32) / 1.8

Significant figures indicate the number of digits known exactly for a measured (or calculated) quantity within a degree of uncertainty.

For example, if a thermometer indicates a boiling point of 36.2 °C, and has an uncertainty of ± 0.2 °C, all three figures in 36.2 are significant (including the 0.2, which is uncertain). The notation 36.2 ± 0.2 °C indicates the known quantity, along with the digits of uncertainty.

For example, convert the mass of a 23 lbs. object to kilograms?

The pound units are converted to grams, and then grams are converted to kilograms.

Use the proper units to set up the problem:

$$23 \text{ lbs.} = (23 \text{ lbs.} / 1) \times (454 \text{ g} / 1 \text{ lbs.}) \times (1 \text{ kg} / 1 \times 10^3 \text{ g})$$

$$23 \text{ lbs.} = (23 \cancel{\text{lbs.}} / 1) \times (454 \cancel{\text{g}} / 1 \cancel{\text{lbs.}}) \times (1 \text{ kg} / 1 \times 10^3 \cancel{\text{g}})$$

$$23 \text{ lbs.} = 10 \text{ kg}$$

A conversion problem may include multiple units.

For example, convert the pressure 14 lb/in^2 to g/cm^2. For problems that involve multiple conversions, work with one unit at a time.

Convert the pounds to gram units:

$$14 \text{ lb/in}^2 \times 454 \text{ g/lb}$$

Convert the in^2 unit to cm^2 units. Set up the conversion without the exponent first, using the conversion factor 1 in = 2.54 cm.

Since in^2 and cm^2 are needed, raise each value to the second power:

$$= 14 \text{ lb/in}^2 \times 454 \text{ g/lb} \times 1^2 \text{ in}^2/2.54 \text{ cm}^2$$

$$= 9.9 \times 10^2 \text{ g/cm}^2$$

When the units are squared, the numbers associated with them must be squared. For 1 in^2, it does not make a difference because of $1^2 = 1$. In most cases, it is essential to squaring of quantities (numbers) when the units are squared.

Always check the units to verify that the problem has been set up correctly.

Description of the composition by percent mass

Percent mass (% mass) is a unit of concentration that *compares the mass of one part of a substance to the mass of the whole.* The percent mass unit is often used for solutions, especially concentrated acid, or base solution, where the concentration is often expressed as percent by mass on the bottle.

Percent mass is:

% mass = (mass of species of interest / total mass) × 100%

Find the % mass of sodium (Na) in sodium bicarbonate ($NaHCO_3$). Use the periodic table to find that the atomic mass of the elements. Na = 22.99 g/mol, H = 1.01 g/mol, C = 12.01 g/mol and O = 16.00 g/mol. Use the molecular mass of the atoms to calculate the molecular weight of $NaHCO_3$:

1Na (22.99 g/mol) + 1 H (1.01 g/mol) + 1 C (12.01 g/mol) + 3 O (16.00 g/mol)

MW $NaHCO_3$ = 84.01 g/mol

Find the percent mass of Na:

% mass Na = (mass of species of interest / total mass) × 100%

% mass Na = (22.99 g/mol / 84.01 g/mol) × 100%

% mass Na = 27.4%

Concentration can be expressed as % volume, which is calculated like % mass.

Notes

Mole Concept, Avogadro's Number *N*

Stoichiometry uses quantitative relationships between products and reactants; the number of molecules is expressed in the mole unit.

Avogadro's number (N_A) is the number of particles in 1 mole of a substance and is equal to 6.02×10^{23} particles.

For a reaction equation, the mass of the substance and the moles of substance are determined for both the reactants and products.

$2\ H_2$	+	$1\ O_2$	→	$2\ H_20$
2 moles = 4 *g*		1 mole = 32 *g*		2 moles = 36 *g*
= 12.04×10^{23} molecules		= 6.02×10^{23} molecules		= 12.04×10^{23} molecules

Most stoichiometry problems follow a set strategy that involves the mole:

Quantity A → Moles A → Moles B → Quantity B

Many stoichiometry problems are solved using this strategy.

Each step is examined and then combined to solve complicated problems.

Converting Quantity A to Moles A:

How many moles of $CaCO_3$ are in a 25.0 g sample?

Calculate the molar mass of $CaCO_3$, using the atomic mass information from the periodic table. The molar mass should be calculated to the same number of significant figures as the quantity converted:

1 Ca = 40.08 g/mol × 1 = 40.08 g/mol

1 C = 12.01 g/mol × 1 = 12.01 g/mol

3 O = 16.00 g/mol × 3 = 48.00 g/mol

$CaCO_3$ = 40.08 g/mol + 12.01 g/mol + 48.00 g/mol

$CaCO_3$ = 100.09 = 100.1 g/mol

Use the molar mass to convert the 25.0 g mass of $CaCO_3$ to moles $CaCO_3$.

$CaCO_3$ = 25.0 g × (1 mol $CaCO_3$ / 100.09 g $CaCO_3$)

$CaCO_3$ = 0.250 mol

Converting moles into grams:

For example: for 0.750 mol $CaCO_3$, the number of grams of $CaCO_3$ is:

$CaCO_3 = 0.750 \text{ mol} \times (100.09 \text{ g } CaCO_3 / 1 \text{ mol } CaCO_3)$

$CaCO_3 = 75.1 \text{ g } CaCO_3$

Moles A to Moles B Conversions:

How many moles of sodium ion (Na^+) does 0.100 mol of sodium carbonate (Na_2CO_3) have?

One mole of sodium Na_2CO_3 contains 2 moles of Na, 1 mole of C, and 3 moles of O.

In a solution, Na_2CO_3 completely dissociates (i.e., electrolyte) into ions, which is why the sodium atoms are referred to as ions.

Compare the moles of Na_2CO_3 and moles of Na^+:

$= 0.100 \text{ mol } Na_2CO_3 \times (2 \text{ mol } Na^+ / 1 \text{ mol } Na_2CO_3)$

$= 0.200 \text{ mol } Na^+$

Relating Moles of Reactants and Products:

In other examples, a reaction may be involved, and a comparison of two different compounds is necessary.

For a balanced chemical reaction, the stoichiometric coefficients represent the mole ratio between all species.

The following is a typical dissociation reaction:

$2\ KClO_3 \rightarrow 2\ KCl + 3\ O_2$

In this reaction, 2 moles of potassium chlorate ($KClO_3$) decomposes into 2 moles of potassium chloride (KCl) and 3 moles of oxygen (O_2). How many O_2 molecules are produced from this reaction?

If there are 0.400 mol of $KClO_3$ and the number of moles of O_2 is needed, use the stoichiometric coefficients to arrange a mole ratio in which the moles of $KClO_3$ cancel and the moles of O_2 remain:

$= (0.400 \text{ mol } KClO_3) \times (3 \text{ mol } O_2 / 2 \text{ mol } KClO_3)$

$= 0.600 \text{ mol } O_2$

Avogadro's number converts the number of moles to the number of particles.

$= 0.600 \text{ mol } O_2 \times (6.02 \times 10^{23} \text{ molecules } O_2 / 1 \text{ mol } O_2)$

$= 3.61 \times 10^{23}\ O_2 \text{ molecules}$

The mole units of O_2 successfully cancel in the answer.

Definition of Density

Density (ρ) is the ratio of mass over the volume of a given substance.

The SI unit of density is kg/m^3.

Specific gravity is a unit-less ratio between the density of some material and the density of a *reference substance* (most times, the reference is water).

Some relevant density units:

- Density of water = 1 g/mL = 1 g/cm^3
- Specific gravity of water = 1 g/cm^3 / 1 g/cm^3 = 1
- Density of lead = 11 g/cm^3
- Specific gravity of lead = 11 g/cm^3 / 1 g/cm^3 = 11

The densest known element is osmium, with a density of 22.59 g/cm^3.

The least dense known element is hydrogen, with a density of 8.99 × 10^{-5} g/cm^3.

Hydrogen is about 250,000 times less dense than osmium.

Molarity (*M*) of a substance is the moles of solute per liter of solution.

Molarity = moles of solute / L of solution

Molarity is the most common concentration unit used for solutions.

The molality (*m*) of a substance is the moles of solute per kilogram of solvent and is described in the Phase Equilibria chapter.

$$\text{Molality} = \frac{\text{moles of solute}}{\text{kg solvent}}$$

Very dilute solutions use the unit of parts per million (ppm).

$$\text{ppm} = \frac{\text{grams of solute}}{\text{grams of solution}} \times 10^6$$

The amount of solute relative to the amount of solvent is typically minimal. Therefore, the density of the overall solution is, to a first approximation, the same as the density of the solvent. For this reason, parts per million may be expressed as:

$$\text{ppm} = \frac{\text{mg solute}}{\text{kg solution}}$$

If the solvent is water, with a density of 1.00 kg/L, then ppm can be expressed as:

$$\text{ppm} = \frac{\text{mg solute}}{\text{L solution}}$$

Another unit of measurement to express the concentration of even more dilute solutions is parts per billion (ppb). The expression for parts per billion is:

$$\text{ppb} = \frac{\text{grams of solute}}{\text{grams of solution}} \times 10^9$$

The density of a very dilute solution is about equal to the density of the solvent.

Thus, parts per billion may be expressed as:

$$\text{ppb} = \frac{\mu\text{g solute}}{\text{kg solution}}$$

$$\text{ppb} = \frac{\mu\text{g solute}}{\text{L solution}}$$

The *mole fraction* (χ) is the ratio between the moles of a specific molecule over the total moles of all components in the mixture.

The *mole percent* (%χ) is the mole ratio multiplied by 100.

$$\chi_{\text{solute}} = \frac{\text{mol solute}}{\text{total moles of all components}}$$

$$\chi_{\text{solute}}\ \% = \frac{\text{mol solute}}{\text{total moles of all components}} \times 100$$

When converting between units, start by choosing an arbitrary amount of solution in the denominator of the concentration to be converted. For example, if converting percent mass to molarity, assume 100 grams of solution. If converting molarity to percent mass, assume one liter of solution.

What is the molarity, molality, and mole fraction of HCl for a concentrated solution of HCl known to be 37.0% HCl by mass and its density is 1.19 g/ml?

Begin with the valid assumption that the HCl solution is 100 g. Since the % mass of HCl is 37%, that means 37.0 g of the solution is HCl (grams of solute), and the remaining 63.0 g is water (grams of solvent).

To find molarity, determine the moles of HCl (solute) per liter of solution. First, convert the mass of HCl to moles:

mol HCl = 37.0 g HCl × 1 mol HCl/36.5 g HCl

mol HCl = 1.01 mol HCl

Then, using the density of the solution, convert the known mass of solution (100 g) to liters of solution.

$$\text{L solution} = 100\text{ g solution} \times \frac{1\text{ mL solution}}{1.19\text{ g solution}} \times \frac{1\text{ L solution}}{1000\text{ mL solution}}$$

$$\text{L solution} = 100\,\cancel{\text{g solution}} \times \frac{1\,\cancel{\text{mL solution}}}{1.19\,\cancel{\text{g solution}}} \times \frac{1\text{ L solution}}{1000\,\cancel{\text{mL solution}}}$$

L solution = 0.0840 L solution

Using the moles of solute (HCl) and volume of solution in liters, calculate the molarity (M) of the solution as moles of solute per liter of solution:

$$M = \frac{1.01\text{ mol HCl}}{0.0849\text{ L solution}}$$

M = 12.0 M HCl/L solution

From the information above, find the molality of the HCl solution. The moles of solute are known (1.01 mol HCl). Determine the mass of the solvent (H_2O) in kilograms:

63.0 g H_2O × (1 kg H_2O / 1,000 g H_2O) = 0.0630 kg H_2O

Using the moles and the mass of solvent, calculate the molality (m or b):

m = 1.01 mol HCl / 0.0630 kg H_2O

m = 16.0 mol HCl/kg H_2O

m = 16.0 m H_2O

Finally, determine the mole fraction of HCl.

From before, there are 1.01 moles of HCl. Calculate moles of H_2O.

$$\text{mol } H_2O = 63.0 \text{ g } H_2O \times (1 \text{ mol } H_2O / 18.0 \text{ g } H_2O)$$

$$\text{mol } H_2O = 63.0 \cancel{\text{g } H_2O} \times (1 \text{ mol } H_2O / 18.0 \cancel{\text{g } H_2O})$$

$$\text{mol } H_2O = 3.50 \text{ mol } H_2O$$

Now that the moles of all molecules are known, calculate the mole fraction of HCl:

$$\chi_{solute} = \frac{\text{mol solute}}{\text{total moles of all components}}$$

$$\chi_{HCl} = 1.01 \text{ mol HCl} / [(1.01 \text{ mol HCl} + 3.50 \text{ mol } H_2O)]$$

$$\chi_{HCl} = 0.244$$

Oxidation Number

The *oxidation number* (or *oxidation state*) is the charge an atom has or appears to have when a set of rules counts the electrons of the compound. Oxidation numbers are typically used in ionic compounds for oxidation-reduction reactions, which involve the transfer of electrons. For compounds that are not ionic, oxidation numbers determine if the chemical formula of the compound has been written correctly.

Assigning the Oxidation States:

1. The oxidation state of an element is always zero.
2. For main group metals (Groups 1-2, 13-18), the charge of the ion is the same as the valence electrons or group number.
   a. Transition metals, however, do not obey this rule: their oxidation states are sometimes not given by their group number. They may also have multiple oxidation states.
3. The oxygen ion in a compound is typically –2 except the peroxide ion (O_2^{2-}), which is –1.
4. For polyatomic ions or compounds, the sum of oxidation states is equal to the overall charge of the polyatomic ion or neutral molecule.

Determine the oxidation states of the elements in potassium permanganate ($KMnO_4$).

Solution: Apply rule two. Since potassium is a Group IA alkali metal, its oxidation state must be +1. Assign x to Mn for now since manganese may exist in several oxidation states. There are four oxygen atoms in the permanganate ion, with oxidation states of –2 per O atom. The overall charge of the neutral compound equals zero:

K	Mn	O_4
+1	x	4(–2)

The algebraic expression is:

$$1 + x - 8 = 0$$

Solving for x gives the oxidation state of manganese:

$$x - 7 = 0$$

$$x = +7$$

K	Mn	O_4
+1	+7	4(–2)

What is the oxidation state of chromium in dichromate ion $Cr_2O_7^{2-}$?

Start by assigning –2 as the oxidation state for oxygen. Since the oxidation state for chromium is not known, and two chromium atoms are present, assign the algebraic value of $2x$ for chromium:

$Cr_2 \quad O_7^{2-}$

$2x \quad 7(-2)$

Use the algebraic equation to solve for x. Since the overall charge of the ion is –2, the expression is set equal to –2, rather than 0:

$$2x + 7(-2) = -2$$

Solve for x:

$$2x - 14 = -2$$

$$2x = 12$$

$$x = +6$$

Each chromium in the ion has an oxidation state of +6.

What are the oxidation states of the elements in the polyatomic compound $Fe_2(CO_3)_3$?

Two elements (iron and carbon) have more than one possible oxidation state. When considering molecules formed from cations (positive ions) and anions (negative ions), start by splitting the molecule into its constituent ions. Determine the charge on each ion.

Iron (Fe) has more than one oxidation state, but a carbonate ion always has an oxidation state of –2 (CO_3^{2-}). With this information, Fe's oxidation state can be determined:

$Fe_2 \quad (CO_3)^3$

$2x \quad 3(-2)$

$$2x - 6 = 0$$

$$2x = 6$$

$x = 3$ Each iron ion in the compound has an oxidation state of +3.

Next, consider the carbonate ion independent of the iron (III) ion:

CO_3^{2-}

$x \quad 3(-2)$

$$x - 6 = -2$$

$x = +4$ The oxidation state of carbon is +4, and each oxygen is –2.

An *oxidation-reduction reaction* is any process whereby a transfer of electrons occurs between two chemicals. For an oxidation-reduction (or redox) reaction to proceed, one substance in a reaction is oxidized, while another substance in the reaction is reduced; reduction or oxidation processes cannot occur separately.

Oxidation involves a loss of electrons. Oxidation is typically associated with metals and may involve the addition of oxygen or removal of hydrogen.

Reduction involves a gain of electrons, which is typically observed in nonmetals and may involve the addition of hydrogen or removal of oxygen.

In biological systems, cells oxidize and reduce metals. The cytochrome c protein plays a vital role in ATP production. The cytochrome c protein contains a Fe^{2+} cation that undergoes oxidation to form Fe^{3+}, followed by a reduction back into Fe^{2+}.

In organic reactions, look for the movement of oxygen and hydrogen. *Combustion* (i.e., burning) is a reaction with oxygen as an example of a redox reaction.

Common oxidizing and reducing agents

An *oxidizing agent* (or *oxidant*) is a substance that oxidizes another substance by removing electrons. The oxidizing agent gains the removed electrons, thus reducing itself. Therefore, the oxidation number of the oxidizing agent becomes less positive.

A *reducing agent* is a substance that reduces another substance by donating its electrons to the substance being reduced. Thus, the reducing agent loses these electrons and is oxidized. Therefore, the oxidation number of the reducing agent becomes more positive (or less negative).

Using a number line shown below to help identify the substances being reduced or oxidized, and whether they are oxidizing agents or reducing agents, respectively.

Oxidation number decreases, substance reduced

⟶

-|---|---|---|---|---|---|---|---|---|---|---

–5 –4 –3 –2 –1 0 +1 +2 +3 +4 +5

⟵

Oxidation number increases, substance oxidized

The image below summarizes the relationship between reducing agents and oxidizing agents. Essentially, atoms that lose electrons during a chemical reaction undergo oxidation. Atoms that gain electrons during a chemical reaction undergo reduction.

The oxidized species is the reducing agent, and the reduced species is the oxidizing agent.

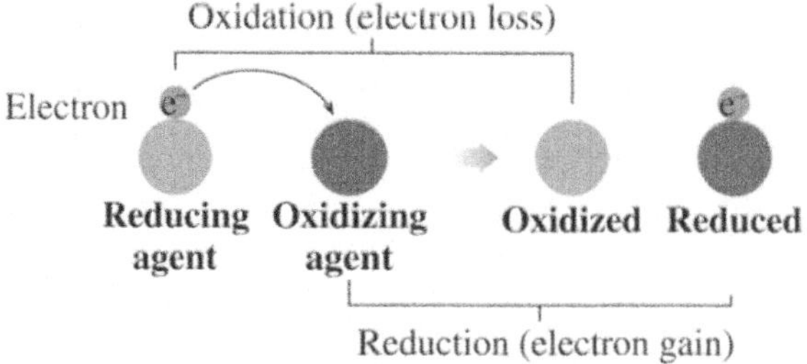

A useful mnemonic for remembering what is happening to the electrons in a redox reaction is "OIL RIG," which stands for "**O**xidation **I**s **L**oss, **R**eduction **I**s **G**ain. Remember that this mnemonic device refers to the loss and gain of electrons.

Identify the substance being oxidized, the substance being reduced, the oxidizing agent, and the reducing agent in the following reaction:

$$Cu\ (s) + 4\ HNO_3\ (aq) \rightarrow Cu(NO_3)_2\ (aq) + 2\ NO_2\ (g) + 2\ H_2O\ (l)$$

Determine the oxidation state for each element in the reactants and products.

Elemental copper, Cu (*s*), has an oxidation state of zero (0).

For nitric acid, HNO_3, set N equal to x:

H N O_3

+1 x 3(–2)

$1 + x - 6 = 0$

$x = +5$

H = +1, N = +5, O = –2

For copper (II) nitrate, $Cu(NO_3)_2$:

$Cu(NO_3)_2$

Nitrates are always –1, while Cu might be +1 or +2.

$Cu(NO_3)_2$

Cu $(NO_3)_2$

$x + 2(-1) = 0$

$x = 2$

In this example, Cu's oxidation state is +2. Oxygen is −2.

For nitrate ion, NO_3^-, set N equal to x:

N $\quad O_3^-$

$x \quad 3(-2)$

$x - 6 = -1$

$x = +5$

so,

Cu = +2, N = +5, O = –2

For nitrogen dioxide, NO_2, set N equal to x:

N $\quad O_2$

$x \quad 2(-2)$

$x - 4 = 0$

$x = +4$

so,

N = +4, O = –2

For water, H_2O:

$H_2 \quad O$

$2(+1) \quad -2$

so,

H = +1, O = –2

Element	**Oxidation State: Reactants**	**Oxidation State: Products**
Cu	0	+2
H	+1	+1
N	+5	+5 (in NO_3^-), +4 (in NO_2)
O	–2	–2

The next step is to identify the molecules that changed the oxidation state. Copper's oxidation number increased from 0 to +2, which means that copper has been oxidized (loses electrons), and therefore, it is the reducing agent. Copper reduced nitric acid by giving away its electrons to the nitrogen in nitric acid. Nitrogen in nitric acid changed from +5 to +4 in nitrogen dioxide, which means that copper has reduced nitric acid. Since nitric acid is reduced, it is the oxidizing agent.

Oxidation cannot occur without also having reduction.

Common oxidizing agents	Common reducing agents
Oxygen O_2, Ozone O_3, Permanganates MnO_4^-, Chromates $CrO4_2^-$, Dichromates $Cr_2O_7^{2-}$, peroxides H_2O_2, Lewis acids, compounds with many oxygens	Hydrogen H_2, metals (such as K), Zn/HCl, Sn/HCl, LAH (Lithium Aluminum Hydride), $NaBH_4$ (Sodium Borohydride), Lewis bases, compounds with many hydrogens

Disproportionation reactions

In *disproportionation reactions*, an atom undergoes both oxidation and reduction to form two different atoms with different oxidation states. Consider the following reaction:

$$2\ Cu^+ \rightarrow Cu + Cu^{2+}$$

The Cu^+ acts as both oxidizing and reducing agents and simultaneously reduces and oxidizes itself. The oxidized Cu^+ becomes Cu^{2+}. The reduced Cu^+ becomes Cu.

Redox titration

Titration, or *volumetric analysis*, is a standard laboratory procedure to determine the unknown concentration of a substance. A reagent of known concentration is slowly added to a sample of unknown concentration until the neutralization reaction (i.e., the reaction between an acid and a base) is complete. The *analyte* is the substance of unknown concentration being analyzed, and the *titrant* is the analytical reagent of known concentration being added to the sample.

To determine the endpoint of a redox titration (i.e., when all the sample molecules have been used), an *indicator* is added to the reaction. A redox indicator undergoes a definite color change at the equivalence point of the neutralization reaction.

An example of a redox reaction is the titration of ascorbic acid (a form of vitamin C) using iodine. Ascorbic acid oxidizes iodine to form iodide ions. Starch is added as an indicator because excess iodine reacts with starch and turn the solution's color from clear to dark blue. Using the initial volume of ascorbic acid (analyte), and the measured volume of iodine (titrant) added, the concentration of ascorbic acid can be calculated.

Description of Reactions by Chemical Equations

Predicting the products that form in each reaction is no easy task. Since many chemical compounds exist, memorizing every possible reaction is not necessary. Most chemical reactions can be classified into several groups.

Being able to identify the type of reaction is a significant step towards predicting the products of a given chemical reaction. The important classes of reactions encountered in general chemistry are shown below:

Reaction Type	**General Reaction Scheme**
Synthesis	A + B → AB
Decomposition	AB → A + B
Replacement	AB + C → AC + B (single) AB + CD → AD + CB (double)

1. *Combination* (Synthesis) *Reactions*: Two (or more) substances react to form a single product.

   The general form of reaction is A + B → AB, in which two substances combine to form one compound.

2. *Decomposition Reactions*: One material reacts to form two or more products. This is the reverse of a combination reaction.

   The general form of this reaction is AB → A + B, in which a compound decomposes to form two (or more) constituent elements.

3. *Replacement Reactions* (also called substitution or displacement reactions): One atom or group replaces another species in a compound.

   - In a *Single Replacement Reaction*, an element replaces the corresponding element in a compound.

     The general form of this type of reaction is AB + C → AC + B, in which element C is replacing element B.

   - In a *Double Replacement* (or *Metathesis Reaction*), two compounds react and switch partners; cations and anions in one compound exchange with counterparts in the other compound.

The general form for this reaction is AB + CD → AD + CB.

- Important Note: In both types of replacement reactions, the number of substances on the reactant side of the equation is the *same* as on the product side.

Some examples of more specific types of reactions.

- Oxidation-Reduction Reaction (also called *redox*): A transfer of electrons occurs between the two reactants. A loss of electrons (increase in oxidation number) is oxidation, while a gain of electrons (or decrease in oxidation number) is a reduction.
    - Many combination and decomposition reactions use oxidation and reduction, along with single displacement reactions.

- *Combustion* involves a carbon-containing molecule (e.g., a hydrocarbon) reacting with oxygen to produce carbon dioxide (CO_2) and water (H_2O).
    - Combustion reactions tend to be highly exothermic and are considered irreversible, and therefore spontaneous (refer to the chapter on thermodynamics).
    - Combustion reactions are a type of decomposition reaction and are also oxidation-reduction reactions. Combustion uses O_2 to produce energy, which dissipates as heat or performs work.
    - The general formula for the combustion of a hydrocarbon:

$$C_xH_y + O_2 \rightarrow CO_2 + H_2O$$

- *Condensation* involves water being produced as part of a double replacement reaction. Condensation reactions occur among functional groups that contain –H and –OH: those two groups break away from their respective compounds and form H_2O.

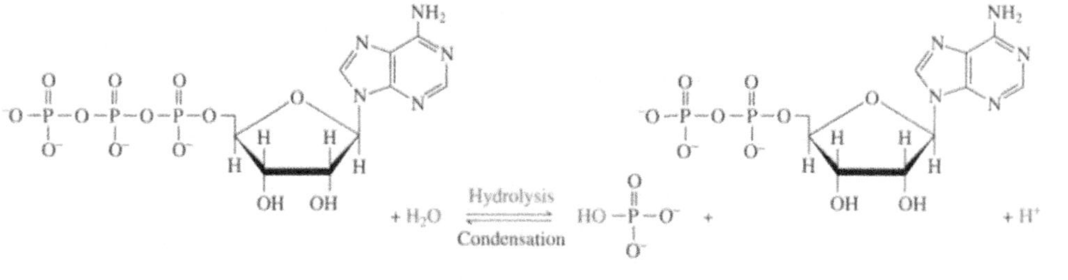

Adenosine triphosphate (ATP)

Adenosine diphosphate (ADP) + Energy

- *Hydrolysis* (hydro- means "water," lysis means "break") is the reverse process of condensation, where water is consumed as a reactant. The other reactant molecule is split into two smaller molecules in the products.

  An example of hydrolysis (and the reverse condensation reaction) is shown above. The more general form is shown in the figure below.

Condensation

Two smaller molecules form a larger one

—OH + HO— —O—

Hydrolysis

One large molecule forms two smaller ones

$+ H_2O$

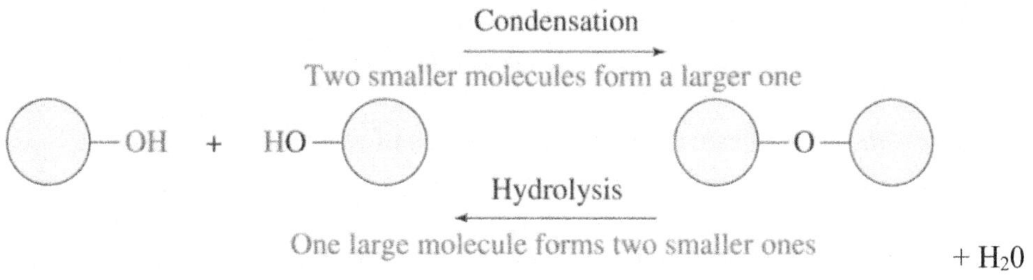

- *Carboxylation* reactions involve the addition of a carboxyl (−C=O) or carboxylic acid (–COOH) group. As carbon dioxide (CO_2) moves through the cell, it is added to and removed from biomolecules by two enzymes (carboxylase to add; decarboxylase to remove).

- An example of carboxylation is shown in the figure below.

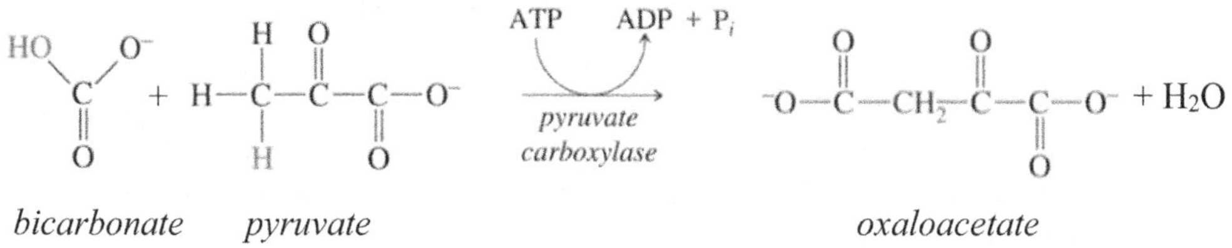

ATP ADP + P_i

pyruvate carboxylase

$+ H_2O$

bicarbonate *pyruvate* *oxaloacetate*

Conventions for writing chemical equations

Chemical equations have several important parts:

$$H_2SO_4\ (aq) + 2\ NaOH\ (aq) \rightarrow 2\ Na^+\ (aq) + SO_4^{2-}\ (aq) + 2\ H_2O$$

phase *coefficient* *direction* *charge*

- The *phase* of a substance is indicated with a letter subscript, such as solid (s), liquid (l), gas (g), or aqueous solution (aq).

- The *coefficient* indicates the number of moles of reactant or product, relative to other components in the reaction.

- The *direction* is represented as a single-headed arrow, which denotes the forward direction of the reaction as written.

  - *Reversible reactions* are at a state of *chemical equilibrium* and are represented with a double-headed arrow. This indicates that the rates of the forward and reverse reactions are equal and constant, and no net products are formed.

  - A double-sided arrow with one side *larger* than the other denotes a *nonequilibrium condition*, which spontaneously favors the direction of the larger arrow.

- The *charge* is indicated with a numerical superscript and +/– sign. It is common practice not to indicate the charge on a neutral compound or substance.

Balancing equations including redox equations

The *law of mass conservation* (refer to the chapter on thermodynamics) states that atoms are neither created nor destroyed in a chemical reaction—they are simply rearranged.

In chemical reactions, the coefficients of the reactants and products must be balanced, which means there are equivalent amounts of all atoms on each side of the reaction.

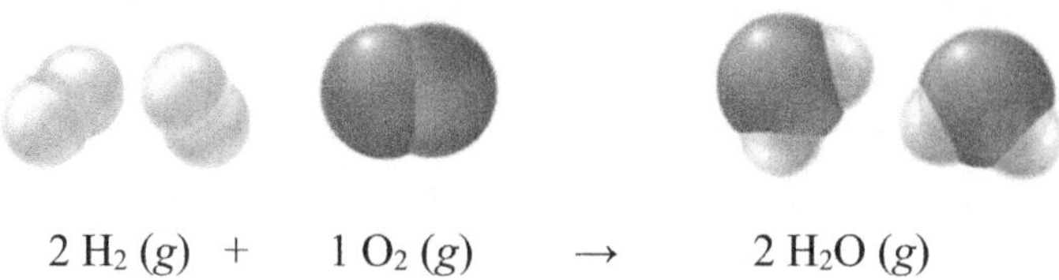

$$2\ H_2\ (g) + 1\ O_2\ (g) \rightarrow 2\ H_2O\ (g)$$

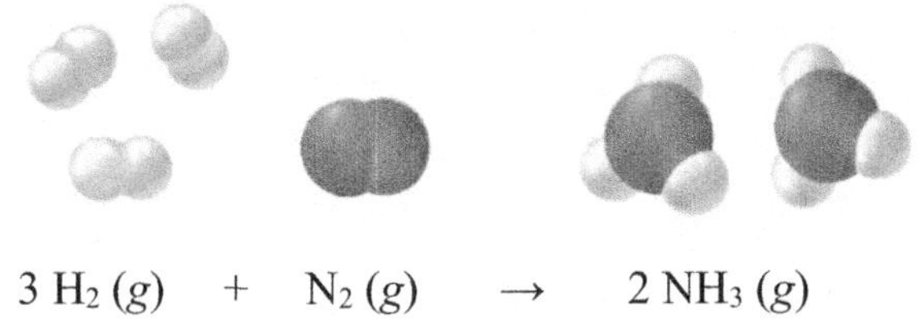

$3\ H_2\ (g)$ + $N_2\ (g)$ → $2\ NH_3\ (g)$

For balancing a chemical equation, coefficients are assigned to each species until all atoms balance on both sides.

Balance the combustion of propanol:

$C_3H_8O + O_2 \rightarrow CO_2 + H_2O$

Select the atom (or ion) that is only present in one species on each side of the equation. Start with carbon. Add the coefficient 3 to CO_2 on the right side since there are three carbons on the left side, indicated by C_3.

$C_3H_8O + O_2 \rightarrow \mathbf{3}\ CO_2 + H_2O$

Hydrogen is the other species present in one molecule on each side so that the hydrogens can be balanced. Add the coefficient 4 to H_2O on the right side because $4 \times 2 = 8$, equals the number of hydrogens on the left side, indicated by H_8.

$C_3H_8O + O_2 \rightarrow 3\ CO_2 + \mathbf{4}\ H_2O$

Starting with hydrogen before balancing carbon yields the same result.

Balance all the elements that are present in one species on each side first, because it prevents needing to go back to change coefficients of species that are already balanced.

Oxygen is present in every term of the equation, so if O were balanced first, it would be more difficult. Because of the adjustments made to balance the other atoms, at this point, there is only one oxygen-containing species that has not yet been balanced.

Count oxygen atoms: there is one from C_3H_8O, six from $3\ CO_2$ (because $3 \times 2 = 6$), and four from $4\ H_2O$.

Set this equation: $1 + 2x = (3 \times 2) + 4$, where x is the coefficient of the last term, O_2. Solve for x, which equals $^9/_2$.

$C_3H_8O + \boldsymbol{x}\ \mathbf{O_2} \rightarrow 3\ CO_2 + 4\ H_2O$

$C_3H_8O + \mathbf{^9/_2\ O_2} \rightarrow 3\ CO_2 + 4\ H_2O$

Remove any fractions, so multiply every term by 2.

$\mathbf{2}\ C_3H_8O + \mathbf{9}\ O_2 \rightarrow \mathbf{6}\ CO_2 + \mathbf{8}\ H_2O$

With all non-integer coefficients removed, the equation is now balanced.

When solving oxidation-reduction equations, a different approach is required. Balancing redox reactions is much more complicated; it involves splitting the redox reaction into half-reactions. There are two methods of balancing half-reactions:

- *The Ion-Electron Method*: balance the elements first, then balance charge by adding electrons.

- *The Oxidation-State Method*: treat the species of interest as a single element (i.e., changes oxidation number) and then balance it.

Learn the following procedure when approaching problems that require the balancing of oxidation-reduction reactions:

1. Separate the equation into two *half-reactions*. A half-reaction contains only the species of interest (i.e., those containing the atom that changes oxidation state).

   a. Each half-reaction corresponds to the oxidation half-reaction or reduction half-reaction. Anything that is not covalently attached to the atom is not part of the species of interest. A species that does not change the oxidation state is a *spectator ion*, (i.e., an ion that is present in solution but is not involved in the reaction).

2. Balance each half-reaction for both the *charge* and *number* of atoms.

   a. Two methods for balancing oxygen elements:

      i. Under acidic conditions: add H_2O to the side that needs the oxygen atom, then add H^+ to the other side.

      ii. Under basic conditions: add 2 OH^- to the side that needs the oxygen atom, then add H_2O to the other side.

3. After the half-reactions are balanced, recombine the half-reactions:

   a. Multiply each half-reaction by a factor, so the electrons cancel.

   b. It is like solving a simultaneous equation where the electron term must be eliminated.

4. Lastly, perform these additional steps:

   a. Combine identical species on the same side of the equation.

   b. Cancel identical species on opposite sides of the equation.

c. Add back in the spectator ions.

d. For the oxidation-state method, now balance the oxygen and hydrogen elements.

e. Check that both sides of the equation have an equal number of atoms and neutral net charge.

Balance the redox reaction below using the ion-electron method:

$$K_2Cr_2O_7\,(aq) + HCl\,(aq) \rightarrow KCl\,(aq) + CrCl_3\,(aq) + H_2O\,(l) + Cl_2\,(g)$$

Solution: Step 1 – Separate into half-reactions:

- Reduction: $Cr_2O_7^{2-} \rightarrow Cr^{3+}$
- Oxidation: $Cl^- \rightarrow Cl_2$

The species of interest for the oxidation reaction is Cl^- (not HCl) because the H^+ is not covalently attached to Cl^- and the ions separate in an aqueous solution.

Similarly, $Cr_2O_7^{2-}$ is used, not $K_2Cr_2O_7$, because K^+ and H^+ are spectator ions.

Step 2 – Balance each of the half-reactions:

For the ion-electron method, balance the elements first, then balance charge.

Balance elements for the reduction half-reaction (ion-electron method):

1. $Cr_2O_7^{2-} \rightarrow Cr^{3+}$
2. $Cr_2O_7^{2-} \rightarrow 2Cr^{3+}$
3. $Cr_2O_7^{2-} + 14\ H^+ \rightarrow 2\ Cr^{3+} + 7\ H_2O$

Balance charge for the reduction half-reaction (ion-electron method):

1. $Cr_2O_7^{2-} + 14\ H^+ + 6\ e^- \rightarrow 2\ Cr^{3+} + 7\ H_2O$

Balance charge for the oxidation half-reaction (ion-electron method):

1. $Cl^- \rightarrow Cl_2$
2. $2\ Cl^- \rightarrow Cl_2$
3. $2\ Cl^- \rightarrow Cl_2 + 2\ e^-$

Step 3 – Recombine the half-reactions:

- $Cr_2O_7^{2-} + 14\ H^+ + 6\ e^- \rightarrow 2\ Cr^{3+} + 7\ H_2O$

- $2\ Cl^- \rightarrow Cl_2 + 2\ e^-$

Multiply each species in the second equation by 3:

- $6\ Cl^- \rightarrow 3\ Cl_2 + 6\ e^-$

Add the two equations:

- $Cr_2O_7^{2-} + 14\ H^+ + 6\ e^- + 6\ Cl^- \rightarrow 2\ Cr_3+ + 7\ H_2O + 3\ Cl_2 + 6\ e^-$

Step 4 – Complete the process:

Except for the electrons, there are no identical species to combine or cancel.

- $Cr_2O_7^{2-} + 14\ H^+ + 6\ Cl^- \rightarrow 2\ Cr^{3+} + 7\ H_2O + 3\ Cl_2$

Step 5 – For the ion-electron method, the equation is now balanced. The spectator ions need to be added to the equation. When adding the spectator ions, add equal numbers of ions to both sides of the reaction.

- To the left side: dichromate ion was paired with K^+, so add 2 K^+ for the dichromate.

- To the right side: match the left side by adding 2 K^+ ions.

$$K_2Cr_2O_7 + 14\ H^+ + 6\ Cl^- \rightarrow 2\ Cr^{3+} + 7\ H_2O + 3\ Cl_2 + \mathbf{2\ K^+}$$

Step 6 – There are 14 H^+ on the left side and 14 on the right, so they are balanced. Referring to the original equation, the H and Cl elements on the left originated from the HCl. Add 8 Cl elements to the product side.

- $K_2Cr_2O_7 + 14\ HCl \rightarrow 2\ Cr^{3+} + 7\ H_2O + 3\ Cl_2 + 2\ K^+ + \mathbf{8\ Cl^-}$

Step 7 – The right side shows that two of the Cl^- need to be combined with the 2 K^+, and the remaining 6 Cl^- goes with the Cr.

Thus, the final balanced redox equation is:

$$K_2Cr_2O_7\ (aq) + 14\ HCl\ (aq) \rightarrow \mathbf{2}\ Cr\mathbf{Cl_3}\ (aq) + 7\ H_2O\ (l) + 3\ Cl_2\ (g) + \mathbf{2\ K}Cl\ (aq)$$

Example: Balance the same reaction using the oxidation-state method:

$$K_2Cr_2O_7\ (aq) + HCl\ (aq) \rightarrow KCl\ (aq) + CrCl_3\ (aq) + H_2O\ (l) + Cl_2\ (g)$$

Step 1 - Separate into half-reactions (same as the ion-electron method):

Reduction: $Cr_2O_7^{2-} \rightarrow Cr^{3+}$

Oxidation: $Cl^- \rightarrow Cl_2$

Step 2 - Balance each half-reaction.

Balance the elements of interest first when using the oxidation-state method.

Balance the elements for the reduction half-reaction (oxidation-state method):

1. $Cr_2O_7^{2-} \rightarrow Cr^{3+}$
2. $Cr_2O_7^{2-} \rightarrow 2\ Cr^{3+}$
3. Each oxygen is 2^- so the 2 Cr on the left must be 6^+
4. $2\ Cr^{6+} \rightarrow 2\ Cr^{3+}$

Balance charge for the reduction half-reaction (oxidation-state method):

1. $2\ Cr^{6+} + 6\ e^- \rightarrow 2\ Cr^{3+}$

Balance charge for the oxidation half-reaction (oxidation-state method):

1. $Cl^- \rightarrow Cl_2$
2. $2\ Cl^- \rightarrow Cl_2$
3. $2\ Cl^- \rightarrow 2\ Cl^\circ$
4. $2\ Cl^- \rightarrow 2\ Cl^\circ + 2\ e^-$

Step 3 - Recombine the half-reactions:

$2\ Cr^{6+} + 6\ e^- \rightarrow 2\ Cr^{3+}$

$2\ Cl^- \rightarrow 2\ Cl^\circ + 2\ e^-$

To cancel the electrons, multiply each term in the second equation by 3:

$2\ Cr^{6+} + 6\ e^- \rightarrow 2\ Cr^{3+}$

$6\ Cl^- \rightarrow 6\ Cl^\circ + 6\ e^-$

Add the two equations:

$2\ Cr^{6+} + 6\ e^- + 6\ Cl^- \rightarrow 2\ Cr^{3+} + 6\ Cl^\circ + 6\ e^-$

Step 4 - Except for the electrons, there are no like terms to combine or cancel.

$2\ Cr^{6+} + 6\ Cl^- \rightarrow 2\ Cr^{3+} + 6\ Cl^\circ$

Convert the elements into species by referring to the original equation.

$$K_2Cr_2O_7 + 6\ HCl \rightarrow 2\ CrCl_3 + 3\ Cl_2$$

Unlike the ion-electron method, where the equation is balanced, and spectator ions are added into the equation, the oxidation-state method requires the equation to be balanced again. This is because after the elements are combined to recreate the molecules, the equation is no longer balanced.

Step 5 - Oxygen: there are seven O atoms on the left, so add 7 H_2O molecules to the right. (Remember that this method applies to acidic reactions – refer to the explanation for basic reactions).

$$K_2Cr_2O_7 + 6\ HCl \rightarrow 2\ CrCl_3 + 3\ Cl_2 + \mathbf{7\ H_2O}$$

Step 6 - Hydrogen: there are 6 H atoms on the left, but 14 H atoms on the right. Eight H atoms should be added to the left for a total of 14 H atoms. All 14 H atoms on the left should be as HCl (refer to the original equation).

$$K_2Cr_2O_7 + \mathbf{14\ HCl} \rightarrow 2\ CrCl_3 + 3\ Cl_2 + 7\ H_2O$$

Important: HCl is both the species of interest and the spectator ion. Some of the HCl contributes to the $Cl^- \rightarrow Cl_2$ oxidation, but the other HCl does not undergo redox. It merely provides the H^+ ions for the water and the Cl^- ions for the KCl and $CrCl_3$.

Step 7 - Chlorine: there are 14 Cl atoms on the left, and 12 Cl atoms on the right. Add 2 Cl atoms to the right. From the original equation, all the right-sided Cl atoms come in the form of KCl (do not modify the Cl_2 since it has already been balanced by the oxidation state method. When balancing equations at this stage, only manipulate the water and spectator species.

$$K_2Cr_2O_7 + 14\ HCl \rightarrow 2\ CrCl_3 + 3\ Cl_2 + 7\ H_2O + \mathbf{2\ KCl}$$

Upon examination, every element is balanced.

The balanced redox equation is:

$$K_2Cr_2O_7\ (aq) + 14\ HCl\ (aq) \rightarrow 2\ CrCl_3\ (aq) + 3\ Cl_2\ (g) + 7\ H_2O\ (l) + 2\ KCl\ (aq)$$

Ten Step for Balancing Redox Half-Reactions

1. Split the equation into two half-reactions.

2. Balance any atom other than O or H.

3. Balance oxygens using H_2O. Add water as needed to balance oxygens on the side deficient in O.

4. Balance hydrogens using H^+ if in acidic solution. Add H^+ ions as needed to the side deficient in H.

5. In basic solution, add OH^- ions equal to the number of H^+ ions to both sides of the chemical equation.

6. The mass should now be balanced. To balance the charge, determine the charge on each side of the reaction and add electrons as needed to the more positive side, so the charge is the same on each side.

7. Repeat steps 1 through 4 for both half-reactions.

8. If the number of electrons lost does not equal the number of electrons gained, multiply each half-reaction by the necessary factor.

9. Sum both half-reactions to obtain the balanced net ionic reaction. In many cases, inspect for H^+ ions and H_2O molecules to cancel.

10. Check the final equation for mass and charge balance.

Important: If expressing H^+ as H_3O^+, change the number of H^+ into the same number of H_3O^+. Add that same number of H_2O to the opposite side of the equation.

Limiting reactants

The *limiting reactant* is the component of a chemical reaction in the lowest stoichiometric quantity, which limits the amount of product formed in a reaction. The limiting reactant is the reagent that is depleted first and stops the reaction.

Consider a 50.6 g sample of magnesium hydroxide $Mg(OH)_2$ that reacts with 45.0 g of hydrogen chloride HCl, according to the reaction given below:

$$Mg(OH)_2 + 2\ HCl \rightarrow MgCl_2 + 2\ H_2O$$

Identify the limiting reactant. Notice how quantities of both reactants are known.

There are three common ways of approaching a limiting reagent problem.

Method 1 for determining the limiting reactant

To identify the limiting reagent, convert each of the grams of reactants given to moles using their molar mass:

$$50.6\ g\ Mg(OH)_2 \times (1\ mol\ Mg(OH)_2 / 58.3\ g\ Mg(OH)_2) = 0.868\ mol\ Mg(OH)_2$$

$$45.0\ g\ HCl \times (1\ mol\ HCl / 36.5\ g\ HCl) = 1.23\ mol\ HCl$$

Select one of these reactants and calculate how many moles of the other reactant is needed to use up the reactant completely.

For example, begin with magnesium hydroxide:

$$0.868\ mol\ Mg(OH)_2 \times (2\ mol\ HCl\ needed / 1\ mol\ Mg(OH)_2)$$

$$= 1.74\ mol\ HCl\ needed$$

Compare the moles of HCl needed to the actual moles of HCl available. In this example, 1.74 moles of HCl is needed, and 1.23 moles of HCl is available.

So, even though it appears that there are more moles of HCl than $Mg(OH)_2$, HCl is the limiting reagent. The HCl is completely consumed before the magnesium hydroxide, thereby limiting the amount of product formed.

Method 2 for determining the limiting reactant

Compare the *theoretical ratio* of reactants to the *actual ratio* of reactants.

First, determine the moles of each reactant using their molar mass:

$$50.6\ g\ Mg(OH)_2 \times (1\ mol\ Mg(OH)_2 / 58.3\ g\ Mg(OH)_2)$$

$$= 0.868\ mol\ Mg(OH)_2\ available$$

$$45.0\ g\ HCl \times (1\ mol\ HCl / 36.5\ g\ HCl)$$

$$= 1.23\ mol\ HCl\ available$$

Consider the balanced reaction:

$$Mg(OH)_2 + 2\ HCl \rightarrow MgCl_2 + 2\ H_2O$$

From the balanced equation, the theoretical mole ratio is:

$$2\ moles\ of\ HCl\ needed / 1\ mol\ Mg(OH)_2$$

The actual mole ratio, based on the actual amounts of reactants present:

$$1.23\ mol\ HCl / 0.868\ Mg(OH)_2 = 1.42\ mol\ HCl\ present / 1\ mol\ Mg(OH)_2$$

From these ratios, there is not enough HCl, so HCl is the limiting reagent.

(2 mol HCl needed / 1 mol $Mg(OH)_2$)

compared to (1.42 mol HCl present / 1 mol $Mg(OH)_2$)

Method 3 for determining the limiting reactant

Calculate the theoretical yield for each reactant and choose the lesser amount:

50.6 *g* $Mg(OH)_2$ × (1 mol $Mg(OH)_2$ / 58.3 *g* $Mg(OH)_2$) × (1 mol $MgCl_2$ / 1 mol $Mg(OH)_2$) × (95.3 *g* $MgCl_2$ / 1 mol $MgCl_2$)

= 82.7 *g* $MgCl_2$

45.0 *g* HCl × (1 mol HCl / 36.5 *g* HCl) × (1 mol $MgCl_2$ / 2 mol HCl) × (95.3 *g* $MgCl_2$ / 1 mol $MgCl_2$)

= 58.6 *g* $MgCl_2$

HCl produced less product (limiting reagent); 58.6 g $MgCl_2$ as theoretical yield.

Theoretical yields

The *theoretical yield* is the amount of product expected from a reaction. Reactions almost always yield less product than the predicted yield for various reasons (e.g., equipment inefficiency or human error).

The *percent yield* is the ratio of experimental (actual) to theoretical yields.

percent yield = (experimental yield / theoretical yield) × 100%

If the reaction of 30.0 grams of calcium carbonate ($CaCO_3$) produces 15.0 grams of calcium oxide (CaO), what is the percent yield for the following reaction?

$CaCO_3 \rightarrow CaO + CO_2$

The experimental yield was 15.0 g, so the theoretical yield needs to be determined. For every mole of $CaCO_3$ reacted, one mole of CaO produced. From the periodic table, the molar mass of $CaCO_3$ is 100.0896, and the molar mass of CaO is 56.0774. Use these conversion factors to perform the stoichiometric calculations and determine the theoretical yield.

30.0 g $CaCo_3$ × (1 mol $CaCO_3$ / 100.0869 *g*) × (1 mol CaO / 1 mol $CaCO_3$)

× (56.0774 *g* CaO / 1 mol CaO) = 16.8 *g* CaO

Theoretical yield = 16.8 *g* CaO

Then, use the formula given above to calculate the percent yield.

% Yield = (Actual Yield / Theoretical Yield) × 100%

% Yield = (15.0 *g* CaO / 16.8 *g* CaO) × 100%

% Yield = 89.3%

Practice Questions

1. Which substance listed below is the strongest reducing agent, given the following spontaneous redox reaction?

$$Mg\ (s) + Sn^{2+}\ (aq) \rightarrow Mg^{2+}\ (aq) + Sn\ (s)$$

A. Sn **B.** Mg^{2+} **C.** Sn^{2+} **D.** Mg **E.** None of the above

2. Which of the following is a guideline for balancing redox equations by the oxidation number method?

A. Verify that the total number of atoms and the total ionic charge is the same for reactants and products
B. In front of the substance reduced, place a coefficient that corresponds to the number of electrons lost by the substance oxidized
C. In front of the substance oxidized, place a coefficient that corresponds to the number of electrons gained by the substance reduced
D. Determine the electrons lost by the substance oxidized and gained by the substance reduced
E. All of the above

3. Which of the following represents the oxidation of Co^{2+}?

A. $Co \rightarrow Co^{2+} + 2\ e^-$
B. $Co^{3+} + e^- \rightarrow Co^{2+}$
C. $Co^{2+} + 2\ e^- \rightarrow Co$
D. $Co^{2+} \rightarrow Co^{3+} + e^-$
E. $Co^{3+} + 2\ e^- \rightarrow Co^+$

4. What is the molecular formula of a compound that has an empirical formula of CHCl with a molar mass of 194 g/mol?

A. $C_4H_4Cl_4$
B. $C_2H_4Cl_3$
C. $C_3H_5Cl_3$
D. CHCl
E. $C_4H_6Cl_4$

5. What is the term for a substance that contains 6.02×10^{23} particles?

A. molar mass
B. mole
C. Avogadro's number
D. formula mass
E. none of the above

6. How many grams are in 0.7 moles of $CaCO_3$?

A. 25 g **B.** 40 g **C.** 70 g **D.** 48 g **E.** 57 g

7. What is the coefficient for CO_2 in the balanced reaction?

$$\_\_C_5H_{12} + \_\_O_2 \rightarrow \_\_CO_2 + \_\_H_2O$$

A. 5 **B.** 7 **C.** 8 **D.** 10 **E.** 12

8. What are the products for this double-replacement reaction?

$$BaCl_2\ (aq) + K_2SO_4\ (aq) \rightarrow$$

A. $BaSO_3$ and $KClO_4$
B. $BaSO_4$ and 2 KCl
C. BaS and $KClO_4$
D. $BaSO_3$ and KCl
E. $BaSO_4$ and $KClO_4$

9. Which coefficients balance the following equation: $\_\_P_4\ (s) + \_\_H_2\ (g) \rightarrow \_\_PH_3\ (g)$?

A. 2, 10, 8
B. 1, 4, 4
C. 1, 6, 4
D. 4, 2, 3
E. 1, 3, 4

10. After balancing the following redox reaction in acidic solution, what is the coefficient of H^+?

$$Mg\ (s) + NO_3^-\ (aq) \rightarrow Mg^{2+}\ (aq) + NO_2\ (aq)$$

A. 1 **B.** 2 **C.** 4 **D.** 6 **E.** None of the above

11. How many atoms are in a sample of phosphorus trifluoride (PF_3) that contains 1.40 moles?

A. 3.37×10^{24}
B. 5.38
C. 3.46
D. 2.218×10^{24}
E. 8.98×10^{23}

12. Which of the following is the percent mass composition of acetic acid (CH_3COOH)?

A. 48% carbon, 8% hydrogen and 44% oxygen
B. 52% carbon, 12% hydrogen and 36% oxygen
C. 32% carbon, 6% hydrogen and 62% oxygen
D. 40% carbon, 7% hydrogen and 53% oxygen
E. 34% carbon, 4% hydrogen and 62% oxygen

13. If one mole of Ag is produced in the following reaction, how many grams of O_2 gas is produced?

$$2\ Ag_2O \rightarrow 4\ Ag + O_2$$

A. 6g **B.** 5g **C.** 2g **D.** 12g **E.** 8g

14. Which metal in the free state has an oxidation number of zero?

A. Mg
B. Na
C. Al
D. Ag
E. All of the above

15. What is the oxidation number of Cr in $K_2Cr_2O_7$?

A. +6 **B.** +5 **C.** +4 **D.** +2 **E.** +1

Detailed Explanations

1. D is correct.

Use the mnemonic OIL RIG: **O**xidation **I**s **L**oss, **R**eduction **I**s **G**ain (of electrons).

Oxidation is the loss of electrons, while reduction is the gain of electrons.

An oxidizing agent undergoes reduction, while a reducing agent undergoes oxidation.

Because the reaction is spontaneous, the reducing agent as a reactant is the strongest reducing agent in the reaction.

Mg is the reducing agent being oxidized because it went from 0 as a reactant to +2 as a product.

2. E is correct.

All statements are correct for balancing redox equations by the oxidation number method.

3. D is correct.

Use the mnemonic OIL RIG: **O**xidation **I**s **L**oss, **R**eduction **I**s **G**ain (of electrons).

Oxidation is the loss of electrons, while reduction is the gain of electrons.

An oxidizing agent undergoes reduction, while a reducing agent undergoes oxidation.

Co^{2+} is the starting reactant because it must lose electrons and produce an ion with a higher oxidation number.

4. A is correct.

Formula mass is a synonym for molecular mass/molecular weight (MW).

Start by calculating the mass of the formula unit (CHCl):

$CHCl = (12.01 \text{ g/mol} + 1.01 \text{ g/mol} + 35.45 \text{ g/mol})$

$CHCl = 48.47 \text{ g/mol}$

Divide the molar mass by the formula unit mass:

$194 \text{ g/mol} / 48.47 \text{ g/mol} = 4.0025$

Round it to the closest whole number: 4

Multiply the formula unit by 4:

$(CHCl)_4 = C_4H_4Cl_4$

5. B is correct.

A mole is a unit of measurement used to express amounts of a chemical substance.

The number of molecules in a mole is 6.02×10^{23}, which is Avogadro's number.

However, Avogadro's number relates to the number of molecules, not the amount of substance.

Molar mass refers to the mass per mole of a substance.

Formula mass is a term that is sometimes used to mean molecular mass or molecular weight, and it refers to the mass of a molecule.

6. C is correct.

MW of $CaCO_3$:

$(Ca = 40.08 \text{ g/mol}) + (C = 12 \text{ g/mol}) + (O = 3 \times 16 \text{ g/mol}) = 100 \text{ g/mol}$

Mass of 0.7 moles of $CaCO_3$: 0.7 mole $CaCO_3 \times 100$ g/mole = 70 g $CaCO_3$

7. A is correct.

The number of oxygens on both sides of the reaction equation must be equal.

On the right side, there are 5 CO_2 molecules; since each CO_2 molecule contains 2 oxygens, multiply $5 \times 2 = 10$.

Also, on the right side are 6 H_2O molecules: $6 \times 1 = 6$.

Add 10 and 6 to get the total number of oxygens on the right side: $10 + 6 = 16$.

Since there are 16 oxygens on the right side, there should be 16 oxygens on the left side.

Each O_2 molecule contains 2 oxygens: $16 / 2 = 8$.

Therefore, the coefficient 8 is needed to balance the equation.

Balanced equation (combustion):

$$C_5H_{12} + 8\ O_2 \rightarrow 5\ CO_2 + 6\ H_2O$$

8. B is correct.

Balanced reaction:

$$BaCl_2\ (aq) + K_2SO_4\ (aq) \rightarrow BaSO_4 \text{ and } 2\ KCl$$

A double replacement reaction indicates an exchange of cations and anions between the reactants.

Separate the reactants into ions and then exchange the cation and anion pairings.

9. C is correct.

Balanced equation (synthesis):

$$P_4\ (s) + 6\ H_2\ (g) \rightarrow 4\ PH_3\ (g)$$

10. C is correct.

Balancing Redox Equations

From the balanced equation, the coefficient for the proton is determined. Balancing a redox equation not only requires balancing the atoms that are in the equation, but the charges must be balanced as well.

The equations must be separated into two different half-reactions. One of the half-reactions addresses the oxidizing component, and the other addresses the reducing component.

Magnesium is oxidized. Therefore the unbalanced oxidation half-reaction is:

$$Mg\ (s) \rightarrow Mg^{2+}\ (aq)$$

Nitrogen is reduced. Therefore the unbalanced reduction half-reaction is:

$NO_3^-\ (aq) \rightarrow NO_2\ (aq)$

Each half-reaction needs to be balanced for each atom, and the net electric charge on each side of the equations is balanced.

Order of operations for balancing half-reactions:

1) Balance all atoms except for oxygen and hydrogen.
2) Balance the oxygen atoms by adding water.
3) Balance the hydrogen atoms by adding protons.
   a) In basic solution, add equal amounts of hydroxide to each side to cancel the protons.
4) Balance the electric charge by adding electrons.
5) If necessary, multiply the coefficients of one half-reaction equation by a factor that cancels the electron count when both equations are combined.
6) Cancel any ions or molecules that appear on both sides of the overall equation.

After determining the balanced overall redox reaction, the stoichiometry indicates the moles of protons that are involved.

For magnesium:

$Mg\ (s) \rightarrow Mg^{2+}\ (aq)$

The magnesium is already balanced with a coefficient of 1. There are no hydrogen or oxygen atoms present in the equation. The magnesium cation has a +2 charge, so to balance the charge, 2 moles of electrons should be added to the right side.

The balanced half-reaction for oxidation:

$Mg\ (s) \rightarrow Mg^{2+}\ (aq) + 2\ e^-$

For nitrogen:

$NO_3^-\ (aq) \rightarrow NO_2\ (aq)$

The equation is balanced for N because one N atom appears on both sides of the reaction.

The nitrate reactant has three O atoms, while the nitrite product has two O atoms.

To balance oxygen, one mole of water is added to the right side of the reaction:

$NO_3^-\ (aq) \rightarrow NO_2\ (aq) + H_2O$

Adding water to the right side of the equation introduces hydrogen atoms to that side. Therefore, the hydrogen atom count needs to be balanced. Water possesses two hydrogen atoms. Therefore, two protons are added to the left side of the reaction:

$NO_3^- (aq) + 2 H^+ \rightarrow NO_2 (aq) + H_2O$

The reaction is occurring in acidic conditions. If the reaction were basic, then OH^- would be added to both sides to cancel the protons.

The net charge needs to be balanced. The left side has a net charge of +1 (+2 from the protons and –1 from the electron), while the right side is neutral.

Therefore, one electron should be added to the left side:

$NO_3^- (aq) + 2 H^+ + e^- \rightarrow NO_2 (aq) + H_2O$

When half-reactions are recombined, the electrons in the overall reaction must cancel.

The reduction half-reaction contributes one electron to the left side of the overall equation, while the oxidation half-reaction will contribute two electrons to the product side.

Therefore, the coefficients of the reduction half-reaction should be doubled:

$2 \times [NO_3^- (aq) + 2 H^+ + e^- \rightarrow NO_2 (aq) + H_2O]$

$= 2 NO_3^- (aq) + \mathbf{4 H^+} + 2 e^- \rightarrow 2 NO_2 (aq) + 2 H_2O$

The answer is 4 at this step.

Combining both half-reactions gives:

$Mg (s) + 2 NO_3^- (aq) + 4 H^+ + 2 e^- \rightarrow Mg^{2+} (aq) + 2 e^- + 2 NO_2 (aq) + 2 H_2O$

Cancel electrons from both sides of the reaction in the following balanced net equation:

$Mg (s) + 2 NO_3^- (aq) + 4 H^+ \rightarrow Mg^{2+} (aq) + 2 NO_2 (aq) + 2 H_2O$

This equation is now fully balanced for mass, oxygen, hydrogen, and electric charge.

11. A is correct.

Number of molecules = moles × Avogadro's constant

Number of molecules = 1.40 mol × 6.02×10^{23} molecules/mol

Number of molecules = 8.428×10^{23} molecules

4 atoms in each PF_3 molecule:

$4 \times 8.428 \times 10^{23} = 3.37 \times 10^{24}$ atoms

12. D is correct.

CH_3COOH:

mass of O atoms = 2 × 16 g/mol = 32 g/mol

mass of H atoms = 4 × 1 g/mol = 4 g/mol

Therefore, the mass % of O is 8 times (32 / 4) greater than the mass % of H.

Calculate the molecular mass (MW) of acetic acid:

MW of CH_3COOH:

= (2 × atomic mass of C) + (4 × atomic mass of H) + (2 × atomic mass of O)

MW of CH_3COOH = (2 × 12.01 g/mole) + (4 × 1.01 g/mole) + (2 × 16.00 g/mole)

MW of CH_3COOH = 60.05 g/mole

To obtain the percent mass composition of each element, calculate the total mass of each element, divide it by the molecular mass, and multiply with 100%.

% mass composition of carbon = [(2 × 12.01 g/mole) / 60.05 g/mole] × 100%

% mass composition of carbon = 40.00%

% mass composition of hydrogen = [(4 × 1.01 g/mole) / 60.05 g/mole] × 100%

% mass composition of hydrogen = 6.71%

% mass composition of oxygen = [(2 × 16.00 g/mole) / 60.05 g/mole] ×100%

% mass composition of oxygen = 53%

13. E is correct.

From the balanced equation:

4 moles of Ag is produced for each mole of O_2.

Therefore, if 1 mole of Ag is produced, ¼ mole of O_2 is produced.

The mass of a ¼ mole of O_2 is: (¼ mol)·(2 × 16 g/mol) = 8 g

14. E is correct.

All metals (and elements) have an oxidation number of zero in their elemental state.

15. A is correct.

Assign oxidation number to each species:

K_2	Cr_2	O_7
2(+1)	$2x$	7(–2)

The sum of charges in a neutral molecule is zero:

$2 + (2x) + (7 \times -2) = 0$

$2 + (2x) + (-14) = 0$

$2x = -2 + 14$

$2x = +12$

+6 = oxidation number of Cr

Chapter 5

Thermochemistry and Thermodynamics

Thermochemistry

- **Thermodynamic System, State Function**
- **Endothermic and Exothermic Reactions**
- **Bond Dissociation Energy as Related to Heats of Formation**
- **Measurement of Heat Changes (Calorimetry), Heat Capacity, Specific Heat Capacity**
- **Gibbs Free Energy (*G*)**
- **Spontaneous Reactions and $\Delta G°$**

Thermodynamics

- **Zeroth Law (Concept of Temperature)**
- **First Law (Conservation of Energy in Thermodynamic Processes)**
- **Equivalence of Mechanical, Chemical, Electrical, and Thermal Energy Units**
- **Second Law (Concept of Entropy)**
- **Third Law (Entropy at Absolute Zero)**
- **Temperature Scales, Conversions**
- **Heat Transfer (Conduction, Convection, Radiation)**
- **Heat of Fusion, Heat of Vaporization**
- **PV Diagram (Work Done = Area Under or Enclosed by Curve)**
- **Coefficient of Expansion**
- **Phase Diagram: Pressure and Temperature**

Thermochemistry:
Energy Changes in Chemical Reactions

Thermochemistry studies energy and heat associated with chemical reactions and physical transformations.

A reaction may release energy (exothermic reactions) or absorb energy (endothermic reactions), and a phase change may either release or absorb energy (e.g., boiling, melting).

Thermochemistry:

explains these energy changes, particularly in the context of the energy exchange between the system and its surroundings.

predicts reactant and product quantities during the reaction.

predicts if a reaction is spontaneous or non-spontaneous, favorable, or unfavorable.

combines the concepts of thermodynamics and energy of chemical bonds.

Thermochemistry commonly includes calculations of such quantities as:

heat capacity

the heat of combustion

the heat of formation

enthalpy, H

entropy, S

Gibbs free energy, G

calories

Thermodynamic System, State Function

Thermodynamics relates to the energy requirements of physical and chemical changes. Thermodynamics defines many variables related to energy.

Some typical thermodynamic variables are listed below:

ΔE or ΔU	measures the total internal energy of a system
q	measures the heat of a system
w	measures the work of a system
ΔH	measures the enthalpy change of a system – whether a process is exothermic or endothermic
ΔG	measures the Gibbs free energy change of a system – whether a process is spontaneous or nonspontaneous
ΔS	measures the entropy change of system or surroundings – whether the randomness of system or surroundings is increasing or decreasing

A *state function* is a quantity whose value does not depend on the path used to measure the value.

State functions such as $\Delta E°$, $\Delta H°$, $\Delta S°$, and $\Delta G°$ depend only on the initial and final states of the system.

Quantities, such as work (w) and heat (q), which are not state functions. These quantities are *path functions* and are generally represented by lowercase letters.

The term *system* defines the part of the universe being studied, such as the substances involved in a chemical reaction or a phase change.

The term *surroundings* are everything other than the system that is being studied.

Different types of systems can be studied:

- An *isolated system* has no exchange of heat, work, or matter between the system and surroundings.

- A *closed system* has an exchange of heat and work but not an exchange of matter between the system and surroundings.

- An *open system* has heat, work, and matter exchanged between the system and surroundings.

Note that heat and temperature are different quantities.

Heat is a specific form of thermal energy that can leave or enter a system.

Temperature is a measure of the average kinetic energy of particles in a system.

Endothermic and Exothermic Reactions

Some chemical reactions produce energy, while others require heat energy. Bonding energy is used to form or break chemical bonds.

The *enthalpy change* (ΔH) describes the energy change associated with the formation or destruction of a chemical bond.

The unit for enthalpy (H) is the Joules (J). The Joule is the same unit used for energy; it can also be expressed as energy per mol (J/mol).

An *exothermic reaction* releases heat.

Energy is released into the surroundings because stronger, more stable bonds are formed in the products compared to the bonds in the reactants.

For exothermic reactions, the energy that is needed to initiate a reaction is less than the energy that is released, leading to the liberation of heat.

An exothermic reaction can be represented in a chemical reaction in the following way, with the heat on the product side:

$$4\ Fe\ (s) + 3\ O_2\ (g) \rightarrow 2\ Fe_2O_3\ (s) + \text{heat}$$

An example of an exothermic reaction is fuel burning.

In an exothermic reaction, heat is released, and the sign of ΔH is (–), meaning that the energy of the products is less than the energy of the reactants.

The amount of energy released can be calculated using bond-dissociation energies, which is explained later.

Exothermic reaction:

energy is released → $\Delta H < 0$ → enthalpy decreases

The chemical reaction for hydrogen gas and chlorine gas (diagram below):

$$H_2\ (g) + Cl_2\ (g) \rightarrow 2\ HCl\ (g) + 185\ \text{kJ}$$

$$\Delta H = -185\ \text{kJ/mol (heat released)}$$

Heat is released, indicating that the reaction is exothermic, and ΔH is negative.

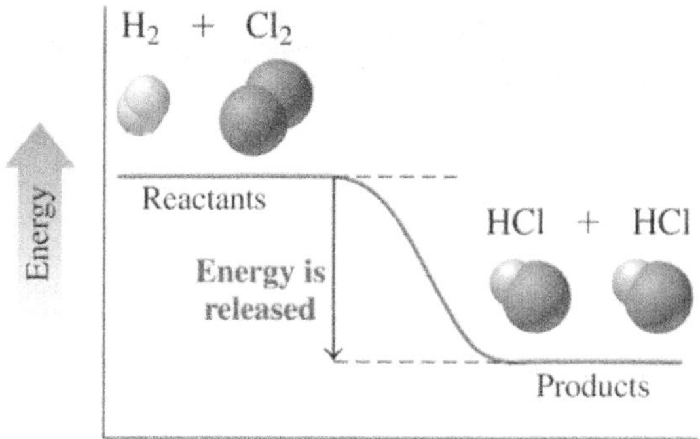

Endothermic reactions require (net) energy to break bonds. In endothermic reactions, weaker bonds form in the products compared to the bonds in the reactants. An endothermic chemical reaction can be represented with the heat on the reactant side:

Heat + $NH_4NO_3\,(s) \rightarrow NH_4NO_3\,(aq)$

The above reaction is the ionic compound ammonium nitrate dissolving in water, which means that it is breaking up into its constituent ions, NH_4 and NO_3.

A few examples of endothermic reactions are melting ice and cooking an egg (heat is required for both processes).

In an endothermic reaction, heat is absorbed, and the sign of ΔH is (+); the energy of the products is greater than the energy of the reactants. The amount of energy absorbed is calculated using bond-dissociation energies.

Endothermic reaction:

energy is absorbed $\rightarrow \Delta H > 0 \rightarrow$ enthalpy increases

The chemical reaction between nitrogen gas (N_2) and oxygen gas (O_2) to form nitrogen monoxide (NO):

$N_2\,(g) + O_2\,(g) + 181 \text{ kJ} \rightarrow 2\,NO\,(g)$

$\Delta H = +181$ kJ (heat added)

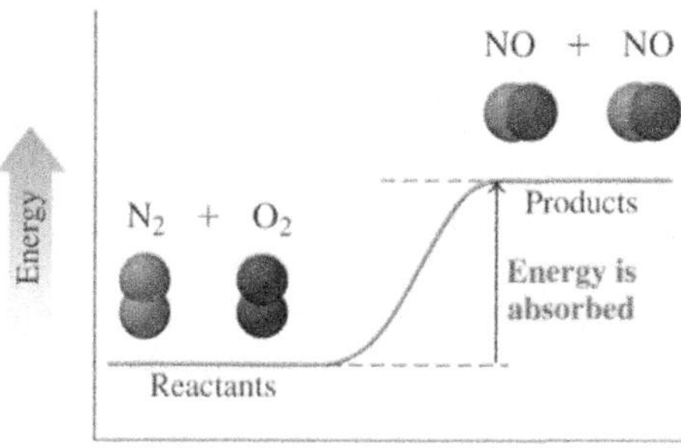

Heat is absorbed, indicating that the reaction is endothermic, and ΔH is positive

Summary of exothermic and endothermic reactions

Reaction	Energy Change	Heat in the Equation	Sign of ΔH
Endothermic	Heat absorbed	Reactant side	Positive (+)
Exothermic	Heat released	Product side	Negative (–)

Is the following reaction endothermic or exothermic?

$$O_2N\text{–}NO_2 \rightarrow O_2N + NO_2$$

This reaction is endothermic (absorbs heat) because it forms two products from a single reactant. The breaking of a chemical bond requires energy.

Enthalpy (*H*), standard heats of reaction and formation

Enthalpy changes can occur at standard temperatures and pressures.

The *standard state* refers to substances in their most stable form (i.e., the lowest-energy state). The standard state in thermodynamics refers to one-atmosphere pressure (1 atm) and a temperature of 25 °C (298 K).

For example, at standard state, oxygen naturally occurs as O_2 (diatomic gas), and carbon exists as C (solid graphite).

The *standard heat of reaction* (*standard enthalpy of reaction* $\Delta H°_{rxn}$) is the change in heat content when one mole of the matter is transformed by a chemical reaction under standard conditions (1 atm and 298 K).

The *standard heat of formation* (*standard enthalpy of formation* $\Delta H°_f$) is the change in heat content associated with the formation of one mole of a compound or molecule from its constituent elements in their standard states. The standard enthalpy of formation has been determined for many molecules, and the $\Delta H°_f$ for the most common molecules are generally listed as tabulated data.

When choosing $\Delta H°_f$ values from a table, choose the value corresponding to the appropriate state of matter (solid, liquid, aqueous, gas). The standard heat of formation for an element in its standard state always has a value of 0 kJ/mol.

The enthalpy change of a chemical reaction can be calculated using both theoretical and experimental methods. The general equation to calculate the heat of a reaction from the difference in the energy of the reactants and products is:

$$\Delta H = H_{\text{products}} - H_{\text{reactants}}$$

For the following balanced chemical equation, calculate the heat of reaction from the heats of formation, and determine if the process is exothermic or endothermic:

$$16\ H_2S\ (g) + 8\ SO_2\ (g) \rightarrow 16\ H_2O\ (l) + 3\ S_8\ (s)$$

From the thermodynamic quantities, use the values for the heats of formation of each component in the reaction, Set the equation to calculate the heat of reaction:

$$\Delta H = H_{\text{products}} - H_{\text{reactants}}$$

$$\Delta H°_{\text{rxn}} = [16\ \text{mol}(\Delta H°_f \text{ of } H_2O\ (l)) + 3\ \text{mol}(\Delta H°_f \text{ of } S_8\ (s))] - [16\ \text{mol}(\Delta H°_f \text{ of } H_2S\ (g)) + 8\ \text{mol}(\Delta H°_f \text{ of } SO_2\ (g))]$$

$$\Delta H°_{\text{rxn}} = [16\ \text{mol}(-285.8\ \text{kJ/mol}) + 3\ \text{mol}(0\ \text{kJ/mol})] - [16\ \text{mol}(-20.2\ \text{kJ/mol}) + 8\ \text{mol}(-296.8\ \text{kJ/mol})]$$

$$\Delta H°_{\text{rxn}} = [-4{,}573\ \text{kJ} + 0\ \text{kJ}] - [-323\ \text{kJ} + (-2{,}374\ \text{kJ})]$$

$$\Delta H°_{\text{rxn}} = -4{,}573\ \text{kJ} + 323\ \text{kJ} + 2374\ \text{kJ}$$

$$\Delta H°_{\text{rxn}} = -1{,}876\ \text{kJ}$$

Since $\Delta H°_{\text{rxn}}$ is negative, the reaction is exothermic, whereby the system releases heat to the surroundings.

Calculate the heat of reaction (ΔH) using these steps:

1) Identify the given and needed quantities.
2) Write an expression using the heat of reaction and molar mass needed.
3) Write the conversion factors, including the heat of the reaction.
4) Set the expression to calculate heat.

In the following reaction, how much heat (kJ) is absorbed when 1.65 grams of nitrogen monoxide gas is produced?

$$N_2\ (g) + O_2\ (g) \rightarrow 2\ NO\ (g)$$

$$\Delta H = +181\ \text{kJ/mol}$$

Step 1: State the given and needed quantities.

Given:

1.65 grams of NO

ΔH = +181 kJ/mol

Solve:

Heat absorbed in kJ.

$N_2(g) + O_2(g) \rightarrow 2\ NO(g)$ $\quad \Delta H = +181$ kJ/mol

1.65 g $\quad$? kJ

Step 2: Write an expression for the heat of reaction and any molar mass needed.

grams of NO → moles of NO → kilojoules of energy

molar mass $\quad$ heat of reaction

Step 3: Write the conversion factors, including the heat of reaction.

1 mole of NO = 30.01 g of NO:

30.01 g NO / 1 mol NO $\quad$ or $\quad$ 1 mol NO / 30.01 g NO

ΔH for 2 moles of NO = +181 kJ/mol

therefore:

2 mol NO / +181 kJ/mol $\quad$ or $\quad$ +181 kJ/mol / 2 mol NO

Step 4: Set the expression to calculate the heat.

1.65 g NO × (1 mol NO/30.01 g NO) × (+181 kJ/mol/2 mol NO)

1.65 ~~g NO~~ × (1 ~~mol NO~~/30.01 ~~g NO~~) × (+181 kJ/mol/2 ~~mol NO~~)

1.65 × (1/30.01) × (+181 kJ/mol/2) = 4.98 kJ/mol

4.98 kJ/mol of nitrogen gas is absorbed when 1.65 g of nitrogen monoxide is produced from nitrogen and oxygen.

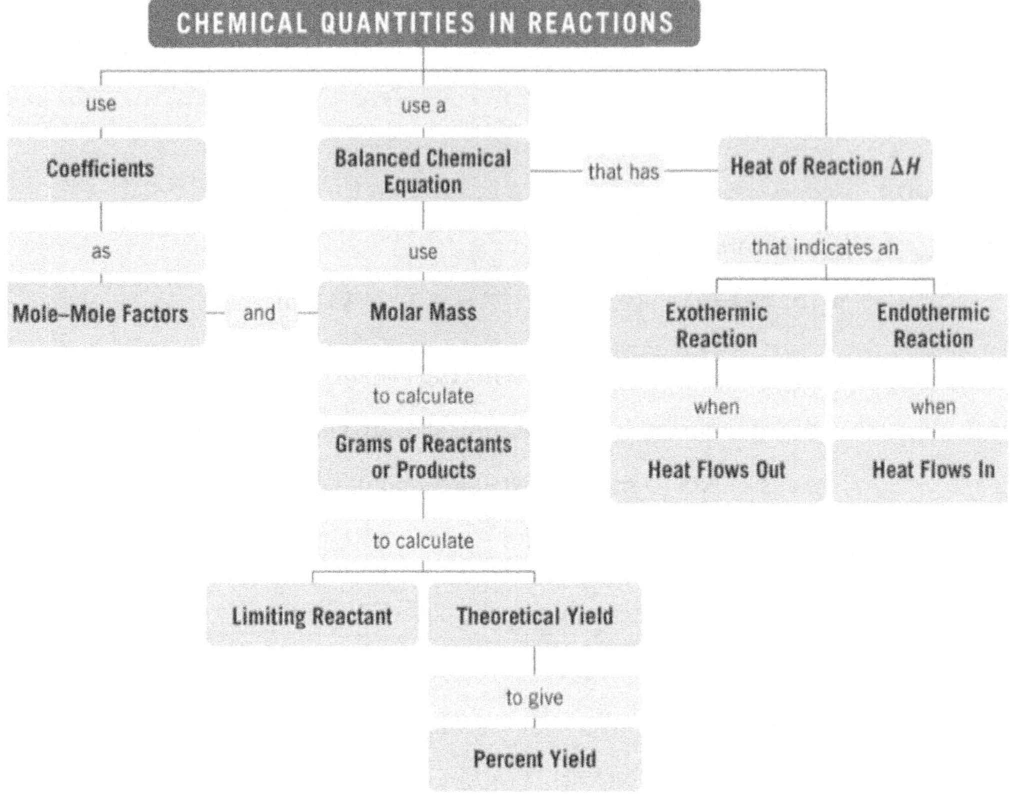

Hess' Law of Heat Summation

Hess' Law of constant heat summation is an essential relationship in physical chemistry. Enthalpy is a *state function* (i.e., depends only on initial and final states; it does not depend on the path of the reaction).

The equation for *Hess' Law of Heat Summation* is:

$$\Delta H^\circ_{rxn} = \Sigma\, n(\Delta H^\circ_f \text{ of products}) - \Sigma\, n(\Delta H^\circ_f \text{ of reactants})$$

where n represents the moles of each reactant or product in the balanced chemical equation. The Greek letter Σ means the sum of the variables.

Essentially, Hess' Law states that the total energy change throughout a reaction is the same, regardless of whether the reaction occurs in one step or several steps.

In addition to applying Hess' Law of Heat Summation, enthalpy can be calculated using alternative means.

For example, if a series of reactions whose enthalpies are known to create an overall reaction that gives the reaction sought, then the sum of those enthalpies is the enthalpy of the entire reaction. This method of calculating the enthalpy change is the *thermochemical systems of equations* method.

From Hess' Law, determine the enthalpy of reaction for the reaction:

$$2\ N_2\,(g) + 5\ O_2\,(g) \longrightarrow 2\ N_2O_5\,(g) \qquad \Delta H = ?$$

Use the following heats of reaction:

Equation (1) $2\ H_2\,(g) + O_2\,(g) \longrightarrow 2\ H_2O\,(l)$ $\qquad \Delta H = -571.7$ kJ

Equation (2) $N_2O_5\,(g) + H_2O\,(l) \longrightarrow 2\ HNO_3\,(l)$ $\qquad \Delta H = -92$ kJ

Equation (3) $N_2\,(g) + 3\ O_2\,(g) \longrightarrow 2\ HNO_3\,(l)$ $\qquad \Delta H = -348.2$ kJ

Manipulate the above equations so that the overall equation sums to the desired equation. First, inspect the compounds in each equation and choose an equation that contains a compound that only appears once in the equations.

Nitrogen, N_2, is found only in equation (3), so start with this equation.

Dinitrogen pentoxide, N_2O_5, is found only in equation (2), so equation (2) could also be used to start solving the problem.

Oxygen, O_2, is found in both equations (1) and (3), so it is not a suitable compound by which to begin solving the problem.

Continue solving the problem using the compound nitrogen and equation (3). Equation (3) is written containing only one mole of nitrogen as a reactant. Two moles of nitrogen are needed as a reactant.

Since nitrogen N_2 is a reactant in both instances, the equation does not need to be reversed, but equation (3) must be multiplied by two for the required two moles of nitrogen.

What is done to manipulate the chemical equation is applied to the heat of reaction; thus, the given heat of reaction is also multiplied by two:

2 × Equation (3):

$$2\ N_2\,(g) + 6\ O_2\,(g) + 2\ H_2O\,(g) \rightarrow 4\ HNO_3\,(l)$$

$$\Delta H = 2(-348.7\text{ kJ})$$

$$\Delta H = -696.4\text{ kJ}$$

Next, dinitrogen pentoxide is introduced into the problem, since only equation (2) contained the compound. In the equation sought, N_2O_5 appears as two moles of product. In equation (2), N_2O_5 appears as one mole of reactant. Reverse the equation and multiply by two. The sign of the heat of reaction is also changed, in addition to the equation being reversed and doubled:

−2 × Equation (2):

4 HNO_3 (*l*) ⟶ 2 N_2O_5 (*g*) + 2 H_2O (*l*)

$\Delta H = -2(-92 \text{ kJ})$

$\Delta H = +184 \text{ kJ}$

The four moles of nitric acid cancel. This is convenient since this compound is not found in the sought equation.

2 × Equation (3):

2 N_2 (*g*) + 6 O_2 (*g*) + 2 H_2O (*g*) ⟶ ~~4 HNO_3 (l)~~

$\Delta H = 2(-348.7 \text{ kJ})$

$\Delta H = -696.4 \text{ kJ}$

−2 × Equation (2):

~~4 HNO_3 (*l*)~~ ⟶ 2 N_2O_5 (*g*) + 2 H_2O (*l*)

$\Delta H = -2(-92 \text{ kJ})$

$\Delta H = +184 \text{ kJ}$

Cancel two moles of H_2 and H_2O, and one mole of O_2 by reversing equation (1).

Remember to reverse the sign for the heat of reaction:

2 × Equation (3):

2 N_2 (*g*) + ~~6~~ O_2 (*g*) + 2 H_2O (*g*) ⟶ ~~4 HNO_3 (*l*)~~

$\Delta H = 2(-348.7 \text{ kJ})$

$\Delta H = -696.4 \text{ kJ}$

−2 × Equation (2):

~~4 HNO_3 (*l*)~~ ⟶ 2 N_2O_5 (*g*) + 2 H_2O (*l*)

$\Delta H = -2(-92 \text{ kJ})$

$\Delta H = +184 \text{ kJ}$

$-1 \times$ Equation (1):

$$2\ H_2O\ (l) \longrightarrow 2\ H_2\ (g) + O_2\ (g)$$

$\Delta H = -1(571.7\ \text{kJ})$

$\Delta H = +571.1\ \text{kJ}$

Summing the chemical equations gives the desired reaction.

Summing the heats of reaction gives the heat of reaction for the overall process:

$$2\ N_2\ (g) + 5\ O_2\ (g) \longrightarrow 2\ N_2O_5\ (g)$$

$\Delta H = +59\ \text{kJ}$

The ΔH values used in calculations need not be for actual, measurable processes.

Values from the table of heats of formation have been calculated via Hess' Law, and can now be used for further calculations.

Notes

Bond-Dissociation Energy as Related to Heats of Formation

Bond-dissociation energy is the standard enthalpy change when a bond is cleaved and is thus a measure of the strength of a chemical bond.

Bond energy is a more specific term that refers to the heat required to break the chemical bonds of one mole of gaseous molecules to give separate gaseous atoms.

Since energy input is required to pull two atoms apart, bond-dissociation energy and bond energy are always positive quantities associated with an endothermic process.

Bond dissociation is the reverse process of *bond formation*, for which the energy is always a negative quantity and is associated with an exothermic process.

Bond energy depends on the types of atoms bonding. The H–H single bond, for example, has a bond energy of 436 kJ per mole, but the H–O single bond has a bond energy of 464 kJ per mole.

Bond energies for some common bonds

Bond	Bond Energy (kJ/mole)	Bond	Bond Energy (kJ/mole)
H–H	436	N–N	159
H–C	414	O–O	138
H–N	389	Cl–Cl	243
H–O	464	C=O	803
H–F	569	N=O	631
H–Cl	431	O=O	498
C–O	351	C≡C	837
C–C	347	N≡N	946

The bond energies in the above chart can be used to estimate ΔH°_f

Calculate ΔH°_f for CH_3OH (*g*) (the bottom reaction in the figure below)

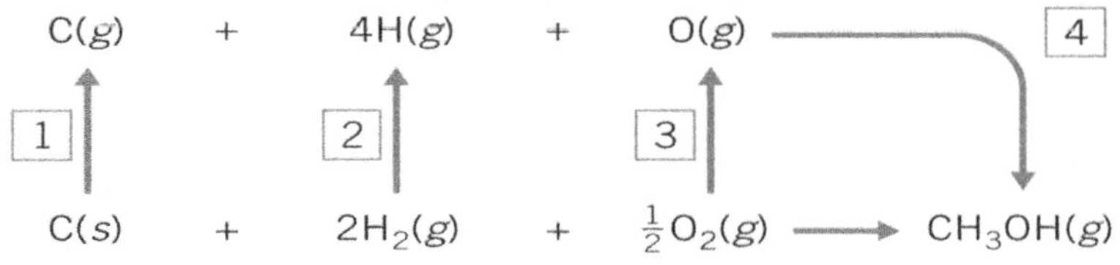

$C(g)$ + $4H(g)$ + $O(g)$ [4]

[1] ↑ [2] ↑ [3] ↑ ↓

$C(s)$ + $2H_2(g)$ + $\frac{1}{2}O_2(g)$ ⟶ $CH_3OH(g)$

C (*g*)	+	4 H (*g*)	+	O (*g*)		Step 4
Step 1 ↑		Step 2 ↑		Step 3 ↑		↓
C (*s*)	+	$2 H_2$ (*g*)	+	½ O_2 (*g*)	→	CH_3OH (*g*)

Use the four-step path to solve:

Step 1: break C–C bonds (positive energy)

Step 2: break H–H bonds (positive energy)

Step 3: break O–O bond (positive energy)

Step 4: form four C–H bonds (negative energy)

$\Delta H^\circ_f CH_3OH (g) = \Delta H^\circ_f C (g) + 4 \times \Delta H^\circ_f H (g) + \Delta H^\circ_f O (g) - \Delta H_{atom} CH_3OH (g)$

$\Delta H^\circ_f CH_3OH (g) = [716.7 + (4 \times 217.9) + 249.2]\ kJ - \Delta H_{atom} CH_3OH (g)$

$\Delta H^\circ_f CH_3OH (g) = 1{,}837.5\ kJ - \Delta H_{atom} CH_3OH (g)$

$\Delta H^\circ_f CH_3OH (g) = 1{,}837.5\ kJ - 3D_{C—H} + D_{C—O} + D_{O—H}$

$\Delta H^\circ_f CH_3OH (g) = 1{,}837.5\ kJ - (3 \times 412) + 360 + 463 = 2{,}059\ kJ$

$\Delta H^\circ_f CH_3OH (g) = 1{,}837.5\ kJ - 2{,}059\ kJ$

$\Delta H^\circ_f CH_3OH (g) = -222\ kJ$

The ΔH°_f of CH_3OH (*g*) is calculated as –222 kJ using bond energies.

According to the chart on the previous page, ΔH°_f of CH_3OH (*g*) was experimentally determined to equal –201 kJ/mol.

Therefore, bond energies give an estimate within 10% of the actual quantity.

Notes

Measurement of Heat Changes, Heat Capacity, Specific Heat Capacity

Calorimetry is an experimental method to calculate enthalpy change. The basic principle behind calorimetry involves heat flow. For an *exothermic* process, the heat loss by the system equals the heat gain by the surroundings. For an *endothermic* process, the heat gained by the system from the surroundings equals the heat lost by the surroundings.

A *calorimeter* is used in calorimetry experiments to measures the amount of energy in food. The food is placed inside a closed container surrounded by water and burned in the presence of oxygen. The change in the temperature of the water is related to the amount of energy released. The energy is commonly reported in units of nutritional Calories, or kilojoules.

Nutrient Molecule in Food	Example	Cal/g	kJ/g
Carbohydrate	Table sugar, potatoes, flour	4	17
Protein	Meats, fish, beans	4	17
Fat	Oil, butter	9	38

A nutritional Calorie (Cal) is an energy unit equal to 1,000 calories (cal) and a kilocalorie (kcal). Even though cells within the body combust the molecules differently, the calorimeter provides an accurate *caloric value* because the end products of the two reactions are the same. Different foods contain varying amounts of energy, as shown in the table.

Heat capacity (*thermal capacity*) is a physical property that relates to the ability of a substance to hold heat. Heat capacity is *the amount of heat energy required to raise a substance by one degree Celsius.*

Specific heat capacity is often used instead of heat capacity. The specific heat capacity differs from heat capacity in that it relates *the amount* of *heat energy required to raise one gram of a substance by one degree Celsius.*

A comparison of heat capacities between metals and water demonstrates how heat capacities of substances can vary over a wide range.

Metals tend to have low specific heat capacities. Water has a relatively high specific heat capacity; it takes 4.2 J of heat energy to raise the temperature of one gram of water by 1 °C. It is this property of water that helps control global temperatures.

Temperature fluctuations during day and night would be much larger if the heat capacity of water were low. This is evident in a desert where water is scarce. Daytime temperatures are high, yet nighttime temperatures can drop to near freezing.

The heat capacity (*mC*) of an object is the product of the mass *m* of the object and the specific heat capacity (*c*) of the substance.

The equation that relates specific heat to heat and temperature is:

$$q = mc\Delta T$$

where *q* is the heat flow, *m* is the mass measured in units of grams, *c* is the specific heat capacity of the substance, and ΔT is the change in temperature.

There are several types of heat capacities measured in chemistry, including:

Molar heat capacity = heat capacity per mole = J/mol·°C

Specific heat capacity (*c*) = heat capacity per mass = J/g·°C

In solving calorimetry problems, know that one-unit degree Celsius (1 °C) is the same as a change in one Kelvin (1 K).

Some useful conversion factors:

1 calorie = 4.2 J

1 Calorie (with capital C) = 1,000 calorie = 1 kilocalorie

1 Calorie = 4,200 J

For water, 1 gram = 1 cubic centimeter = 1 mL

In a typical calorimetry experiment, a hot substance (the system) is introduced to a cold substance (the surroundings). The system and surroundings reach thermal equilibrium at some final temperature. Heat dissipates from the hotter substance to the cooler substance. The heat lost by the hot substance (the system) equals in magnitude to the heat gained by the cooler substance (the surroundings) but opposite in sign.

$$q_{\text{system}} = -q_{\text{surrounding}}$$

There are two types of calorimeters. The *constant-pressure calorimeter* is where heat flow is measured at constant atmospheric pressure. A perfect calorimeter does not exist; no calorimeter can completely insulate heat flow. The calorimeter itself, as part of the surroundings, absorbs some heat. Therefore, the calorimeter must be calibrated to obtain the most accurate result.

The simplest type of constant-pressure calorimeter is a Styrofoam cup, which is often used in introductory chemistry classes. Although the temperature insulating properties of a Styrofoam cup are reasonably good, the cup still absorbs some heat. Calibration is often performed as an experiment in which hot water of known mass and temperature is poured into a coffee cup containing cold water of known mass and temperature. The combined water samples equilibrate to a final maximum temperature, and a calorimeter constant can be determined, as shown in the example below.

Suppose 60.1 g of water at 97.6 °C is poured into a coffee cup calorimeter containing 50.3 g of water at 24.7 °C. The final temperature of the combined water samples reaches 62.8 °C. What is the calorimeter constant? (Use c_{water} = 4.184 J/g°C)

Solution: Consider the basic principle of calorimetry:

heat lost by hot water = heat gained by cool water + heat gained by calorimeter

Determine the mathematical equations corresponding to the premise described:

heat lost by hot water = heat gained by cool water + heat gained by calorimeter

$$-m_{hot}C_{water}(T_{final} - T_{intial, hot}) = m_{cool}\ c_{water}(T_{final} - T_{initial, cool}) + K_{cal}\ (T_{final} - T_{initial, cool})$$

Notice the negative sign for the heat lost by the hot water because heat loss is negative. Substitute the proper masses and temperatures into the expression.

Since the cool water was contained in the cup, the initial temperature of the calorimeter and the cool water are the same.

The specific heat capacity of water:

$$c_{water} = 4.184 \text{ J/g°C}$$

$$-(60.1 \text{ g})\cdot(4.184 \text{ J/g°C})\cdot(62.8 \text{ °C} - 97.6 \text{ °C})$$

$$= (50.3 \text{ g})\cdot(4.184 \text{ J/°C})\cdot(62.8 \text{ °C} - 24.7 \text{ °C}) + K_{cal}\ (62.8 \text{ °C} - 24.7 \text{ °C})$$

Combining terms:

$$8.75 \times 10^3 \text{ J} = 8.02 \times 10^3 \text{ J} + K_{cal}\ (38.1 \text{ °C})$$

Solving for K_{cal} :

$$K_{cal} = 8.75 \times 10^3 \text{ J} = 8.02 \times 10^3 \text{ J} / 38.1 \text{ °C}$$

$$K_{cal} = 19.2 \text{ J/°C}$$

Now that the calorimeter has been calibrated, use it to determine the specific heat capacity of a metal. In this experiment, a known quantity of metal is heated to a known temperature. The heat metal sample is placed in a coffee cup calorimeter containing a known quantity of water at a known temperature. The temperature is allowed to equilibrate, and the final temperature is measured.

Suppose 28.2 g of an unknown metal is heated to 99.8 °C and placed into a coffee cup calorimeter containing 150.0 g of water at a temperature of 23.5 °C. The temperature of the water equilibrates at a final temperature of 25.0 °C. The calorimeter constant has been determined to be 19.2 J/°C. What is the specific heat capacity of the metal?

This problem is based on heat flow:

heat lost by hot metal = heat gained by water + heat gained by calorimeter

$-(28.2\ g)\cdot(C_{metal})\cdot(25.0\ °C - 99.8\ °C)$

$= (150.0\ g)\cdot(4.184\ J/g°C)\cdot(25.0\ °C - 23.5\ °C) + 19.2\ J/°C(25.0\ °C - 23.5\ °C)$

Combining terms:

$2.11 \times 10^3\ g°C\ (C_{metal}) = 9.4 \times 10^2\ J + 29\ J$

$2.11 \times 10^3\ J\ (C_{metal}) = 969\ J$

Solving for specific heat capacity:

$C_{metal} = 969\ J / 2.11 \times 10^3\ g°C$

$C_{metal} = 0.459\ J/\ g°C$

The *constant-volume calorimeter* is the second type of calorimeter, where heat flow is measured at constant volume. The most used constant-volume calorimeter is the bomb calorimeter, also known as a decomposition vessel. Combustion is the process where oxygen reacts with a substance to produce carbon dioxide, water, and energy.

Bomb calorimeters are most frequently used for the combustion of hydrocarbons.

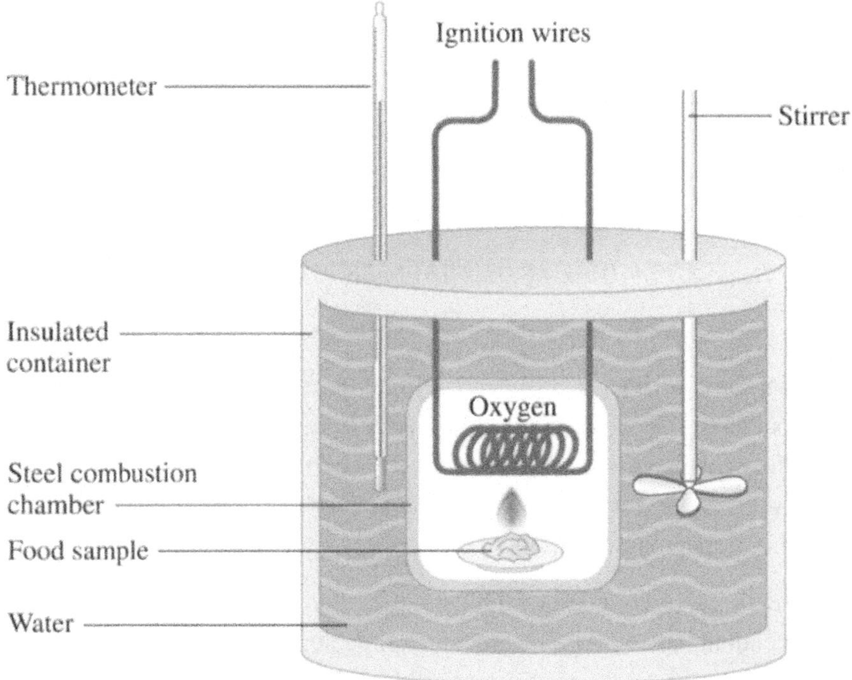

A typical setup for a bomb calorimeter; constant-volume calorimeter

Gibbs Free Energy *G*

The *Gibbs free energy* (G) is the amount of energy present in molecules available to do work. The *free energy change* ΔG is the energy difference between the free energy present in the product and reactant molecules. When molecules react, some chemical bonds may break in the reactants, and others form in the products.

Breaking a bond is a process that requires the input of energy, and making a bond releases energy, so free energy exchanges during a chemical reaction. Not all chemical bonds have the same strength, and therefore do not require the same amount of energy.

The Gibbs free energy change ΔG is equal to the maximum amount of energy produced during a reaction that can theoretically be harnessed as work. More specifically, the Gibbs free energy change equals the amount of work that a *reversible reaction* can produce.

Recall that reactions that give off heat are exothermic, and those that absorb heat are endothermic. For a chemical reaction to occur, the reactants must collide with enough energy to react. The energy necessary to align the reactant molecules and to cause them to collide with enough energy to form products is the *activation energy*.

If the energy in the reactant molecules is less than the activation energy, the molecules do not form products, and a chemical reaction does not occur.

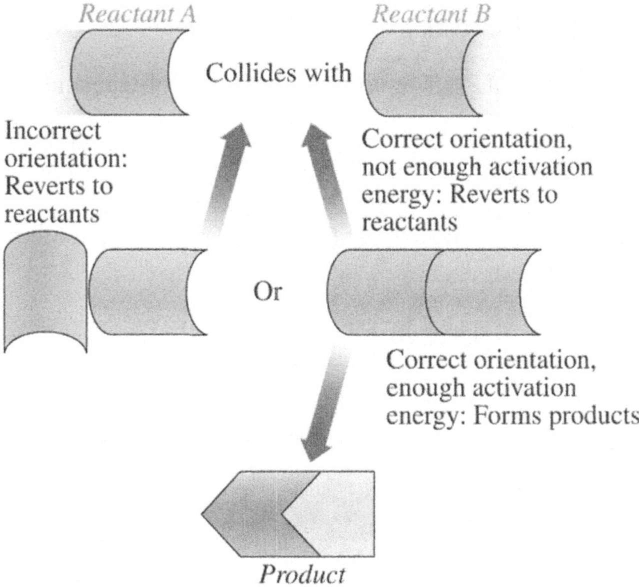

The reactants must be 1) appropriately aligned and 2) possess sufficient energy to overcome the energy of activation for the formation of products

The thermodynamics of a chemical reaction can be represented on a *reaction energy diagram* showing the energy of the reactants, products, the activation energy, and the free energy change ΔG.

Energy appears on the y-axis, and the reaction progress (time) from the starting reactants to the final products appears on the x-axis.

Each reaction has an energy of activation (E_{ac}) hill or *energy barrier* before the reactants transform into products.

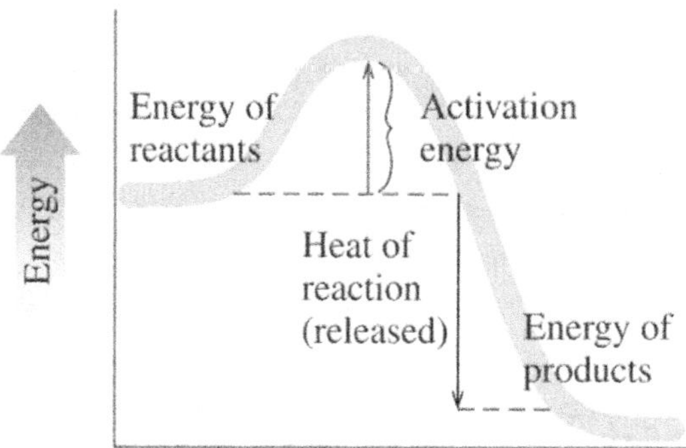

Reaction diagram of energy vs. time (reaction progress)

There are several methods to calculate Gibbs free energy ΔG. Since Gibbs free energy is a state function, one method is analogous to previous methods discussed for enthalpy and entropy.

The Gibbs free energy equation is:

$$\Delta G^\circ_{rxn} = \Sigma\, n(\Delta G^\circ_f \text{ of products}) - \Sigma\, n(\Delta G^\circ_f \text{ of reactants})$$

where n represents the moles of each product or reactant and is given by the coefficient in the balanced chemical equation.

As with enthalpy, the standard Gibbs free energy of formation of elements in their standard states is 0 kJ/mol.

Notice the similarity of this equation to Hess' Law of heat summation.

Spontaneous Reactions and ΔG

Gibbs free energy (ΔG) determines a reaction as spontaneous or nonspontaneous.

If $\Delta G < 0$, then the reaction is spontaneous in the forward reaction

If $\Delta G = 0$, the reaction is at equilibrium.

If $\Delta G > 0$, then the reaction is spontaneous in the reverse direction

The term *spontaneous* describes a process that occurs independently without any input of energy from outside the system, related to the Second Law of Thermodynamics.

If ΔG of the reaction is negative, the reaction is *spontaneous* (or exergonic).

If ΔG of the reaction is positive, the reaction is *nonspontaneous* (or endergonic).

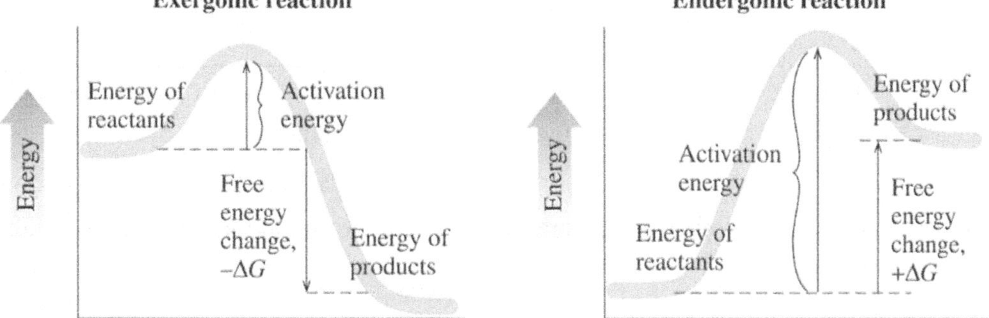

Entropy (S) is a thermodynamic quantity that measures the spreading of energy or disorder of a system.

An increase in entropy of a system corresponds to an increase in the randomness or disorder of the system.

$\Delta S° < 0$ represents a decrease in entropy.

$\Delta S° > 0$ represents an increase in entropy.

For instance, crystals form naturally within a supersaturated solution of sugar water. The formation of these sugar crystals increases entropy. An increase in entropy occurs because the crystals form "on their own" in a spontaneous process and thus must result in an entropy increase.

Enthalpy, temperature, and entropy are related to free energy by the following:

$$\Delta G° = \Delta H° - T\Delta S°$$

where $\Delta G°$ is the change in free energy, $\Delta H°$ is the change in enthalpy, T is the temperature, and $\Delta S°$ is the change in entropy.

The table displays how ΔH, ΔS and T affect the spontaneity of a reaction.

It is not correct to assume that an endothermic reaction is nonspontaneous, because a large, positive ΔS can make it spontaneous.

ΔH	ΔS	Sign of ΔG	Spontaneous?
–	+	$\Delta G = (-) - [T(+)] = -$	Always, regardless of T
+	–	$\Delta G = (+) - [T(-)] = +$	Never, regardless of T
+	+	$\Delta G = (+) - [T(+)] = ?$	Depends; spontaneous at high T, $-\Delta G$
–	–	$\Delta G = (-) - [T(-)] = ?$	Depends; spontaneous at low T, $-\Delta G$

It is also not correct to assume that an exothermic reaction is spontaneous, because a large, negative ΔS can cause it to become nonspontaneous.

Temperature-controlled reactions are the reactions that are spontaneous at one temperature and not at another. These are the reactions in the bottom left and top right quadrants of the pie chart.

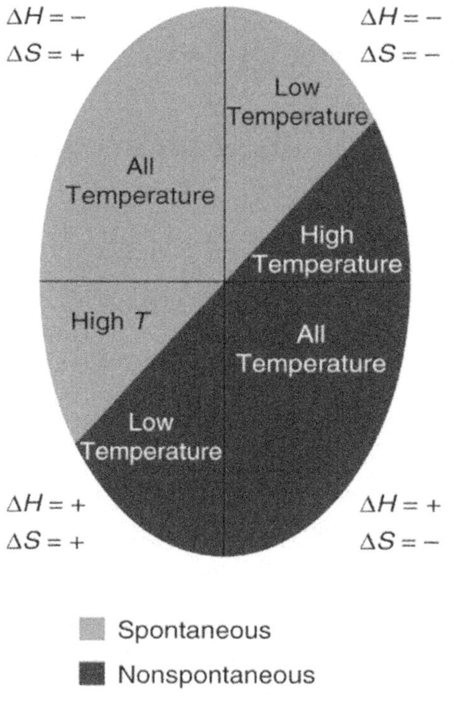

If the enthalpy change of a reaction is –1,876 kJ, the entropy change is –3.375 kJ/K, and the temperature has a value of 298 K (standard temperature), is the reaction spontaneous? The units must be consistent.

Entropy is often expressed in units of J/K, that has been converted to kJ/K.

Substituting these quantities into the equation for Gibbs Free Energy:

$$\Delta G^\circ = \Delta H - T\Delta S$$

$$\Delta G^\circ = -1876 \text{ kJ} - [(298 \text{ K})\cdot(-3.375 \text{ kJ/K})]$$

$$\Delta G^\circ = -870.2 \text{ kJ}$$

Since ΔG° is negative, this is a spontaneous reaction.

There are two other equations to calculate ΔG°.

The first of these equations is related to *chemical equilibrium*:

$$\Delta G^\circ = -RT \ln K$$

$$\Delta G^\circ = -2.303RT \log K$$

where R is the ideal gas law constant ($R = 8.314$ J/mol·K), T is the temperature in Kelvin, and K is the equilibrium constant for the chemical equilibrium.

The ΔG only implies the direction and extent of a chemical reaction; it does NOT imply a rate for the reaction. Do not assume that a spontaneous reaction occurs quickly because it may take a million years for it to happen, depending on its kinetics.

Another equation to calculate ΔG° in calculations for *electrochemistry*:

$$\Delta G^\circ = -nF\text{E}^\circ$$

where n represents the moles of electrons transferred, F is Faraday's constant (96,485 C/mol), and E° is the standard potential for an electrochemical cell.

Calculations of the Gibbs free energy from these equations involve the substitution of the appropriate quantities into the expression.

Thermodynamics

Thermodynamics answers several fundamental questions regarding whether a given reaction occurs, to what extent, and whether the reaction releases or absorbs heat.

Thermodynamics states four laws that characterize thermodynamic systems by describing the behavior of physical quantities such as temperature, energy, and entropy.

Notes

Zeroth Law: Concept of Temperature

The *Zeroth Law* of Thermodynamics states that *when two bodies are placed into contact, heat flows from the hotter body to the colder body until thermal equilibrium is reached.*

After thermal equilibrium is achieved, there is no more heat transfer. This law helps to define the concept of temperature.

If two systems are each in thermal equilibrium with a third system, then each is in thermal equilibrium with the other.

Thermal equilibrium is expressed by the equation:

$$T_A = T_B, \text{ and } T_C = T_B$$

Then

$$T_A = T_C$$

Two systems are in thermal equilibrium when they are separated by a barrier permeable to heat, and no heat is transferred.

The law is crucial for the mathematical formulation of thermodynamics.

The Zeroth Law of thermodynamics establishes that *temperature is a core, measurable property of matter.*

Notes

First Law: Conservation of Energy

The *First Law of Thermodynamics* states that *energy can neither be created nor destroyed, and the total energy in a system remains constant.*

Energy is conserved in a chemical reaction. This principle is the foundation for the law of conservation of energy.

Essentially, the First Law of Thermodynamics states that the change in total internal energy of a system is equal to the contributions from heat and work:

$$\Delta E = q - w$$

where ΔE is the change in internal energy (also represented as ΔU), q is the contribution from heat, and w is the contribution from work.

For the above expression, use the following sign conventions:

- q is positive if heat is *absorbed into* the system (i.e., heating it).
- q is negative if heat is *released from* the system (i.e., cooling it).
- w is positive if work is done *by* the system (i.e., compression).
- w is negative if work is done *on* the system (i.e., expansion).

An alternative expression for the First Law of Thermodynamics is:

$$\Delta E = q + w$$

where work is positive if done *on* the system, or negative if done *by* the system.

For this form of the equation, use the sign conventions given in the chart below:

q	+	Heat is **absorbed** by the system	E_{system} increases
q	–	Heat is **released** by the system	E_{system} decreases
w	+	Work is done **on** the system	E_{system} increases
w	–	Work is done **by** the system	E_{system} decreases

Note: Relate the sign to what is happening to the system. Heat or energy flowing out of the system is negative; heat or energy flowing into the system is positive.

Unfortunately, some other disciplines of science choose to follow the opposite sign convention. It is essential to note which sign convention is used.

Consider a problem involving internal energy, heat, and work. Suppose there is a process in which 3.4 kJ of heat flows out of the system while 4.8 kJ of work is done *by* the system on the surroundings. What is the internal energy?

Apply the equation $\Delta E = q + w$ to calculate the internal energy:

$$\Delta E = q + w$$

$$\Delta E = -3.4 \text{ kJ} + (-4.8 \text{ kJ})$$

$$\Delta E = -8.2 \text{ kJ}$$

Use the appropriate signs for heat and work.

Energy has flowed out of the system to the surroundings by –8.2 kJ.

Equivalence of Mechanical, Chemical, Electrical and Thermal Energy Units

The units for mechanical, chemical, electrical, and thermal energy are equivalent. If it is energy, it can be measured in Joules (J).

For example, 1 J of mechanical energy can be converted into 1 J of electrical energy (ignoring heat loss) – no more, no less.

Second Law: Concept of Entropy

The *Second Law of Thermodynamics* states that *the natural order of the universe favors a net increase in the overall entropy*. An isolated system increases in entropy over time and never decreases. An open system, however, can decrease in entropy, but only at the expense of a greater increase in entropy of its surroundings.

There are two types of thermodynamic processes: reversible and irreversible. The reversible process is idealized, while the irreversible process frequently occurs in nature.

For reversible processes $\Delta S = q / T$

For irreversible processes $\Delta S > q / T$

Real processes in the world are irreversible, so entropy change is always greater than the heat transfer over temperature. Due to the irreversible nature of real processes, the entropy of the universe is always increasing.

Reactions at equilibrium are *reversible processes* since there is no net change to the system or surroundings, and they proceed in both directions. Reversible processes are those that reverse direction whenever some infinitesimally small change is made to some property of the system. This can be written as follows:

$$\Delta S_{total} = \Delta S_{system} + \Delta S_{surroundings}$$

The change in entropy of the system is always equal in magnitude, but *opposite in sign*, of the change in entropy of the surroundings.

The change in entropy is:

$$\Delta S_{surr} = -q / T$$

$$\Delta S_{surr} = -\Delta H / T \text{ (at constant pressure)}$$

Combining these two equations:

$$\Delta S_{total} = \Delta S_{system} - (\Delta H_{system} / T)$$

One method to calculate entropy change for a reaction is to use tabulated values for the standard molar entropies. This calculation follows the same format as the calculations for the enthalpy change and Gibbs free energy change for the heats of formation is shown below.

The equation to calculate ΔS from tabulated values:

$$\Delta S^\circ_{rxn} = \Sigma\, n(\Delta S^\circ_f \text{ of products}) - \Sigma\, n(\Delta S^\circ_f \text{ of reactants})$$

where n represents the moles of each reactant or product, as a balanced chemical equation.

Determine the entropy change for this chemical equation. It is the same chemical equation for which the change in enthalpy ΔH was calculated in a previous section.

$$16\ H_2S\ (g) + 8\ SO_2\ (g) \rightarrow 16\ H_2O\ (l) + 3\ S_8\ (s)$$

Set the equation:

$$\Delta S^\circ_{rxn} = [(16\text{ mol})\cdot(S^\circ_f \text{ of } H_2O\ (l)) + (3\text{ mol})\cdot(S^\circ_f \text{ of } S_8\ (s))]$$

$$-[(16\text{ mol})\cdot(S^\circ_f \text{ of } H_2S\ (g)) + (8\text{ mol})\cdot(S^\circ_f \text{ of } SO_2\ (g))]$$

From the table of standard entropies of formation, substitute the quantities for the standard entropies of formation:

$$\Delta S^\circ_{rxn} = \Sigma\, n(\Delta S^\circ_f \text{ of products}) - \Sigma\, n(\Delta S^\circ_f \text{ of reactants})$$

$$\Delta S^\circ_{rxn} = [(16\text{ mol})\cdot(69.91\text{ J/mol}\cdot\text{K}) + (3\text{ mol})\cdot(31.80\text{ J/mol}\cdot\text{K})]$$

$$-[(16\text{ mol})\cdot(205.79\text{ J/mol}\cdot\text{K}) + (8\text{ mol})\cdot(161.92\text{ J/mol}\cdot\text{K})]$$

Collect terms, and account for the mathematical signs:

$$\Delta S^\circ_{rxn} = [1{,}118\text{ J/K} + 95.40\text{ J/K}] - [3{,}292.6\text{ J/K} - 1{,}295.4\text{ J/K}]$$

$$\Delta S^\circ_{rxn} = -3{,}375\text{ J/K}$$

In this example, ΔS°_{rxn} is a measure of the entropy change of a system. The entropy change of the surroundings is related to the heat flow and the temperature as shown:

$$\Delta S_{surr} = -q\,/\,T \text{ or } -\Delta H\,/\,T \text{ (at constant pressure)}$$

The negative sign in the equation accounts for the heat of the surroundings that is opposite in sign to the heat of the system.

From the previous example, the enthalpy of the reaction was calculated at –1,876 kJ at standard temperature, 298 K. The entropy change of this system was calculated to be –3,375 J/K (or –3.375 kJ/K). What is the entropy change of the surroundings?

Using the above equation:

$$\Delta S_{surr} = -(-\ 1{,}876\text{ kJ})\,/\,298\text{ K}$$

$$\Delta S_{surr} = 6.30\text{ kJ/K or } 6.30 \times 10^3\text{ J/K}$$

The entropy of the surroundings increased. The total entropy change is the sum of the entropy change of the system and the entropy change of the surroundings:

$\Delta S_{total} = \Delta S_{sys} + \Delta S_{surr}$

$\Delta S_{total} = -3.375 \times 10^3$ J/K $+ 6.30 \times 10^3$ J/K

$\Delta S_{total} = 2.92 \times 10^3$ J/K

Entropy as a measure of "disorder"

Entropy (S) measures randomness or disorder. More specifically, entropy measures the number of ways in which a thermodynamic system can be arranged.

To understand entropy, imagine the following situation. Someone decides to take a piece of paper, tear it up into small pieces, and throw the pieces into the air. The pieces of paper will randomly scatter about them. The pieces would not fly back together into the original piece of paper. They just increased the entropy of the universe. It did not take much energy to tear off the piece of paper and toss the pieces in the air.

Now, how much energy would it take to restore this piece of paper into its original form? First, some energy must be expended to go around the room and pick up the pieces of paper. The paper is still not restored to its original state. Next, they must take the pieces of paper to a pulp mill and have the paper turned back into pulp and pressed again into the original piece. These processes would take much more energy than the energy required to rip up the paper and toss it in the air. Imagine tearing up the paper as a spontaneous process and putting it back together as a nonspontaneous process.

Thus, the entropy (or disorder) of the universe is always increasing.

The higher the entropy and lower the energy, the more stable the system.

Higher entropy + lower energy → more stability

Lower entropy + higher energy → less stability

Entropy has the dimension of energy divided by temperature, and the unit is Joules per Kelvin (J/K). However, it can also be expressed as entropy per unit mass (J/K·kg) or entropy per unit amount of substance (J/K·mol).

Relative entropy for gas, liquid, and crystal states

As the temperature increases, entropy increases as follows:

$T = 0$ K, particles are in equilibrium lattice positions and S relatively low

$T > 0$ K, molecules vibrate, S increases

T increases further, more violent vibrations occur and S higher

This may seem analogous to the molecules of a substance in different phases (i.e., solid, liquid, gas). The relative differences in entropies between the phases of matter can be understood regarding the relative degrees of freedom of forms of kinetic energy.

Solids have low entropy because the particles forming a solid are packed in an orderly matrix in the crystal. Particles of a solid are confined to vibrating in fixed positions and are limited in their freedom of forms of kinetic energy.

Liquids are intermediate in entropy. Intermolecular forces still hold the particles in a liquid together but have increased freedom to slide past one another.

Gases are high in entropy because the gas particles have complete freedom to randomly move around in their container with very few particles arranged in an orderly fashion. A low-pressure, a gas particle experiences little to no intermolecular forces of attraction or repulsion.

The entropy changes accompanying phase changes can be estimated, at least qualitatively. A decrease in entropy accompanies freezing as the relatively disordered liquid becomes a well-ordered solid.

Boiling is accompanied by a substantial increase in entropy as the liquid becomes a much more highly disordered gas.

For any substance, sublimation (phase change from a solid to a gas) is the phase transition with the highest entropy change.

The figure displays gas, liquid, and crystalline states ordered by increasing entropy.

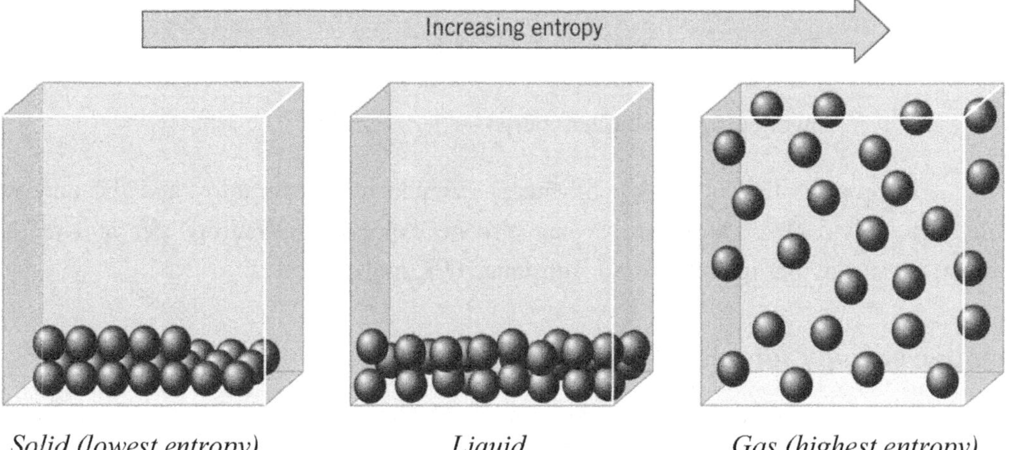

Solid (lowest entropy) *Liquid* *Gas (highest entropy)*

For gases, entropy increases as volume increases.

Take the example in the image below:

The gas is separated from a vacuum by a partition (figure on the left).

When the partition is removed (figure in the center), there are more ways to distribute energy.

The gas expands to achieve higher particle distribution (figure on the right). The configuration of gas molecules is more random and a more positive *S* (more disorder).

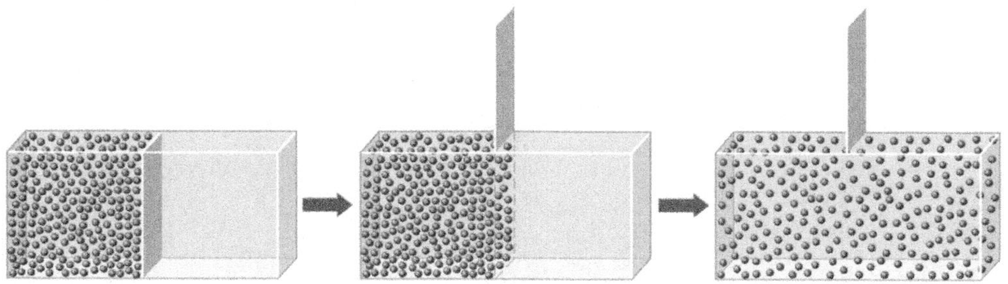

Gas particles are highly ordered on the left (low S) and random on the right

Entropy is also affected by the number of particles.

A reaction that produces more particles has more positive ΔS.

By adding particles to a system (shown below on the right), there are increased numbers of ways that energy can be distributed in that system.

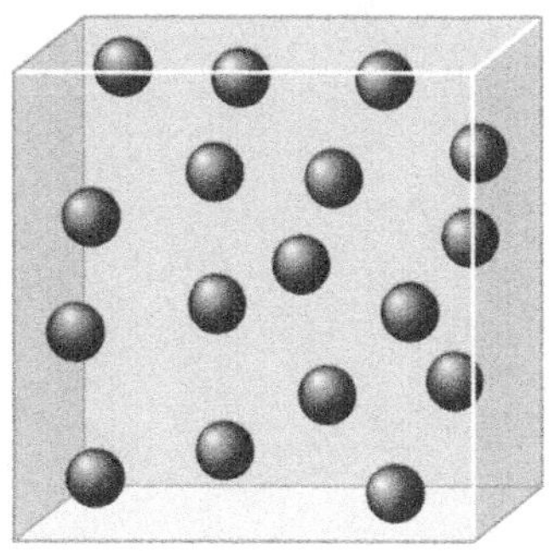

Low entropy

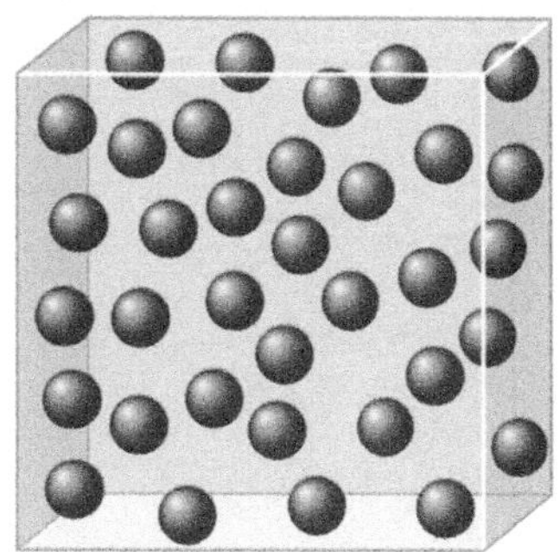

High entropy: more disorder

Notes

Third Law: Entropy at Absolute Zero

The *Third Law of Thermodynamics* states that *the entropy of a perfectly ordered, pure crystalline solid is zero at the absolute temperature of zero.*

At *absolute zero* (coldest temperature possible on the Kelvin scale), the average kinetic energy of the particles equals zero; they essentially possess no kinetic energy.

As the entropy increases, the average kinetic energy of the particles increases. The substance transitions to liquid and, ultimately, a gaseous state.

For a given substance, the gas state always has the highest entropy.

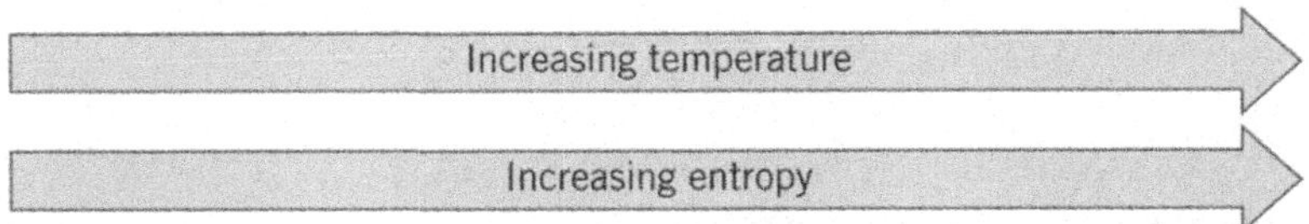

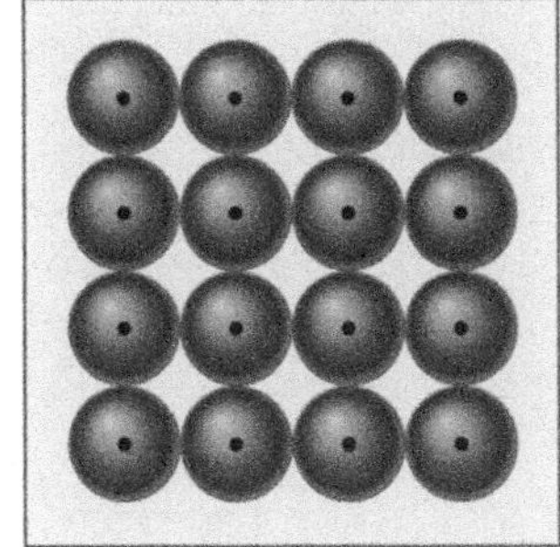

As a crystalline solid, there is minimal kinetic energy, and the particles are arranged orderly.

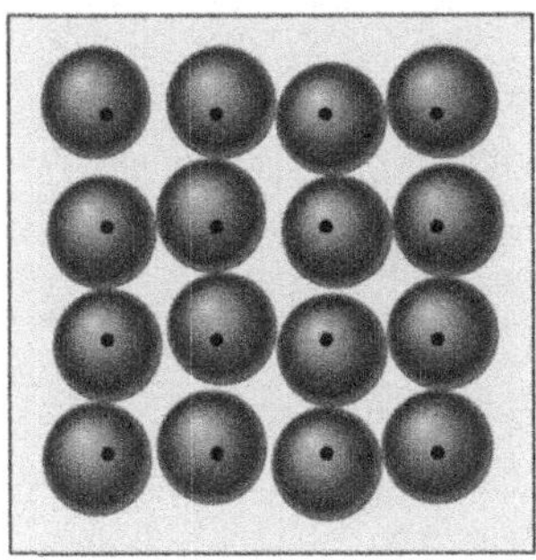

At high temperatures, the particles vibrate as they have higher KE and more ways to distribute, so entropy is high.

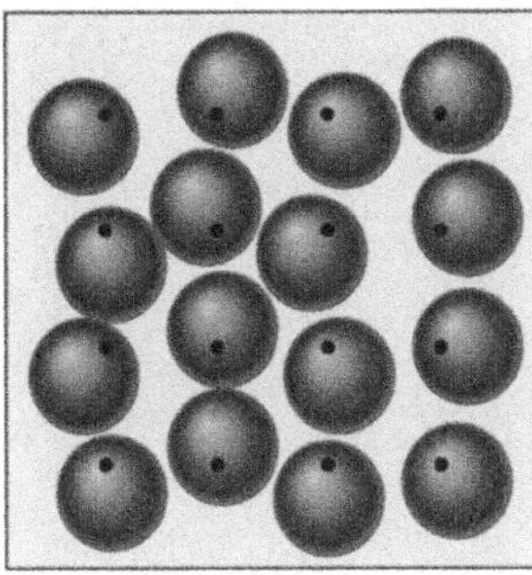

At a higher temperature, the particles have even more kinetic energy and higher entropy *S*.

For the following reaction, determine the sign of ΔS.

$$CaCO_3\,(s) + 2\,H^+\,(aq) \rightarrow Ca^{2+}\,(aq) + H_2O + CO_2\,(g)$$

$$\Delta n_{gas} = 1\text{ mol} - 0\text{ mol}$$

$$\Delta n_{gas} = 1\text{ mol}$$

since Δn_{gas} is positive, ΔS is positive.

For the following reaction, determine the sign of ΔS.

$$2\ N_2O_5\,(g) \longrightarrow 4\ NO_2\,(g) + O_2\,(g)$$

$\Delta n_{gas} = 4$ mol + 1 mol – 2 mol

$\Delta n_{gas} = 3$ mol

since Δn_{gas} is positive, ΔS is positive.

For the following reaction, determine the sign of ΔS.

$$OH^-\,(aq) + H^+\,(aq) \longrightarrow H_2O$$

$\Delta n_{gas} = 0$ mol

$\Delta n_{gas} = 1$ mol – 2 mol

$\Delta n_{gas} = -1$ mol

since Δn is negative, ΔS is negative.

Temperature Scales, Conversions

There are three commonly used temperature scales.

In 1724, Daniel Fahrenheit (1686-1736) proposed the classic English system of measuring temperature. Fahrenheit developed this scale by dividing the difference between the boiling and freezing point of water into 180 equal degrees.

Celsius is the modern metric way of measuring temperature; it divides the difference between the boiling and freezing point of water into 100 equal degrees.

The expression to convert between Fahrenheit and Celsius is:

$$°F = (1.8 \times °C) + 32$$

Kelvin is the third temperature scale; unlike the Fahrenheit and Celsius scales, it is not commonly used daily (e.g., when talking about the weather), but it is the most important scale in science. The Kelvin scale is based on the concept of absolute zero.

The size of the units on the Kelvin scale is the same as for degrees on the Celsius scale. The way to convert between Kelvin and Celsius is:

$$K = °C + 273$$

Comparison of temperatures among the three different scales

	K	°C	°F
Absolute zero	0	–273	–460
Freezing point of water / melting point of ice	273	0	32
Room temperature	298	25	77
Body temperature	310	37	99
Boiling point of water / condensation of steam	373	100	212

Notes

Heat Transfer: Conduction, Convection, Radiation

As described by the First Law of Thermodynamics, *heat always flows from a hotter system to a colder system until the two systems are in thermal equilibrium.*

There are three ways in which heat can be transferred:

conduction

convection

radiation

Conduction is heat transfer between substances that are in direct contact with each other (i.e., they must be touching). When particles of a hotter substance are vibrating, these molecules bump into nearby particles and transfer some energy to them. Conduction is the most important means of heat transfer for solids. Metals are especially conductive.

Convection occurs when warmer areas of a liquid or gas rise to colder areas of the liquid or gas, and the cooler liquid or gas takes the place of the warmer areas. It refers to the physical flow of matter. Convective heat transfer takes place by diffusion (the random Brownian motion of particles) and by advection (larger-scale flowing currents transport matter or heat).

Radiation is a method of heat transfer that does not require contact between the heat source and the heated object. Radiation consists of electromagnetic waves traveling at the speed of light; no mass is exchanged, and no medium is required (i.e., it can occur in a vacuum). An example is a thermal radiation emitted by the sun, which warms Earth.

Notes

Heat of Fusion, Heat of Vaporization

Energy is released when a gas changes phase into a liquid (*condensation*), when a liquid transforms into a solid (*freezing*), or when a gas changes phase directly into a solid (*deposition*).

The energy released is the same as the energy of their reverse processes.

Vaporization is the reverse of condensation.

Melting is the reverse of freezing, and *sublimation* is the reverse of deposition.

The *heat of fusion* (ΔH_{fus}) is the energy input needed to melt a given quantity of a solid substance to the liquid phase at a constant temperature.

The energy it takes to melt a solid is ΔH_{fus} times the number of moles of that solid, represented by the equation:

$$q = n \times \Delta H_{fus}$$

where q is heat energy, and n is the number of moles.

The *heat of vaporization* (ΔH_{vap}) is the energy input needed to vaporize a given quantity of a liquid substance to the gas phase at a constant temperature.

The energy it takes to vaporize a liquid is ΔH_{vap} times the number of moles of that liquid, represented by the equation:

$$q = n \times \Delta H_{vap}$$

where q is heat energy, and n is the number of moles.

The heat of fusion and heat of vaporization generally have the unit kJ/mol, but can also be expressed as J/g, where energy can be obtained by multiplying the latent heats by the mass of the substance, rather than the moles.

Example: 31.5 g of H_2O is being melted at its melting point of 0 °C. How many kJ is required? The molar heat of fusion for water is 6.02 kJ/mol.

Find the number of moles by using the molar mass = 18 g/mol.

$$31.5\ \text{g} / 18\ \text{g/mol} = 1.75\ \text{mol}$$

Use the equation given above.

$$q = 1.75\ \text{mol} \times 6.02\ \text{kJ/mol}$$

$$q = 10.54\ \text{kJ}$$

10.54 kJ heat is required to melt 31.5 g of water.

Notes

Pressure vs. Volume (*PV*) Diagram

A *pressure vs. volume diagram* (*PV* diagram) describes corresponding changes in the pressure and volume in a system.

Two *pressure vs. volume* (*PV*) graphs are shown below.

The area under the curve on a *PV* diagram is equal to the work done.

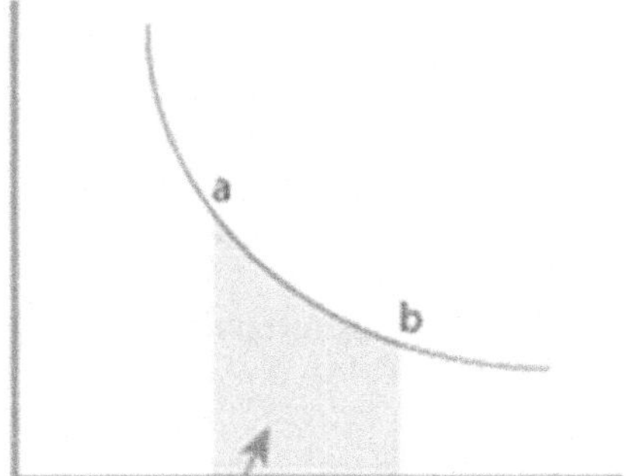

Work is done from point a to point b

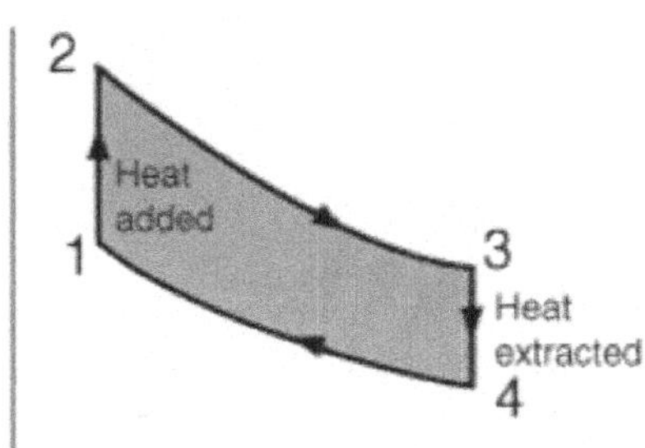

Points 4 to 1 show work done ***on*** *the gas; points 2 to 3 work done* ***by*** *the gas*

The figure above illustrates that:

expansion is work done *by* the system ($\Delta E = q - w$)

compression is work done *on* the system ($\Delta E = q + w$)

For a gas phase that consists of the gas and its container, work can be calculated.

Since gases may be expanded or compressed, work may be related by the pressure of the gas and the change in volume of the gas:

$$w = -P\Delta V$$

The change in volume ΔV is the final volume minus the initial volume of the gas:

$$\Delta V = V_{final} - V_{initial}$$

To expand a gas, the volume of the gas is increased (ΔV is positive).

The gas (of the system) must do work on the surroundings

Thus, the work must be negative

To compress a gas, the volume of the gas is decreased (ΔV is negative).

The surroundings must do work on the system

Thus, the work must be positive

There are four types of thermodynamic processes often depicted by PV diagrams:

- *Adiabatic process*:

  no heat exchanges

  $q = 0$

  $\Delta E = w$

- *Isovolumetric* (*isochoric*) *process*:

  volume is constant

  $w = 0$

  $\Delta E = q$

- *Isothermal process*:

  temperature is constant

  $\Delta T = 0$

- *Isobaric process*:

  pressure is constant

  $w = P\Delta V$

Coefficient of Expansion

Thermal expansion is the tendency of matter to change in volume in response to changes in temperature. This process occurs through heat transfer. In general, when a substance is heated, its molecules increase kinetic energy and begin moving more; thus, greater average separation is maintained, leading to an increase in volume (expansion).

The degree of expansion divided by the change in temperature is the material's *coefficient of thermal expansion*, α.

There are different expansion relationships for linear expansion, area expansion, and volume expansion.

Linear expansion: $\Delta L / L_0 = \alpha\Delta T$

Area expansion: $\Delta A / A_0 = 2\alpha\Delta T$

Volume expansion: $\Delta V / V_0 = 3\alpha\Delta T$

The value of the coefficient α depends on the type of expansion.

A metal rod of length 7.00 m is being heated to 35 °C. If the length of the rod expands to 7.12 m after some time, calculate the linear expansion coefficient (the temperature of the room is 27 °C).

Calculate the change in length ΔL from the initial length L_0 = 7.00 m and expanded length L = 7.12 m.

$\Delta L = 7.12 \text{ m} - 7.00 \text{ m}$

$\Delta L = 0.12 \text{ m}$

Calculate the change in temperature ΔT from the initial room temperature T_0 = 27 °C and the heated temperature T = 35 °C.

$\Delta T = 35\ °\text{C} - 27\ °\text{C}$

$\Delta T = 8\ °\text{C}$

Convert ΔT to Kelvin.

$\Delta T = 8 + 273$

$\Delta T = 281 \text{ K}$

Manipulate the formula for linear thermal expansion to isolate α.

$\alpha = \Delta L / (L_0 \times \Delta T)$

$\alpha = 0.12 \text{ m} / (7 \text{ m} \times 281 \text{ K})$

$\alpha = 6.1 \times 10^{-5}$

Notes

Phase Diagram: Pressure and Temperature

A phase diagram indicates the relationship between temperature, pressure, and associated phases. The phase of a substance (i.e., solid, liquid, or gas) mostly depends on temperature and pressure. If the pressure is exceptionally high, a substance is less likely to be in a gas phase because the pressure brings the molecules so close together that the substance either liquefies or sublimes.

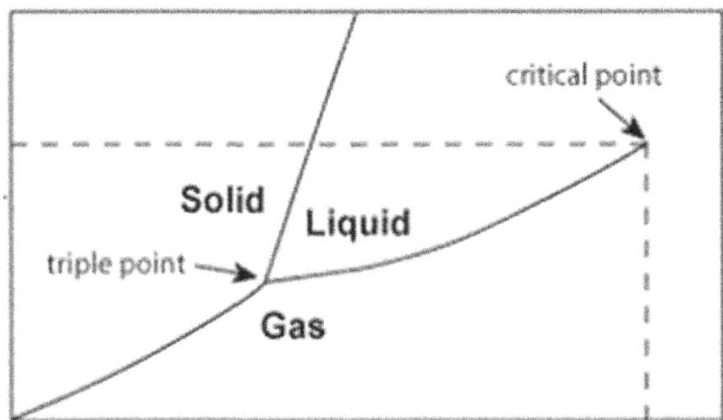

Phase diagram of pressure vs. temperature with triple point and critical point labeled. The dashed horizontal line represents the critical pressure, and the vertical dashed line represents the critical temperature.

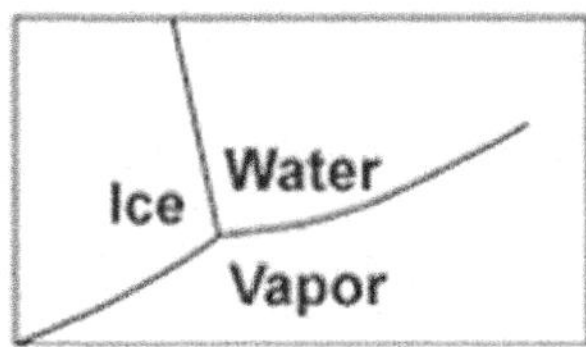

Phase Diagram of pressure vs. pressure for water is different because the solid-liquid boundary is slanted to the left. This is because water (liquid) is denser than ice (solid), and if the pressure at a given temperature is increased, ice melts into water.

The *solid-liquid boundary* is where solid and liquid exist in equilibrium.

The *solid-gas boundary* is where solid and gas exist in equilibrium.

The *liquid-gas boundary* is where liquid and gas exist in equilibrium.

The *triple point* is the temperature and pressure at which all three phases of matter coexist in equilibrium.

The *critical point* is the temperature and pressure at which liquids and gases become indistinguishable.

The *critical temperature* is the temperature above which a liquid cannot form, no matter how much pressure is put on it.

Notes

Practice Questions

1. If a stationary gas has kinetic energy of 500 J at 25 °C, what is its KE at 50 °C?

A. 125 J
C. 540 J
B. 450 J
D. 1,120 J
E. 270 J

2. What is the purpose of the hollow walls in a closed, hollow-walled container that is effective at maintaining the temperature inside?

A. Traps air trying to escape from the container, which minimizes convection
B. Act as an effective insulator, which minimizes convection
C. Acts as an effective insulator, which minimizes conduction
D. Provides an additional source of heat for the container
E. Reactions occur within the walls that maintain the temperature within the container

3. How much heat is energy (in Joules) required to heat 21.0 g of copper from 21.0 °C to 68.5 °C? (Use specific heat c of Cu = 0.382 J/g·°C)

A. 462 J
B. 188 J
C. 522 J
D. 662 J
E. 381 J

4. The species in the reaction $KClO_3$ (*s*) → KCl (*s*) + $3/2O_2$ (*g*) have the values for standard enthalpies of formation at 25 °C. At constant physical states, assume that the values of $\Delta H°$ and $\Delta S°$ are constant throughout a broad temperature range. Which of the following conditions may apply for the reaction? (Use $KClO_3$ (*s*) with $\Delta H°_f = -391.2$ kJ mol^{-1} and KCl (*s*) with $\Delta H°_f = -436.8$ kJ mol^{-1})

A. Nonspontaneous at low temperatures, but spontaneous at high temperatures
B. Spontaneous at low temperatures, but nonspontaneous at high temperatures
C. Nonspontaneous at all temperatures over a broad temperature range
D. Spontaneous at all temperatures over a broad temperature range
E. No conclusion can be drawn about spontaneity based on the information

5. Determine the value of $\Delta E°_{rxn}$ for this reaction, whereby the standard enthalpy of reaction ($\Delta H°_{rxn}$) is –311.5 kJ mol^{-1}:

$$C_2H_2\,(g) + 2\,H_2\,(g) \rightarrow C_2H_6\,(g)$$

A. –306.5 kJ mol^{-1}
C. +346.0 kJ mol^{-1}
B. –318.0 kJ mol^{-1}
D. +306.5 kJ mol^{-1}
E. +466 kJ mol^{-1}

6. The process of H_2O (*g*) → H_2O (*l*) is nonspontaneous under a pressure of 760 torr and temperatures of 378 K because:

A. $\Delta H = T\Delta S$
B. $\Delta G < 0$
C. $\Delta H > 0$
D. $\Delta H < T\Delta S$
E. $\Delta H > T\Delta S$

7. A fuel cell contains hydrogen and oxygen gas that react explosively, and the energy converts water to steam, which drives a turbine to turn a generator that produces electricity. The fuel cell and the turbine represent which forms of energy, respectively?

A. Electrical and mechanical energy
B. Electrical and heat energy
C. Chemical and mechanical energy
D. Chemical and heat energy
E. Nuclear and mechanical energy

8. The thermodynamic systems that have high stability tend to demonstrate:

A. maximum ΔH and maximum ΔS
B. maximum ΔH and minimum ΔS
C. minimum ΔH and maximum ΔS
D. minimum ΔH and minimum ΔS
E. none of the above

9. Which is valid for the thermodynamic functions *G*, *H*, and *S* in $\Delta G = \Delta H - T\Delta S$?

A. *G* refers to the universe, *H* to the surroundings and *S* to the system
B. *G*, *H*, and *S* refer to the system
C. *G* and *H* refers to the surroundings and *S* to the system
D. *G* and *H* refer to the system and *S* to the surroundings
E. *G* and *S* refers to the system and *H* to the surroundings

10. Whether a reaction is endothermic or exothermic is determined by:

A. an energy balance between bond breaking and bond forming, resulting in a net loss or gain of energy
B. the presence of a catalyst
C. the activation energy
D. the physical state of the reaction system
E. none of the above

11. What is the term for a reaction that proceeds by releasing heat energy?

A. Endothermic reaction
B. Isothermal reaction
C. Exothermic reaction
D. Nonspontaneous
E. Endergonic

12. The bond dissociation energy is:

I. useful in estimating the enthalpy change in a reaction
II. the energy required to break a bond between two gaseous atoms
III. the energy released when a bond between two gaseous atoms is broken

A. I only
B. II only
C. I and II only
D. I and III only
E. I, II and III

13. Which of the following statements is true for the following reaction? (Use the change in enthalpy, $\Delta H° = -113.4$ kJ/mol and the change in entropy, $\Delta S° = -145.7$ J/K mol)

$2\ NO\ (g) + O_2\ (g) \rightarrow 2\ NO_2\ (g)$

A. The reaction is at equilibrium at 25 °C under standard conditions
B. The reaction is spontaneous at only high temperatures
C. The reaction is spontaneous only at low temperatures
D. The reaction is spontaneous at all temperatures
E. $\Delta G°$ becomes more favorable as temperature increases

14. Which law explains the observation that the amount of heat transfer accompanying a change in one direction is equal in magnitude but opposite in sign to the amount of heat transfer in the opposite direction?

A. Law of Conservation of Mass
B. Law of Definite Proportions
C. Avogadro's Law
D. Boyle's Law
E. Law of Conservation of Energy

15. Calculate the value of $\Delta H°$ of reaction using provided bond energies.

$H_2C = CH_2\ (g) + H_2\ (g) \rightarrow H_3C–CH_3\ (g)$

C–C: 348 KJC≡C: 960 kJ

C=C: 612 kJC–H: 412 kJ

H–H: 436 kJ

A. –348 kJ
B. +134 kJ
C. –546 kJ
D. –124 kJ
E. –238 kJ

Detailed Explanations

1. C is correct.

Temperature is a measure of the average kinetic energy of the molecules.

Kinetic energy (*KE*) is proportional to temperature.

In most thermodynamic equations, temperatures are expressed in Kelvin, so convert the Celsius temperatures to Kelvin:

$$25\text{ °C} + 273.15 = 298.15\text{ K}$$

$$50\text{ °C} + 273.15 = 323.15\text{ K}$$

Calculate the kinetic energy using simple proportions:

$$KE = (323.15\text{ K} / 298.15\text{ K}) \times 500\text{ J}$$

$$KE = 540\text{ J}$$

2. C is correct.

An insulator reduces conduction (i.e., transfer of thermal energy through matter).

Air and vacuum are excellent insulators. Storm windows, which have air wedged between two glass panes, work by utilizing this principle of conduction.

3. E is correct.

Heat capacity is the amount of heat required to increase the temperature of the *whole sample* by 1 °C.

Specific heat is the heat required to increase the temperature of *1 gram* of a sample by 1 °C.

Heat = mass × specific heat × change in temperature:

$$q = \text{m} \times c \times \Delta T$$

$$q = 21.0\text{ g} \times 0.382\text{ J/g·°C} \times (68.5\text{ °C} - 21.0\text{ °C})$$

$$q = 21.0\text{ g} \times 0.382\text{ J/g·°C} \times (47.5\text{ °C})$$

$$q = 381\text{ J}$$

4. D is correct.

Calculate the value of $\Delta H_f = \Delta H_{f\,product} - \Delta H_{f\,reactant}$

$$\Delta H_f = -436.8 \text{ kJ mol}^{-1} - (-391.2 \text{ kJ mol}^{-1})$$

$$\Delta H_f = -45.6 \text{ kJ mol}^{-1}$$

To determine spontaneity, calculate the Gibbs free energy:

$$\Delta G = \Delta H° - T\Delta S$$

The reaction is spontaneous if ΔG is negative:

$$\Delta G = -45.6 \text{ kJ} - T\Delta S$$

Typical ΔS values are around 100–200 J.

If the $\Delta H°$ value is –45.6 kJ or –45,600 J, the value of ΔG would still be negative unless $T\Delta S$ is less than –45,600.

Therefore, the reaction would be spontaneous over a broad range of temperatures.

5. A is correct.

Relationship between enthalpy (ΔH) and internal energy (ΔE):

Enthalpy = Internal energy + work (for gases, work = PV)

$$\Delta H = \Delta E + \Delta(P\text{V})$$

Solving for ΔE:

$$\Delta E = \Delta H - \Delta(P\text{V})$$

According to ideal gas law:

$$P\text{V} = nRT$$

Substitute ideal gas law to the previous equation:

$$\Delta E = \Delta H - \Delta(nRT)$$

R and T are constant, which leaves Δn as the variable.

The reaction is $C_2H_2\,(g) + 2H_2\,(g) \rightarrow C_2H_6\,(g)$.

There are three gas molecules on the left and one on the right, so $\Delta n = 1 - 3 = -2$.

In this problem, the temperature is not provided. However, the presence of degree symbols ($\Delta G°$, $\Delta H°$, $\Delta S°$) indicates that those are standard values, which are measured at 25 °C or 298.15 K.

Always double-check the units before performing calculations; ΔH is in kilojoules, while the gas constant (R) is 8.314 J/mol K. Convert ΔH to Joules before calculating.

Use the Δn and T values to calculate ΔE:

$$\Delta E = \Delta H - \Delta nRT$$

$$\Delta E = -311{,}500 \text{ J/mol} - (-2 \times 8.314 \text{ J/mol K} \times 298.15 \text{ K})$$

$$\Delta E = -306{,}542 \text{ J} \approx -306.5 \text{ kJ}$$

6. E is correct.

The spontaneity of a reaction is determined by evaluating the Gibbs free energy, or ΔG.

$$\Delta G = \Delta H - T\Delta S$$

A reaction is spontaneous if $\Delta G < 0$ (or ΔG is negative).

For ΔG to be negative, ΔH must be less than $T\Delta S$.

A reaction is nonspontaneous if $\Delta G > 0$ (or ΔG is positive, and ΔH is greater than $T\Delta S$).

7. C is correct.

Potential Energy **Stored energy and the energy of position (gravitational)**	**Kinetic Energy** **Energy of motion: the motion of waves, electrons, atoms, and molecules.**
Chemical Energy Chemical energy is the energy stored in the bonds of atoms and molecules. Biomass, petroleum, natural gas, propane, and coal are stored chemical energy. **Nuclear Energy** Nuclear energy is the energy stored in the nucleus of an atom. It is the energy that holds the nucleus together. The nucleus of the uranium atom is an example of nuclear energy.	**Radiant Energy** Radiant energy is electromagnetic energy that travels in transverse waves. Radiant energy includes visible light, x-rays, gamma rays, and radio waves. Solar energy is an example of radiant energy. **Thermal Energy** Thermal energy (or heat) is the internal energy in substances; it is the vibration and movement of atoms and molecules within substances. Geothermal energy is an example of thermal energy.

Stored Mechanical Energy Stored mechanical energy is energy stored in objects by the application of a force. Compressed springs and stretched rubber bands are examples of stored mechanical energy. **Gravitational Energy** Gravitational Energy is the energy of place or position. Water in a reservoir behind a hydropower dam is an example of gravitational potential energy. When the water is released to spin the turbines, it becomes kinetic energy.	**Motion** The movement of objects or substances from one place to another is motion. Wind and hydropower are examples of motion. **Sound** Sound is the movement of energy through substances in longitudinal (compression/rarefaction) waves. **Electrical Energy** Electrical energy is the movement of electrons. Lightning and electricity are examples of electrical energy.

8. C is correct.

Gibbs free energy:

$$\Delta G = \Delta H - T\Delta S$$

Stable molecules are non-spontaneous because of negative (or relatively low) ΔG.

ΔG is most negative (most stable) when ΔH is small, and ΔS is large.

9. B is correct.

All thermodynamic functions in $\Delta G = \Delta H - T\Delta S$ refer to the system.

10. A is correct.

ΔH refers to enthalpy (or heat).

Endothermic reactions have heat as a reactant.

Exothermic reactions have heat as a product.

Endothermic reactions absorb energy to break strong bonds to form a less stable state (i.e., positive enthalpy).

Exothermic reaction release energy during the formation of stronger bonds to produce a more stable state (i.e., negative enthalpy).

The reaction is nonspontaneous (i.e., endergonic) if the products are less stable than the reactants, and ΔG is positive.

The reaction is spontaneous (i.e., exergonic) if the products are more stable than the reactants, and ΔG is negative.

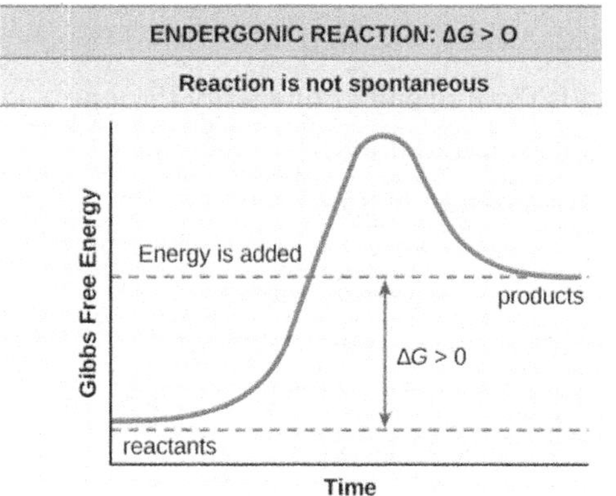

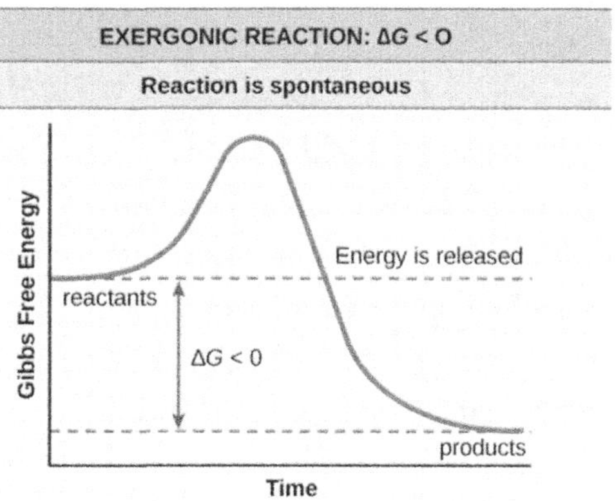

Exothermic reactions release heat and cause the temperature of the immediate surroundings to rise (i.e., a net loss of energy).

Endothermic process absorbs heat and cools the surroundings (i.e., a net gain of energy).

11. C is correct.

ΔH refers to enthalpy (or heat).

Endothermic reactions have heat as a reactant.

Exothermic reactions have heat as a product.

Exothermic reactions release heat and cause the temperature of the immediate surroundings to rise (i.e., a net loss of energy).

Endothermic process absorbs heat and cools the surroundings (i.e., a net gain of energy).

Endothermic reactions absorb energy to break strong bonds to form a less stable state (i.e., positive enthalpy).

Exothermic reaction release energy during the formation of stronger bonds to produce a more stable state (i.e., negative enthalpy).

The reaction is nonspontaneous (i.e., endergonic) if the products are less stable than the reactants, and ΔG is positive.

The reaction is spontaneous (i.e., exergonic) if the products are more stable than the reactants, and ΔG is negative.

12. C is correct.

Bond dissociation energy is the energy required to break a bond between two gaseous items and is useful in estimating the enthalpy change in a reaction.

13. C is correct.

To predict spontaneity of reaction, use the Gibbs free energy equation:

$$\Delta G = \Delta H^\circ - T\Delta S$$

The reaction is spontaneous if ΔG is negative.

Substitute the given values to the equation:

$$\Delta G = -113.4 \text{ kJ/mol} - [T \times (-145.7 \text{ J/K mol})]$$

$$\Delta G = -113.4 \text{ kJ/mol} + (T \times 145.7 \text{ J/K mol})$$

Important: note that the units are not identical, ΔH° is in kJ and ΔS° is in J. Convert kJ to J (1 kJ = 1,000 J):

$$\Delta G = -113{,}400 \text{ J/mol} + (T \times 145.7 \text{ J/K mol})$$

It can be predicted that the value of ΔG would be negative if the value of T is small.

If T increases, ΔG approaches a positive value, and the reaction is nonspontaneous.

14. E is correct.

The Law of Conservation of Energy states that the total energy (e.g., potential, or kinetic) of an isolated system remains constant and is conserved.

Energy can be neither created nor destroyed but is transformed from one form to another.

For instance, chemical energy can be converted to kinetic energy in the explosion of a firecracker.

The Law of Conservation of Mass states that for any system closed to all transfers of matter and energy, the mass of the system must remain constant over time.

Law of Definite Proportions (Proust's Law) states that a chemical compound always contains the same proportion of elements by mass.

The law of definite proportions forms the basis of stoichiometry (i.e., from the known amounts of separate reactants, the amount of the product can be calculated).

Avogadro's Law is an experimental gas law relating the volume of a gas to the amount of substance of gas present. It states that equal volumes of all gases, at the same temperature and pressure, have the same number of molecules.

Boyle's Law is an experimental gas law that describes how the pressure of a gas tends to increase as the volume of a gas decreases. It states that the absolute pressure exerted by a given mass of an ideal gas is inversely proportional to the volume it occupies if the temperature and amount of gas remain unchanged within a closed system

15. D is correct.

In a chemical reaction, bonds within reactants are broken down, and new bonds are formed to create products.

Therefore, for bond dissociation,

$$\Delta H_{reaction} = \text{sum of bond energy in reactants} - \text{sum of bond energy in products}$$

for

$$H_2C{=}CH_2 + H_2 \rightarrow CH_3{-}CH_3$$

$$\Delta H_{reaction} = [(C{=}C) + 4(C{-}H) + (H{-}H)] - [(C{-}C) + 6(C{-}H)]$$

$$\Delta H_{reaction} = [612\text{ kJ} + (4 \times 412\text{ kJ}) + 436\text{ kJ}] - [348\text{ kJ} + (6 \times 412\text{ kJ})]$$

$$\Delta H_{reaction} = -124\text{ kJ}$$

Remember that this is the opposite of ΔH_f problems, where:

$$\Delta H_{reaction} = (\text{sum of } \Delta H_f \text{ products}) - (\text{sum of } \Delta H_f \text{ reactants})$$

Chapter 6

Rate Processes in Chemical Reactions: Kinetics and Equilibrium

- **Reaction Rates**
- **Dependence of Reaction Rate on Concentrations of Reactants**
- **Rate-Determining Step**
- **Dependence of Reaction Rate on Temperature**
- **Kinetic Control vs. Thermodynamic Control of a Reaction**
- **Catalysts, Enzyme Catalysis**
- **Equilibrium in Reversible Chemical Reactions**
- **Relationship of the Equilibrium Constant and $\Delta G°$**
- **Relationship of the Equilibrium Constant and $\Delta H°$, $\Delta S°$**

Reaction Rates

The *reaction rate* is the rate of change in the concentration of reactants or products (i.e., how fast a reactant is consumed and how fast a product is formed).

The reaction rates are shown in the formulas below:

$$\text{rate of reaction} = \frac{\text{change in concentration of reactant or product}}{\text{change in time}}$$

$$\text{rate of reaction} = \frac{\Delta\ \text{concentration}}{\Delta\ \text{time}}$$

The rate is given by concentration divided by time; the unit is molarity per second (M/s).

The rate of reaction can be written for both the forward and reverse directions of a given chemical reaction.

A subscript "fwd" or "rev" is often used to specify the direction in which the reaction rate is being given.

Forward Rate of Reaction:

$$\text{rate}_{\text{fwd}} = \frac{-\Delta\text{reactan}}{\Delta\text{time}} = \text{how fast a reactant disappears}$$

$$\text{rate}_{\text{fwd}} = \frac{\Delta\text{product}}{\Delta\text{time}} = \text{how fast a product forms}$$

Reverse Rate of Reaction:

$$\text{rate}_{\text{rev}} = \frac{-\Delta\text{product}}{\Delta\text{time}} = \text{how fast a product disappears}$$

$$\text{rate}_{\text{rev}} = \frac{\Delta\text{reactant}}{\Delta\text{time}} = \text{how fast a reactant reforms}$$

A reaction between two molecules occurs if the molecules 1) have sufficient *kinetic energy,* and 2) both molecules are *appropriately oriented* to start a reaction, shown in the image below. The *collision theory* focuses on gas-phase chemical reactions.

Three collision-related factors affect the rate of a chemical reaction:

1. *Collision frequency:* An increase in the frequency at which molecules collide, increases the rate of reaction.

   The more collisions, the higher the probability that a collision produces the product.

2. *Collision energy:* The molecules must collide with enough energy to form new bonds for a reaction to occur.

3. *Collision orientation:* The reactants must have the correct orientation (i.e., adequately aligned) for products to be formed.

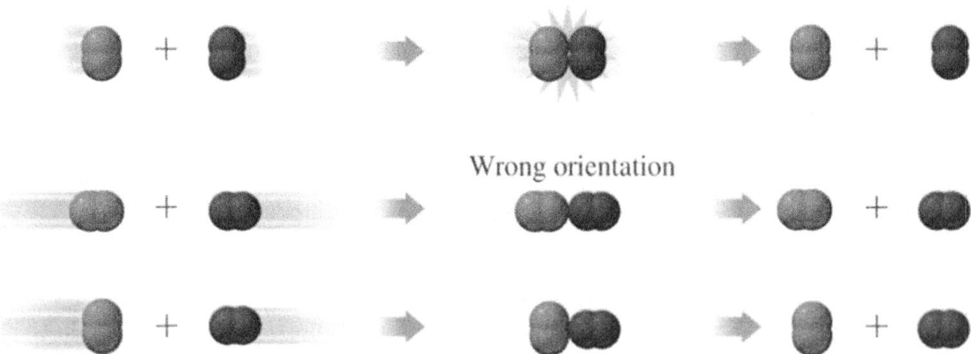

A reaction forming a product from two molecules occurs if the molecules have sufficient kinetic energy, and the molecules are appropriately oriented. These two requirements are satisfied when $N_2 + O_2$ *transforms into NO as shown in the top example above*

Dependence of Reaction Rate on Concentrations of Reactants

If the initial concentration of the reactants is increased, there are more reactant molecules; thus, the molecules are closer and collide more frequently. A higher collision frequency results in a higher rate of reaction. As the reaction proceeds, the concentration of reactant molecules decreases, and the forward rate of the reaction decreases. Thus, the concentration of product molecules increases, and the reverse rate of reaction increases.

Rate law, the rate constant

The *rate law expression* describes how the rate of reaction changes with the concentration of reactants and products.

Consider the general expression for a homogeneous reaction:

$$a\mathbf{A} + b\mathbf{B} \longrightarrow c\mathbf{C} + d\mathbf{D}$$

Its rate law expression is:

$$\text{rate} = -\Delta[\text{A}] / \Delta t = k[A]^x[B]^y$$

where $[A]$ and $[B]$ are initial concentrations of the reactants, x and y are partial rate orders determined experimentally, and k is the rate constant.

The rate constant k quantifies the rate of a chemical reaction; it accounts for other factors influencing reaction rate, including temperature and solvent used. Therefore, the constant is specific to the experimental conditions.

There is no correlation between the stoichiometric coefficients (a, b) and the rate exponents (x, y).

Exponents tend to be small integers, but may be simple fractions (e.g., ½) or zero.

Reaction order

The *partial rate order* is the exponent associated with each concentration term.

- For a *zero-order* reaction, the rate is constant and does not depend on the concentration of that reactant. In this case, the reactant's exponent is zero. Thus the term is equal to one (since anything to the zeroth power is one). Since it is equal to one, it disappears from the rate law expression, as the reaction rate does not depend on a zero-order reactant.

- For a reaction that is *first-order* for a reactant, the rate of reaction is directly proportional to the concentration of that reactant. The exponent is one.

- For a reaction that is *second-order* in a reactant, the rate of reaction is directly proportional to the square of the concentration of that reactant. The exponent is two.

The *overall order of the reaction* is the sum of the partial orders.

Most differential rate laws for reactions are zero-, first-, or second-order overall:

- *Zero-order overall*: For a zero-order overall reaction, the differential rate law is in the general form $r = k$. The units of the rate constant k are mole × L^{-1} × sec^{-1} (or M/s).

- *First-order overall*: For a first-order overall reaction, the differential rate law is in the general form $r = k[A]$. The units of the rate constant k are sec^{-1}.

- *Second-order overall*: For a second-order overall reaction, the differential rate law is in the general form $r = k[A]^2$. The units of the rate constant k are L × $mole^{-1}$ × sec^{-1} (or $M^{-1} \cdot s^{-1}$).

The example below illustrates how to sum the partial orders to obtain the overall order for an equation.

Consider the reaction:

$$2\ NO\ (g) + O_2\ (g) \leftrightarrow 2\ NO_2\ (g)$$

The rate law was experimentally determined to be:

$$r = k[NO]^2[O_2]$$

In the rate law, the partial order for NO is two.

In the rate law, the partial order for O_2 is one.

The overall order is three; therefore, this reaction is third order overall.

Note that in this reaction, the orders match the coefficients from the reaction, but that rate orders and coefficients are *not necessarily* correlated.

Coefficients only equal rate orders for elementary-step reactions. In most problems, it is unknown if the reaction is an elementary or multi-step reaction unless it is explicitly stated.

In defining the overall order of a given reaction, it is important to discuss its *molecularity*. The molecularity of a reaction mechanism is determined by the number of molecules (typically the coefficients) or ions that participate in the rate-determining step.

A *unimolecular* reaction is when a single reactant makes up the transition state.

A *bimolecular* reaction is a mechanism that consists of two reacting species.

A *termolecular* reaction is a mechanism when three independent molecules collide and react. Termolecular reactions are rare as the factors necessary for a successful reaction are so specific that termolecular reactions are unlikely.

Reaction Type	Overall Reaction Order	Rate Law(s)
Unimolecular	1	$r = k[\mathrm{A}]$
Bimolecular	2	$r = k[\mathrm{A}]^2$, $r = k[\mathrm{A}]\cdot[\mathrm{B}]$
Termolecular	3	$r = k[\mathrm{A}]^3$, $r = k[\mathrm{A}]^2[\mathrm{B}]$, $r = k[\mathrm{A}]\cdot[\mathrm{B}]\cdot[\mathrm{C}]$
Zero order	0	$r = k$

The *differential rate law* and *integrated rate law* are methods to express rate law.

The *differential rate* law is the form that is expressed above.

Integrated rate laws express the change in concentration of component *vs*. time,

Differential rate laws express the rate of reaction *vs*. concentration.

The integrated rate laws are listed below:

- Integrated rate law for 0th order:

  $[A]_t = -kt + [A]_0$

- Integrated rate law for 1st order:

  $\ln[A]_t = -kt + \ln[A]_0$

- Integrated rate law for 2nd order:

  $1 / [A] = 1 / [A]_0 + kt$

The *method of initial rates* described below is a method to determine the rate law of a reaction by using experimental data to determine the order of each reactant.

The results of an experiment show the reaction rates relative to the initial concentrations of reactants:

	Rate (M/s)	Initial Concentrations (M)	
		[NO]	$[O_2]$
Trial 1	1.2×10^{-8}	0.10	0.10
Trial 2	2.4×10^{-8}	0.10	0.20
Trial 3	1.08×10^{-7}	0.30	0.10

Based on the data above, determine:

a) reaction orders of the reactants

b) value and units of k

To determine the rate law, compare two trials in which the concentration of one reactant changes, while the concentration of the other reactant is held constant. This is necessary in determining the partial order for the reactant for changed concentrations. By keeping the concentration of the other reactant constant, the change in rate is due to the change in concentration of only one reactant.

Consider the results for Trial 1 and Trial 2.

A comparison of these two trials shows that the concentration of NO is constant, while the concentration of O_2 is doubled.

To determine the partial order of O_2, note that doubling the concentration of O_2 also doubles the rate of the reaction.

This may be expressed as:

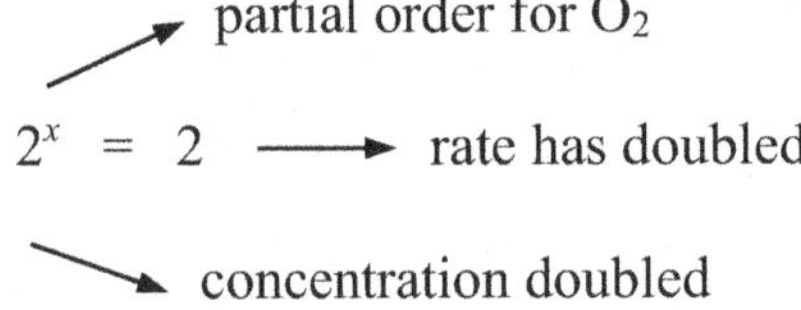

For equality, x must equal one. Therefore, the partial order for O_2 is first-order.

Use the same procedure to determine the partial order of NO. Select two trials in which the concentration of O_2 is constant while the concentration of NO is varied.

These criteria are satisfied with Trial 1 and Trial 3.

The concentration of O_2 is held constant, while the concentration of NO has tripled. Under these conditions, the rate of the reaction increased nine-fold.

This may be expressed as:

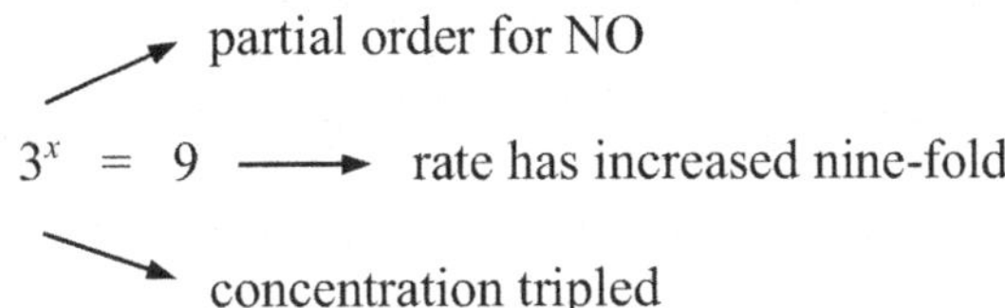

For equality, x must be two.

Therefore, the partial order for NO is second-order.

The calculated rate orders can be substituted into the rate law:

$$\text{rate} = k[NO]^2[O_2]$$

Partial first orders are typically omitted from the rate law of most reactions, which is why $[O_2]$ does not have an exponent; the exponent is assumed to be one.

With partial rate orders, it is also possible to solve for the rate constant, k.

For example, choose an experimental trial, and substitute the values of rate and concentrations of NO and O_2.

Using data from Trial 1:

$$k = \text{rate} / [NO]^2 \cdot [O_2]$$

$$k = 1.2 \times 10^{-8}\ M^{-2} \cdot s^{-1} / [(0.10\ M)^2 \cdot (0.10\ M)]$$

$$k = 1.2 \times 10^{-5}\ M^{-2} \cdot s^{-1}$$

Thus, the rate law becomes:

$$\text{rate} = 1.2 \times 10^{-5}\ M^{-2} \cdot s^{-1}[NO]^2 \cdot [O_2]$$

The reaction is third order overall, and the units are $M^{-2} \times s^{-1}$.

Notice that the units of k cancel to yield M/s for the rate of reaction.

Notes

Rate-Determining Step

Most chemical reactions occur in several steps, known as a *reaction mechanism*. The reaction mechanism is a sequence of events that take place as reactant molecules transform into products.

Each step in a multiple-step mechanism is an *elementary step*, which has an individual rate constant and rate law.

The *rate-determining step* has the slowest rate and is the most significant contributor to the overall rate of the reaction. In most cases, the overall rate of reaction is almost equal to the rate-determining step's rate because the other steps are fast (almost instantaneous), and they do not affect the overall rate significantly.

For this reaction, these steps were experimentally observed:

Overall:

$$(CH_3)_3CCl\ (aq) + OH^-\ (aq) \rightarrow (CH_3)_3COH\ (aq) + Cl^-\ (aq)$$

Step 1:

$$(CH_3)_3CCl\ (aq) \rightarrow (CH_3)_3C^+\ (aq) + Cl^-\ (aq) \quad \text{(slow)}$$

$$r_1 = k[(CH_3)_3CCl]$$

Step 2:

$$(CH_3)_3C^+\ (aq) + OH^-\ (aq) \rightarrow (CH_3)_3COH\ (aq) \quad \text{(fast)}$$

$$r_2 = k[(CH_3)_3C^+]\cdot[OH^-]$$

The first step, which is the formation of the charged carbonium ion $(CH_3)_3C^+$, is slow compared to the second step, where the carbocation immediately reacts with the OH^- ion to form the neutral product $(CH_3)_3COH$.

Therefore, the first step is the rate-determining step, and the rate law of this step determines the overall rate:

$$r_1 = k[(CH_3)_3CCl]$$

There are also cases when the rate-determining step, or slowest step, is not the first step in the reaction mechanism. In these cases, an intermediate may appear in the rate-determining step. The intermediate must be substituted for so that it is not in the rate law. This requires a slightly more detailed approach to determining the rate law of the reaction mechanism.

Notes

Dependence of Reaction Rate on Temperature

An increase in temperature increases the average kinetic energy of the colliding molecules; thus, the molecules move faster. These faster molecules are more likely to possess the energy necessary to overcome the activation energy and successfully collide and react. Faster molecules also collide with each other at an increased frequency. Therefore, using collision theory, an increase in temperature increases the reaction rate.

However, some reactions might slow or even stop as temperature increases. This is quite common among reactions in biological systems because enzyme proteins involved in the reaction may be damaged by higher temperatures, reducing the enzyme's capability to participate in reactions.

Activation energy

For a reaction to start, the reactants must have enough energy to break existing bonds and create new ones. This minimum energy threshold is the *activation energy* (E_a).

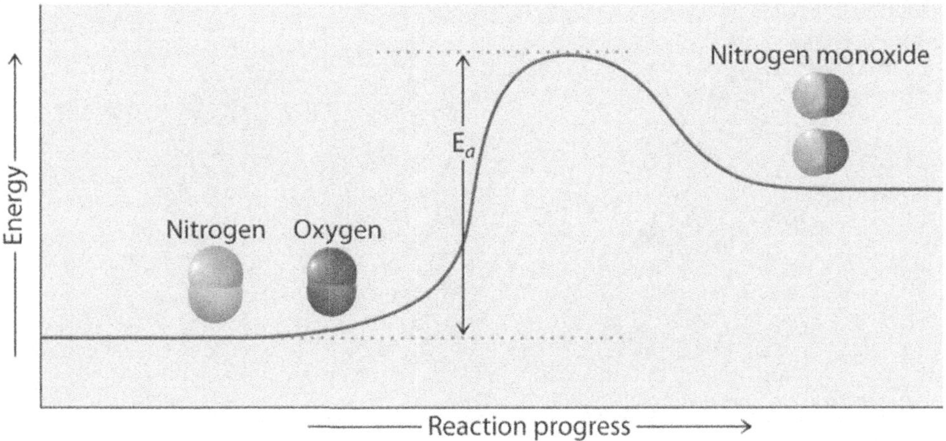

The *activation energy barrier* (E_a as the hill on the reaction profile graph) is the difference between the energy of the reactants and the highest point on the reaction profile in progress toward product formation. The reacting molecules need to increase their energy and proceed over the energy barrier before the reaction can transform into products.

The activation energy E_a is the difference between the energy of reactants and the top (transition state) of the reaction profile.

Activated complex or transition state

The transition state is located at the energy peak (y-axis) of the energy profile. In this state, bonds are beginning to break (reactants), and other bonds are beginning to form (products). A molecule in the transition state can go either way; it can revert to reactants, or it can form products.

Transition states form if the reactants have enough energy to achieve the needed activation energy (highest energy location on the reaction progress graph). Molecules in the transition state cannot be isolated.

Intermediates (e.g., carbocations, radicals) are produced in unimolecular reactions (two peaks on the reaction progress diagram). The formation of an intermediate of the first step is often the rate-determining step. The intermediate is then consumed in the next step as the reaction proceeds toward product formation.

The image below on the left shows a completed bimolecular (E_2 or Sn_2) reaction.

The image below on the right shows a bimolecular reaction whereby the activation energy (energy barrier) is not overcome: no transition state form and no products generated.

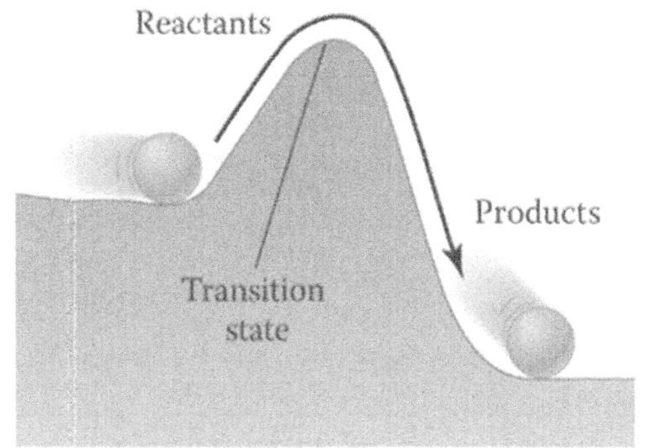

Exergonic reaction forms product

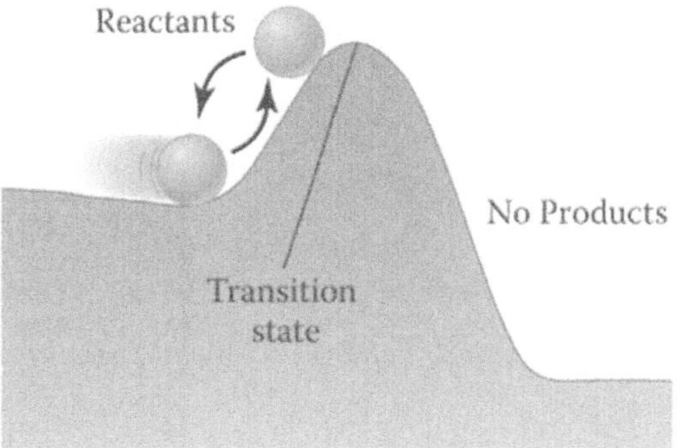

Exergonic reaction without transition state

Transition states signify bond breaking and bond making events are shown below. Transition states are denoted by square brackets [ ] and a double dagger (‡) around the fleeting (i.e., unable to be isolated) molecule. Dashed lines symbolize temporary bonds being simultaneously formed and broken, as shown below.

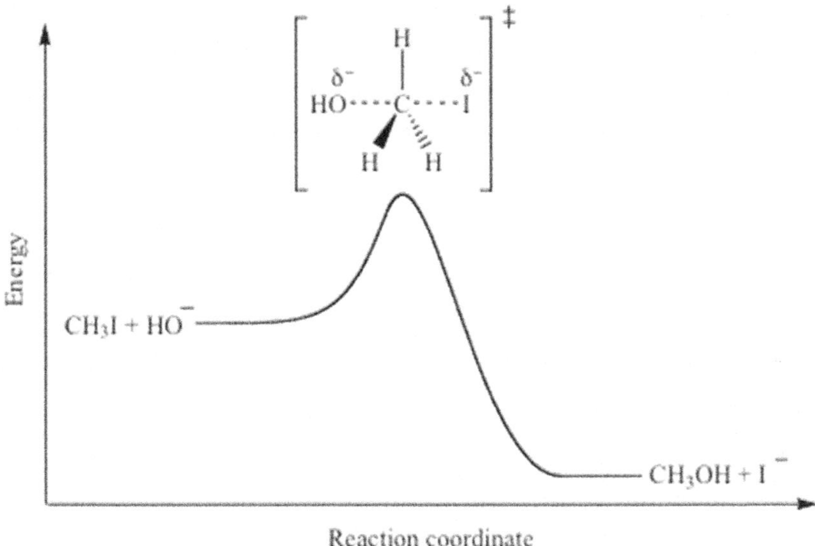

Interpretation of energy profiles showing energies of reactants and products, activation energy (E_{ac}), enthalpy (ΔH) for the reaction

In a reaction, the products and reactants often have different energy levels. The energy differential is released or absorbed from the system as heat. This heat is the heat of reaction (or enthalpy) of the reaction (ΔH).

Reactions are classified as *endothermic* or *exothermic,* based on the enthalpy.

In *endothermic* reactions, the products are at a higher energy level than reactants. To balance the energy levels, heat is absorbed by the system from its surroundings ($+\Delta H$). Energy is absorbed to break bonds. This is an energetically unfavorable reaction, and the temperature of the surroundings decreases.

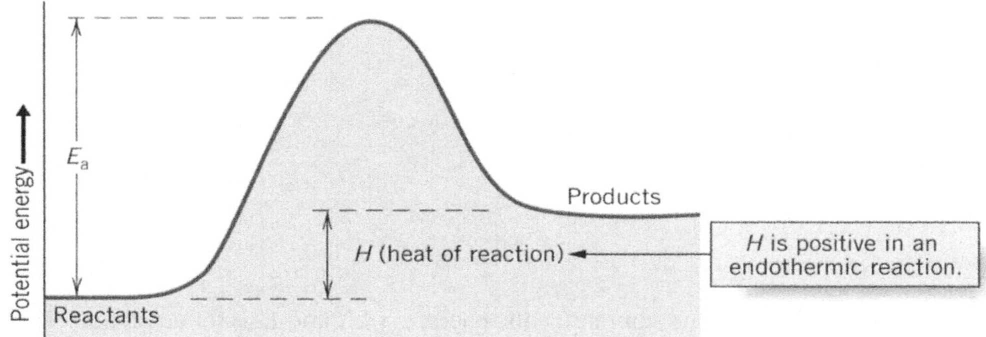

Endothermic reactions have energy of the products higher than reactants ($+\Delta H$). Heat was required as a reactant to drive the reaction forward. The high point of the graph indicates the energy of activation E_{ac} barrier that must be overcome during the reaction.

In *exothermic reactions*, the products have lower energy (i.e., more stable) than the reactants. Excess heat formed by the reaction is released by the system towards its surroundings ($-\Delta H$). Energy is released when bonds are formed. This is an energetically favorable reaction, and the temperature of the surroundings increases.

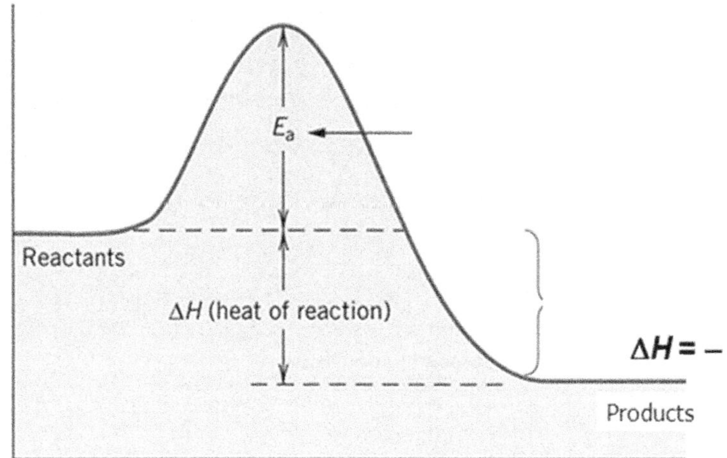

Exothermic reactions form products with lower energy than reactants

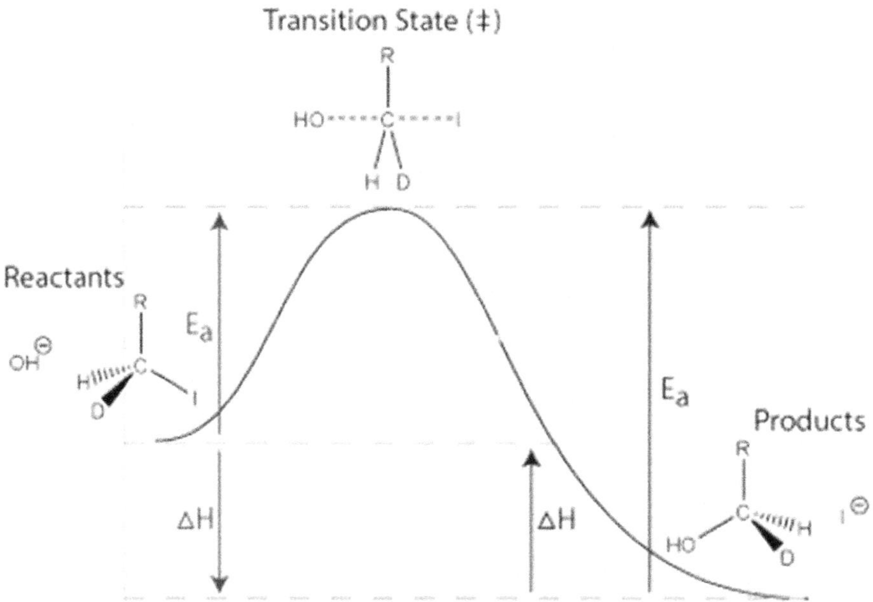

Energy profile diagrams summarize the energies (ΔH and E_{ac}) for a reaction. The forward reaction has a $+\Delta H$ in the forward direction and a $-\Delta H$ in the reverse direction. The E_{ac} is lower (measured from energy of reactants) to the peak of the energy barrier compared to the E_{ac} in the reverse reaction (measure from the energy of the products).

Arrhenius equation

Chemical reactions generally occur more rapidly at higher temperatures. In 1889, Swedish Nobel laureate Svante Arrhenius (1859-1927) described this observation.

Arrhenius equation describes the relationship between the rate constant (k), temperature (T), and activation energy (E_a):

$$k = \mathrm{A}e^{-Eac\,/\,RT}$$

where k is the rate constant, R is the ideal gas constant ($R = 8.314$ J mol^{-1} K^{-1}), E_{ac} is the activation energy, T is the temperature, and A is the *Arrhenius frequency factor* (the pre-exponential factor or the collision factor) that represents the frequency of collisions.

For the Arrhenius equation, ensure that the units for the different quantities are proper. T is in Kelvin, E_{ac} is in J/mol, and A has the same units as the rate constant k.

Taking the natural logarithm of both sides of the Arrhenius equation yields the point-slope form for the equation:

$$\ln k = -\,(E_a\,/\,RT) + \ln A$$

If experimental data is graphed, where ln k is on the y-axis, and $1\,/\,T$ is on the x-axis, a straight-line slope is produced.

On a plot of ln k vs. $1\,/\,T$, the slope of the line is $E_a\,/\,R$, and the y-intercept corresponds to ln A. By determining the slope of the line mathematically, the E_a can be determined. Similarly, by extrapolation, the line to the y-intercept, the ln A collision factor, can be determined.

Since the slope embodies the activation energy (E_{ac}), a simplified form of the Arrhenius equation determines the E_{ac}. The information required is the rate constants of the reaction at two different temperatures. The slope of a line is determined by the change in y over the change in x ("rise" over "run").

The rate constants relationship is expressed by:

$$\ln k_2 - \ln k_1 = \left(\ln A - \frac{E_a}{RT_2}\right)\left(\ln A - \frac{E_a}{RT_1}\right) = \frac{E_a}{R}\left(\frac{1}{T_1} - \frac{1}{T_2}\right)$$

Eliminate the ln A term by subtracting the expressions for the two ln-k terms.

Rearrange the expression on the right to isolate the activation energy.

$$E_a = \frac{R \ln \dfrac{k_2}{k_1}}{\dfrac{1}{T_1} - \dfrac{1}{T_2}}$$

The equation can be used to calculate the rate constant k of a reaction at two different temperatures (in Kelvin). The y-intercept ($\ln A$) can be ignored because it is a constant and not part of the slope calculation.

Example: A rate of reaction for a chemical process is investigated at two different temperatures. The rate of a reaction at 25 °C is 1.55×10^{-4} s^{-1}. At 50 °C, the rate of reaction is 3.88×10^{-4} s^{-1}.

Based on this data, what is the energy of activation for the chemical process expressed in J/mol?

Inspect the units of the quantities given in the problem. The units for both rates, s^{-1}, cancel. Since the energy of activation is to be expressed in J/mol, the logical choice of the gas constant R is 8.314 J/mol·K. This choice of the gas constant R dictates the units for temperature. Since the R constant has temperature units expressed in K, the given temperatures must be converted to Kelvin:

$$T_1 = 25\ °\text{C} + 273 = 298\ \text{K}$$

$$T_2 = 50\ °\text{C} + 273 = 323\ \text{K}$$

$$E_a = \frac{R \ln \frac{k_2}{k_1}}{\frac{1}{T_1} - \frac{1}{T_2}}$$

Substitute the rate constants and temperatures into the Arrhenius equation:

$$E_{ac} = \frac{8.314\ \text{J/mol·K} \times \ln(3.88 \times 10^{-4}\ \text{s}^{-1} / 1.55 \times 10^{-4}\ \text{s}^{-1})}{(1/298\ \text{K}) - (1/323\ \text{K})}$$

$$E_{ac} = \frac{8.314\ \text{J/mol·K} \times \ln(2.503)}{(2.6 \times 10^{-4}\ \text{K})}$$

$$E_{ac} = 2.93 \times 10^4\ \text{J/mol}$$

A modification of the Arrhenius equation replaces E_{ac} with ΔH.

This modified Arrhenius equation is the van't Hoff equation:

$$\ln\left(\frac{K_{T_2}}{K_{T_1}}\right) = \frac{\Delta H^o}{R}\left(\frac{1}{T_1} - \frac{1}{T_2}\right)$$

The above equation is *not* the same as the Arrhenius equation. The van't Hoff equation calculates the equilibrium constant K (uppercase K for equilibrium; lowercase k for rate) of a reaction at two different temperatures in Kelvin.

Kinetic Control Versus Thermodynamic Control of a Reaction

Several reactions in chemistry have two possible products, and the reaction conditions affect the selectivity of the reaction. Thermodynamics explains how ΔG and other variables affect if a reaction occurs spontaneously. Kinetics describes how fast a reaction occurs (based on the activation energy), and kinetics has no impact on the spontaneity or the thermodynamics of the reaction.

The *kinetic product* requires lower activation energy and is formed preferentially (according to transition state being of lower energy) at a lower temperature.

The *thermodynamic product* has a greater negative ΔG (favorable) and is formed preferentially at a higher temperature.

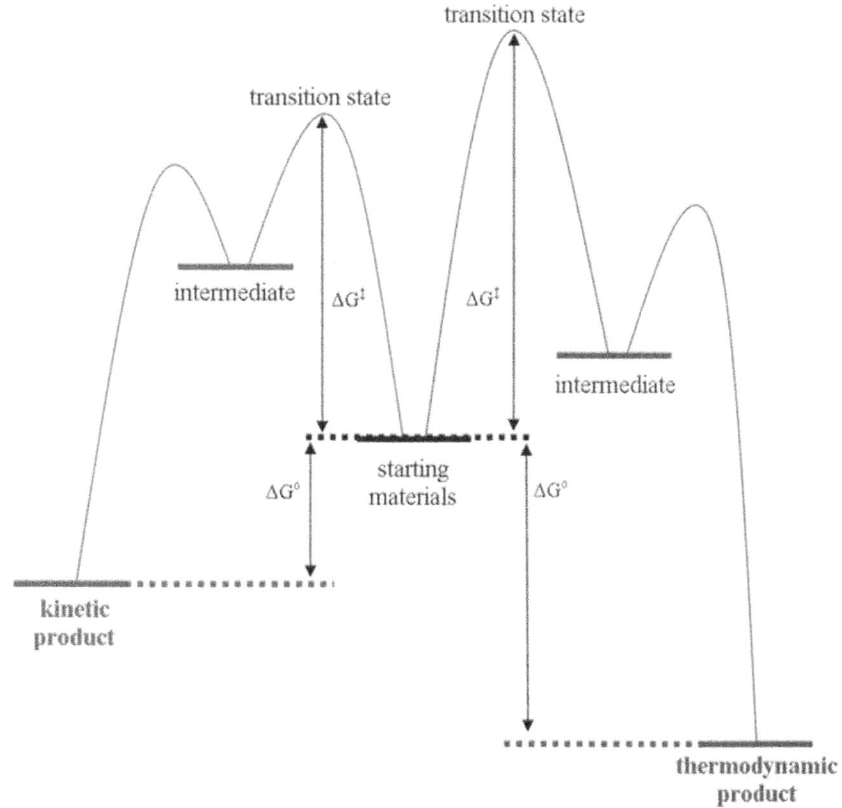

The reactants (starting material) are in the center. On the left is an intermediate with lower energy than on the right. The intermediate with lower energy forms faster (kinetic control). However, the final product from this lower energy intermediate yields a product with higher energy than the more stable product on the right (thermodynamic product). The product on the right is the thermodynamic product (overall more stable than the kinetic product) but requires a higher transitions state to overcome before the product is formed. The kinetic product forms fast because of a lower energy transition state.

The energy profile diagram displays a reaction under kinetic control vs. thermodynamic control. The starting materials (in the center) are the same for both. However, the ΔG, the transition states, and the energy of the intermediates are different. The product of the reaction is also different, depending on whether it is under kinetic control (left) or thermodynamic (right).

When strong acids are added to certain conjugated dienes, such as butadiene, there are two possible products: the 1,2-product and the 1,4-product, in the image below.

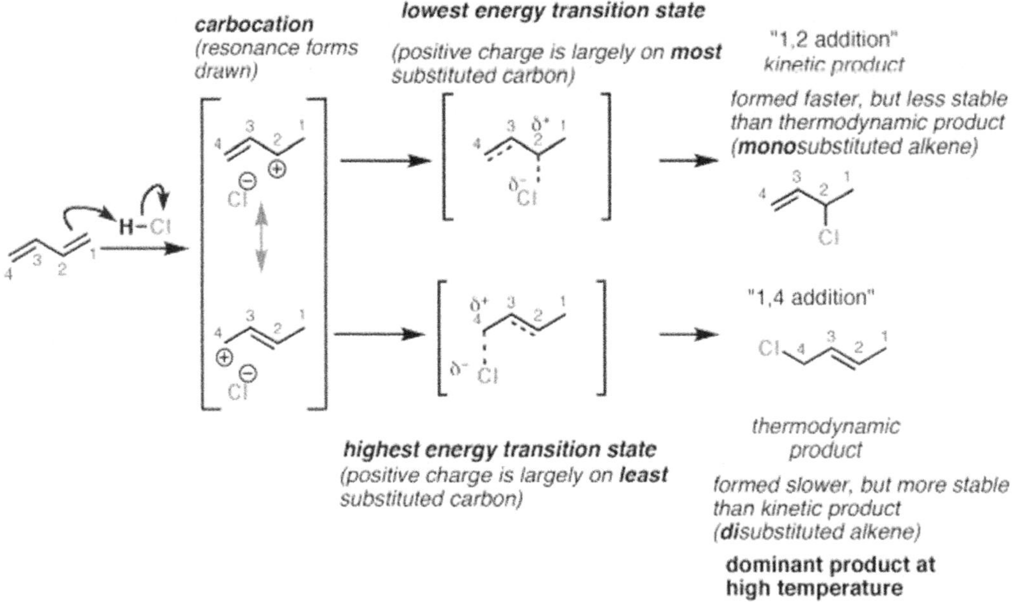

In the top transition state on the diagram above, the positive charge of the carbocation is localized on the most substituted (3°) carbon. Therefore, this transition state has a lower energy. This intermediate indicates that the top reaction has lower activation energy. The 1,2 *kinetic product* is preferentially formed at lower temperatures.

The bottom transition state is higher in energy because the positive charge of the carbocation is localized on the less substituted (2°) carbon. Therefore, this transition state has higher energy. This intermediate indicates that the bottom reaction has higher activation energy. The 1,4 *thermodynamic product* is preferentially formed at higher temperatures needed to proceed through the less stable (2°) carbon intermediate.

The 1,4-product is more stable than the 1,2-product because it is a disubstituted alkene, as opposed to a monosubstituted alkene. Therefore, the 1,4-addition product is more thermodynamically stable (as a product) but proceeds more slowly through the less stable transitions state. The 1,4-addition product is the major product formed at higher temperatures needed to proceed through the less stable transitions state.

The energy profile diagram for an E_1 vs. E_2 reaction is shown below.

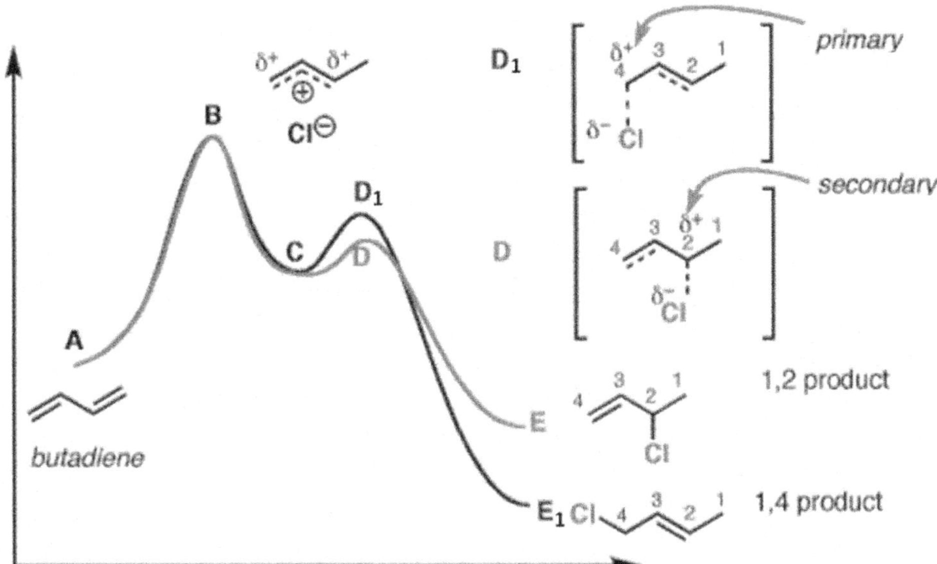

Energy coordinate for HCl addition to 1,2-addition vs. 1,4- additions to butadiene

The height of transition states D and D_1 determines the reaction rate once the carbocation C has formed.

Lower energy transition states (D *vs* D_1) proceed faster to form products.

Therefore, E is formed faster from the primary carbocation intermediate C since the energy of transition state D is less than D_1.

The energy of E and E_1 is related to the greater stability of the 1,4-alkene (disubstituted *vs*. monosubstituted).

E_1 has a more substituted (internal) double bond than E, so it is more stable.

E_1 represents the thermodynamic (more stable) product that is slower to form but is more stable than the kinetic product E which formed faster, but is less stable than E_1.

Notes

Catalysts and Enzyme Catalysis

Catalysts are substances in a chemical reaction that increase the reaction rate. Catalysts lower the activation energy (often through an alternative pathway) needed for the reaction to proceed.

A catalyst participates in a chemical reaction but remains chemically unchanged and can be used repeatedly. For a reaction mechanism separated into elementary steps, a catalyst appears at the start and reappears at the end of the reaction. A catalyst is classified as neither a reactant nor a product in the reaction.

Acids, bases, and metal ions often act as catalysts.

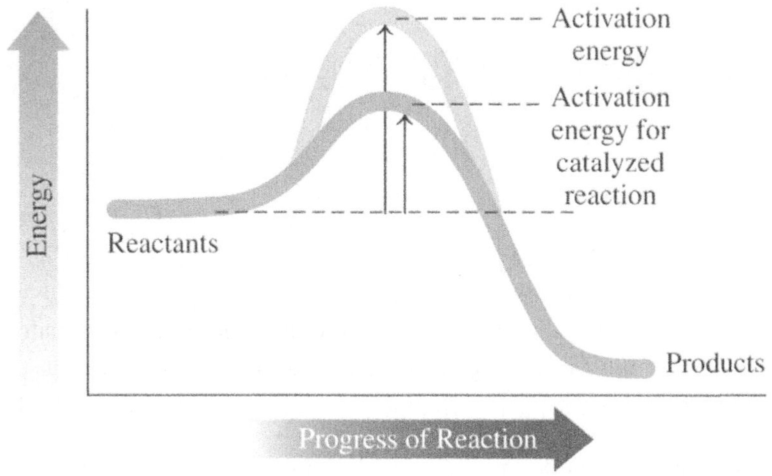

There are homogeneous catalysts and heterogeneous catalysts.

Homogeneous catalysts are in the same phase as the reactants (e.g., the reactant and catalyst are in the gas phase).

A *heterogeneous catalyst* is in a different phase than the reactants. These catalysts often immobilize reactants near each other, increasing the likelihood of a collision and, therefore, the rate of reaction. An example of a heterogeneous catalyst is a dissolved acid that catalyzes the hydrolysis of esters in an aqueous solution. Most heterogeneous catalysts are solids, and the reactants are generally gases or liquids.

For example, iron catalyzes the synthesis of NH_3 from gaseous N_2 and H_2.

Modern cars are equipped with *catalytic converters* in their exhaust systems. These heterogeneous catalysts provide a solid surface for automobile exhaust molecules to bind and react. Catalytic converters help to reduce pollutants such as carbon monoxide

(CO), nitrogen oxide (NO), and hydrocarbons (e.g., octane as C_8H_{18}). The catalyst usually consists of solid particles, including platinum (Pt) and palladium (Pd).

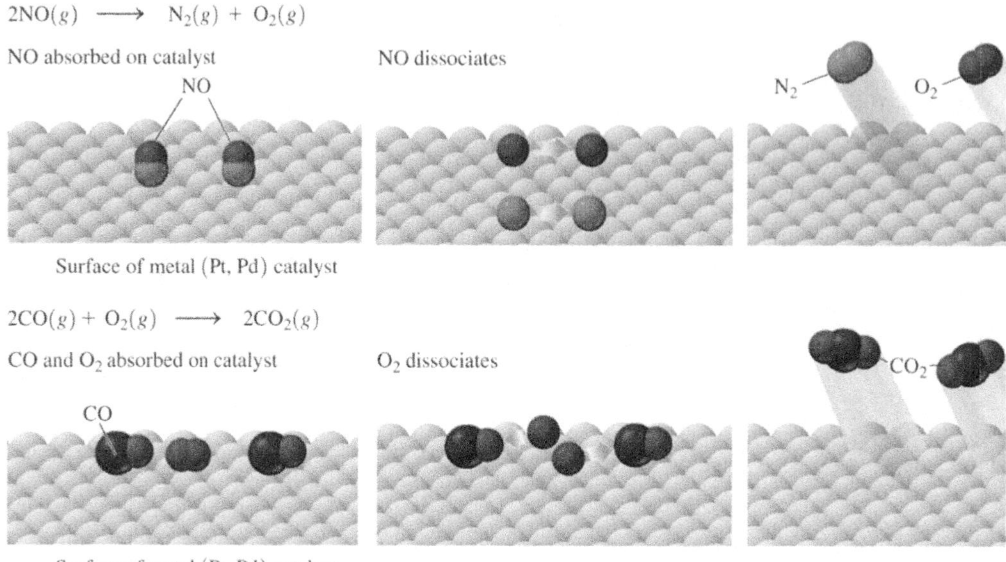

Enzymes are biological catalysts (e.g., proteins) that increase the reaction rates of biochemical reactions. Biological enzymes are more efficient than chemical catalysts; enzymes can enhance the rate of a chemical reaction by a factor of more than ten million (1×10^7). Enzymes are highly specific; they form enzyme-substrate complexes with the reactants. A *substrate* is a molecule on which the enzymes act.

When hydrogen peroxide (H_2O_2) is applied to a wound, oxygen gas is produced by the decomposition of hydrogen peroxide by the catalase enzyme found in the blood.

$$2\,H_2O_2\,(l) \xrightarrow{\text{catalase}} 2\,H_2O\,(l) + O_2\,(g)$$

Catalysts are written either above or below the arrow in the reaction equation.

Changing variables (i.e., reactant concentration, temperature) could increase the reaction rate. Catalyst has a similar effect on the reaction rate but uses a different method.

Factor	Reason
Increasing reactant concentration	More collisions
Increasing temperature	More collisions exceeding the energy of activation
Adding a catalyst	Lowers energy of activation

Equilibrium in Reversible Chemical Reactions

Some chemical reactions are reversible; they can form products in either direction, resulting in an equilibrium mixture of reactants and products. If a chemical bond can be formed, it can also be broken with the input of sufficient enough. Reversible chemical reactions are indicated by either a double-headed arrow or two arrows facing opposite directions.

$$A + B \leftrightarrow AB$$

$$A + B \leftrightarrows AB$$

The forward reaction occurs when A and B form AB. When AB accumulates, the reverse reaction occurs from the decomposition of AB, and A and B are reformed.

At the point when the forward and the reverse reactions occur at the *same rate*, the reaction is in chemical *equilibrium*. Equilibrium does not mean that there are equal amounts of reactants and products but is the point where the number of reactants and products remain constant: the forward and reverse reactions proceed at equal rates.

Consider the following reaction:

$$2\ SO_2\,(g) + O_2\,(g) \rightleftarrows 2\ SO_3\,(g)$$

Starting with the reactants SO_2 and O_2, the reaction proceeds in the forward direction, and the SO_3 product forms until equilibrium (i.e., rate) is reached.

Starting with the product of SO_3, the reaction proceeds in the reverse direction and forms SO_2 and O_2 until equilibrium is reached.

As shown below, the equilibrium concentrations of SO_2, O_2, and SO_3 are the same for both the forward and reverse reactions.

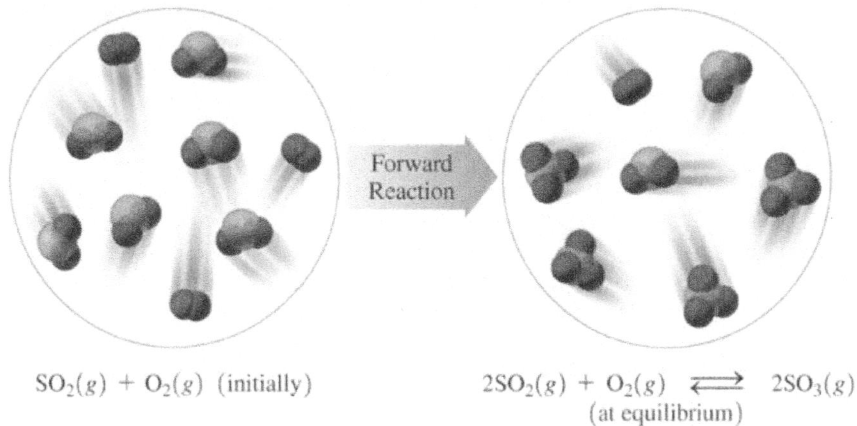

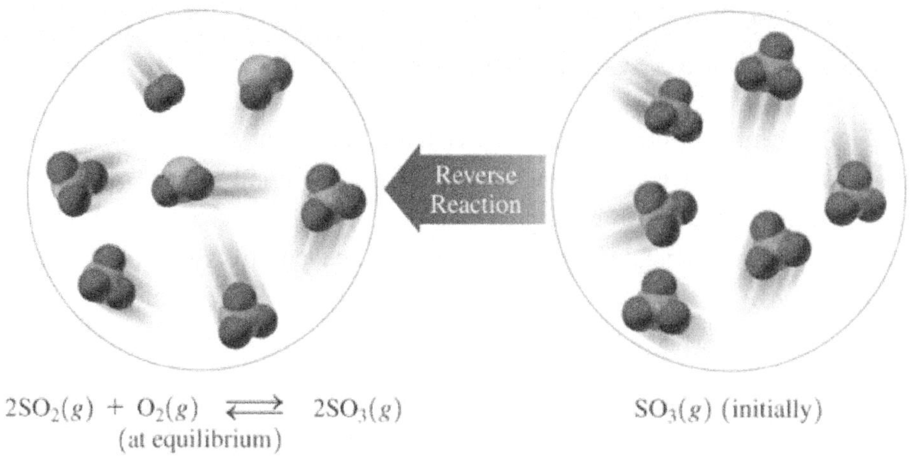

Consider another reversible reaction: the evaporation and condensation of water. If water is placed over a gas-filled chamber, separated by a moving piston as diagramed below, work *on* or *by* the system is observed by measuring the compression or expansion of the gas. As water evaporates, the water weighs less, and it allows the gas to expand. When water vapor condenses, it increases the weight of water and compresses the gas.

However, once the system reaches equilibrium, the liquid-to-vapor ratio is constant, and the gas will not compress or expand, which leads to the conclusion that there is no work being done *on* or *by* the system at equilibrium.

Water molecules are still constantly transforming between liquid and gas phases, but the rate of evaporation matches the rate of condensation, so there is not a noticeable shift in composition.

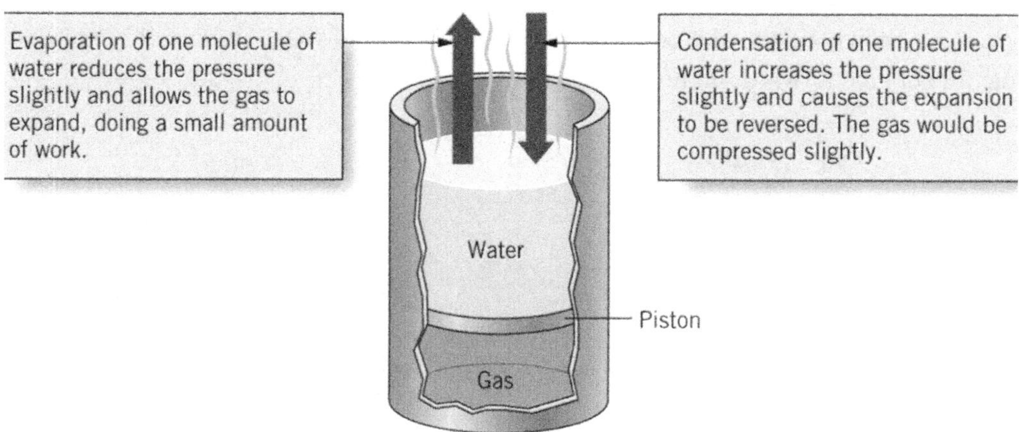

Piston system to detect equilibrium during the condensation and evaporation of water

Law of mass action

The *law of mass action* states that the *reaction rate depends on the concentration of the participating substances in the reaction.*

$$aA + bB \rightleftarrows cC + dD$$

$$\text{rate}_{\text{rxn}} = \left(-\frac{1}{a}\right)\frac{\Delta A}{\Delta t} = \left(-\frac{1}{b}\right)\frac{\Delta B}{\Delta t} = \left(+\frac{1}{c}\right)\frac{\Delta C}{\Delta t} = \left(+\frac{1}{d}\right)\frac{\Delta D}{\Delta t}$$

For the law of mass action, the equilibrium constant is derived by setting the forward reaction rate equal to the reverse reaction rate, which happens at equilibrium.

For the reaction:

$$aA + bB \rightleftarrows cC + dD$$

The following relationships are true:

- $r_{\text{forward}} = r_{\text{reverse}}$
- $k_{\text{forward}}{\cdot}[A]^a{\cdot}[B]^b = k_{\text{reverse}}{\cdot}[C]^c{\cdot}[D]^d$
- $K_c = k_{\text{forward}} / k_{\text{reverse}} = [C]^c{\cdot}[D]^d / [A]^a{\cdot}[B]^b$

The law of mass action is based on the equilibrium constant K_c, which is the ratio of the concentration of products [products] to the concentration of reactants [reactants].

When writing the mass action expression of equilibrium, only include the components in the equilibrium that are in the same phase. Pure solids and liquids do not appear in the mass action expression.

The absence of pure solids and liquids from the equation is seen:

$$CaCO_3\ (s) \leftrightarrows CaO\ (s) + CO_2\ (g)$$

$$K_c = [CO_2]$$

$$2\ HCl\ (aq) + CaCO_3\ (s) \leftrightarrows CaCl_2\ (aq) + CO_2\ (g) + 2\ H_2O\ (l)$$

The net ionic equation for the above reaction is:

$$2\ H^+\ (aq) + CaCO_3\ (s) \leftrightarrows Ca^{2+}\ (aq) + CO_2\ (g) + 2\ H_2O\ (l)$$

Cl^- ions were present on both sides of the equation; therefore, they are spectator ions and can be removed from the net equation.

$$K_{c1} = [Ca^{2+}] / [H^+]^2 \qquad \text{or} \qquad K_{c2} = [CO_2]$$

There are two forms of mass action equilibrium constants.

K_{c1} is expressed as the concentrations of the aqueous species.

K_{c2} is expressed regarding the concentration of the gaseous species.

The K_{c1} and K_{c2} constants have different values; the commonly used K values are based on the most accessible phase to measure experimentally. Do not mix phases, and do not include the solid in either expression. When solving mass action problems, always check the phases of the molecules in the reaction.

Equilibrium Constant, K_c

The *equilibrium constant* K_c represents the ratio of the molar concentrations of products divided by reactants at equilibrium. The K_c can be abbreviated as K_{eq}, which is a more general term for an equilibrium constant that refers to either the equilibrium for concentration or the equilibrium for partial pressures.

For the reaction $a\text{A} + b\text{B} \leftrightarrows c\text{C} + \text{dD}$, the equilibrium constant is:

$$K_c = \frac{[\text{C}]^c[\text{D}]^d}{[\text{A}]^a[\text{B}]^b} = \frac{[\text{products}]}{[\text{reactants}]}$$

The equilibrium constant K_c is a temperature-specific constant. It always has the same value, regardless of the molar concentrations, if the temperature remains constant.

The units of K_c depend on the specific reaction, and K_c is without units.

Guidelines for calculating the K_c value:

1. State the given and needed qualities.
2. Write the K_c expression for the equilibrium.
3. Substitute equilibrium (molar) concentrations and calculate K_c.

What is the value of K_c at 443 °C for the equilibrium concentrations:

$H_2\,(g) + I_2\,(g) \leftrightarrows 2\,HI\,(g)$

$[H_2] = 1.2$ mol/L

$[I_2] = 1.2$ mol/L $[HI] = 0.35$ mol/L

Step 1: State the given and needed quantities.

Given values		**Needed quantities**
Reactants	*Products*	
$[H_2] = 1.2$ mol/L	$[HI] = 0.35$ mol/L	
$[I_2] = 1.2$ mol/L		K_c

Step 2: Write the K_c expression for the equilibrium.

$K_c = [HI]^2 / [H_2]\cdot[I_2]$

Step 3: Substitute equilibrium (molar) concentrations and calculate K_c.

$K_c = [0.35]^2 / [1.2]\cdot[1.2]$

$K_c = 8.5 \times 10^{-2}$

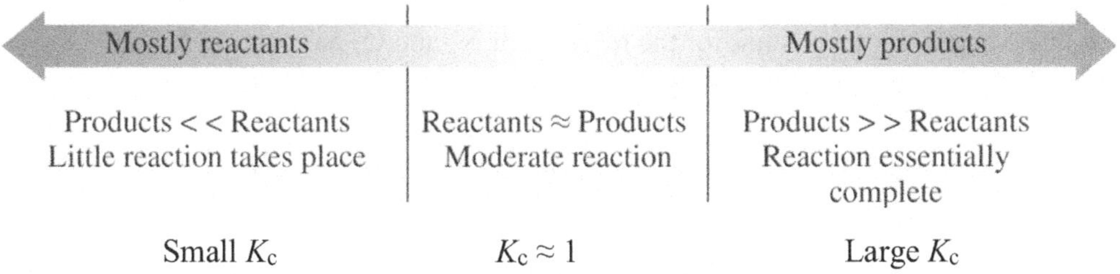

Mostly reactants		Mostly products
Products < < Reactants Little reaction takes place	Reactants ≈ Products Moderate reaction	Products > > Reactants Reaction essentially complete
Small K_c	$K_c \approx 1$	Large K_c

The value of K_c depends on whether the equilibrium is reached with a greater quantity of products or reactants. However, the size of the equilibrium constant does not affect how fast equilibrium is reached.

Reactions with a large K_c have more products created from the forward reaction at equilibrium, and these products predominate at equilibrium.

Reactions with a small K_c have more reactants that predominate at equilibrium.

The equilibrium constant for the reaction of SO_2 and O_2 has a large K_c.

$2\ SO_2\,(g) + O_2\,(g) \leftrightarrows 2\ SO_3\,(g)$

At equilibrium, the reaction contains mostly products and few reactants.

$K_c = [SO_3]^2 / [SO_2]^2\cdot[O_2]$

K_c = [many products] / [few reactants]

$K_c = 3.4 \times 10^2$

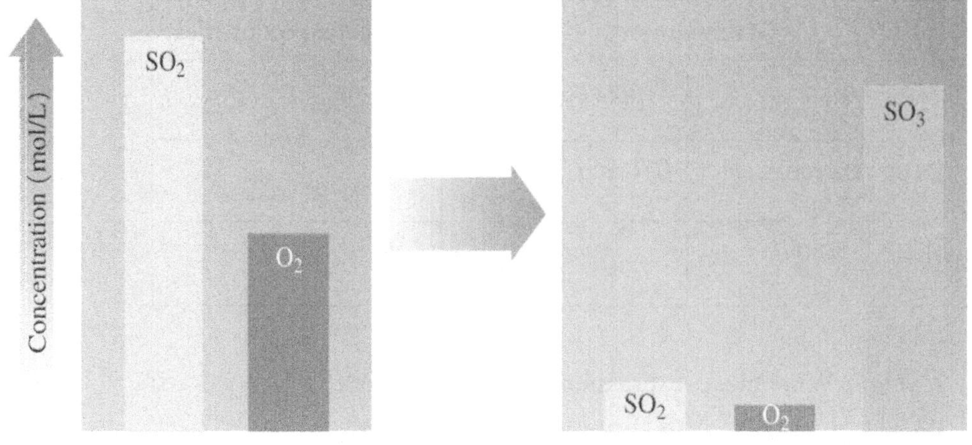

$2\ SO_2\ (g) + O_2\ (g) \leftrightarrows 2\ SO_3\ (g)$

Meanwhile, reactions with a small K_c have an equilibrium mixture with a low concentration of [products] and a high concentration of [reactants].

The equilibrium constant for the reaction of N_2 and O_2 has a small K_c.

$N_2\ (g) + O_2\ (g) \leftrightarrows 2\ NO\ (g)$

At equilibrium, the reaction mixture contains few products and mostly reactants.

$K_c = [NO] / [N_2]{\cdot}[O_2]$

K_c = [few products] / [mostly reactants]

$K_c = 2 \times 10^{-9}$

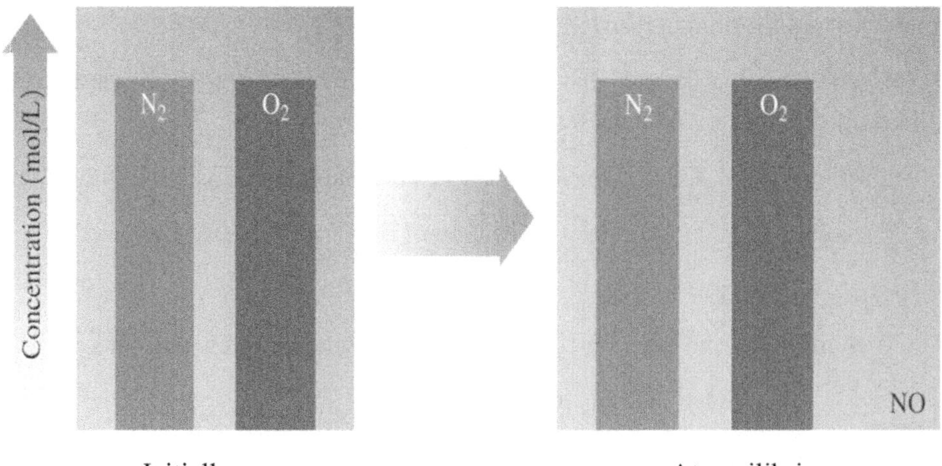

A *homogeneous equilibrium* is a reaction in which all products and reactants are in the same physical state, such as the following reaction:

$2\ SO_2\ (g) + O_2\ (g) \rightleftarrows 2\ SO_3\ (g)$

$K_c = [SO_3]^2 / [SO_2]^2 \cdot [O_2]$

A *heterogeneous equilibrium* is a reaction where one of the substances is in a different physical state, such as in the following reaction:

$C\ (s) + H_2O\ (g) \rightleftarrows CO\ (g) + H_2\ (g)$

$K_c = [CO] \cdot [H_2] / [H_2O]$

The concentrations of liquids and solids do not change and are omitted from the equilibrium constant K_c expression. This omission is like the mass-action equilibrium.

The reaction quotient Q_c is calculated using the same method as K_c, but the calculation of Q_c can be performed at any time during the reaction.

Q_c is a "snapshot" of the system, representing the ratio between products and reactants at a given time.

A comparison of Q_c and K_c can determine the direction of the reaction:

If $Q_c > K_c$, the reaction creates more reactants

If $Q_c < K_c$, the reaction creates more products

If $Q_c = K_c$, the reaction is at equilibrium, $\Delta G = 0$

Application of Le Châtelier's principle

Le Châtelier's principle states that when a reversible reaction at equilibrium is stressed by a change in concentration, pressure, volume, or temperature, the equilibrium shifts to relieve the effect of that change.

For the equilibrium between colorless N_2O_4 and brown NO_2:

$N_2O_4\ (g) \rightleftarrows 2\ NO_2\ (g)$

Effect of Concentration

If the amount of N_2O_4 reactant is increased, the reaction shifts to the right to produce more NO_2 products.

If the amount of NO_2 product is increased, the reaction shifts to the left to produce more N_2O_4 reactants.

Effect of Pressure

In a gaseous equilibrium, increasing the pressure shifts the reaction to the side with fewer gas molecules.

For the reaction:

$N_2O_4\ (g) \rightleftarrows 2\ NO_2\ (g)$

Increasing the pressure shifts the reaction to the left, producing more N_2O_4.

If there are the same moles of gases on both the reactant and product side, then changing the pressure does not shift the equilibrium.

Effect of Decreasing Volume on Equilibrium

A change in the volume of a gas mixture at equilibrium changes the concentration of the gases in the mixture.

Decreasing volume, shifts toward fewer moles

$2\ CO_2\ (g) + O_2\ (g) \rightleftarrows 2\ CO_2\ (g)$

Decreasing the volume increases the concentration of the gases. The system shifts in the direction of the smaller number of moles from the decrease in volume.

1.00 L at equilibrium *0.750 L at equilibrium*

Effect of Increasing Volume on Equilibrium:

A change in the volume of a gas mixture changes the concentration of the gases.

Increasing volume, shifts toward more moles

$$2\ CO\ (g) + O_2\ (g) \rightleftarrows 2\ CO_2\ (g)$$

Increasing the volume decreases the concentration of the gases. Therefore, the system shifts in the direction of the larger number of moles to compensate.

The effect of temperature on equilibrium depends on the ΔH of a reaction (i.e., endothermic, or exothermic).

Endothermic Reaction Equilibrium and Temperature

Decreasing the temperature of an endothermic reaction ($+\Delta H$; heat is a reactant) causes the system to respond by shifting the reaction toward more heat; in this case, it is toward the reactants, increasing heat in the system.

Decrease temperature

$$N_2O_4\ (g) + \textit{heat} \rightleftarrows 2\ NO_2\ (g)$$

Increasing the temperature of an endothermic reaction ($+\Delta H$; heat is a reactant) causes the system to respond by shifting the reaction to remove heat; in this case, it is towards the products, consuming the heat.

Increase temperature

$$N_2O_4\ (g) + \textit{heat} \rightleftarrows 2\ NO_2\ (g)$$

Exothermic Reaction Equilibrium and Temperature

Decreasing the temperature of an exothermic reaction ($-\Delta H$; heat is a product) causes the system to respond by shifting the reaction toward more heat; it shifts the reaction toward the products, increasing heat in the system.

Decrease temperature

$$2\ SO_2\ (g) + O_2\ (g) \leftrightarrows 2\ SO_3\ (g) + \text{heat}$$

Increasing the temperature of an exothermic reaction ($-\Delta H$; heat is a product) causes the system to respond by shifting the reaction toward removing heat; it shifts the reaction toward the reactants, decreasing heat in the system.

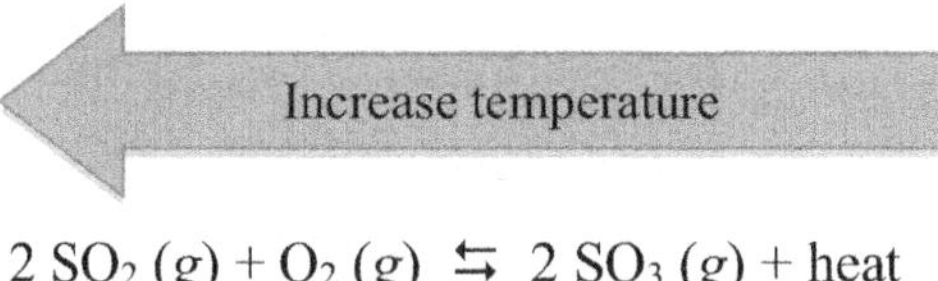

$$2\ SO_2\ (g) + O_2\ (g) \leftrightarrows 2\ SO_3\ (g) + \text{heat}$$

Changes in conditions and effect on equilibrium

Condition	**Change (Stress)**	**Equilibrium Shifts Towards…**
Concentration	Add a reactant	Products (forward reaction)
	Remove a reactant	Reactants (reverse reaction)
	Add a product	Reactants (reverse reaction)
	Remove a product	Products (forward reaction)
Volume (container)	Decrease volume	Side with lower total coefficients (*aq* and *g*)*
	Increase volume	Side with higher total coefficients (*aq* and *g*)
Pressure	Decrease pressure	Side with higher total coefficients (*aq* and *g*)
	Increase pressure	Side with lower total coefficients (*aq* and *g*)
Temperature	**Endothermic Rxn**	
	Raise Temperature	Products (forward reaction to remove heat)
	Lower Temperature	Reactants (reverse reaction to add heat)
	Exothermic Rxn	
	Raise Temperature	Reactants (reverse reaction to add heat)
	Lower Temperature	Products (forward reaction to remove heat)
Catalyst	Increases rates equally	No effect

* Pure solids and pure liquids are not included in the equilibrium expression.

Equilibrium is an important concept in many systems, including biological ones. For example, oxygen transport involves an equilibrium between hemoglobin (Hb), oxygen, and oxyhemoglobin (HbO_2).

$$Hb\ (aq) + O_2\ (g) \rightleftarrows HbO_2\ (aq)$$

$$K_c = [HbO_2] / [Hb] \cdot [O_2]$$

If there is a high concentration of O_2 (reactant) in the alveoli of the lungs, the reaction shifts to the right to make more oxyhemoglobin (product). When the concentration of O_2 is low in the tissues, the reverse reaction releases O_2 from oxyhemoglobin.

At normal atmospheric pressure, oxygen diffuses into the blood because the partial pressure of oxygen in the alveoli is higher than that in the blood. At altitudes above 8,000 ft, a decrease in atmospheric pressure results in lower partial pressure of O_2.

Hypoxia may occur at high altitudes where the $[O_2]$ is low. At an altitude of 18,000 ft., a person obtains 29% less oxygen and may experience hypoxia. According to Le Châtelier's principle, a decrease in O_2 shifts the equilibrium in the direction of the reactants and depletes the concentration of HbO_2, which can cause hypoxia.

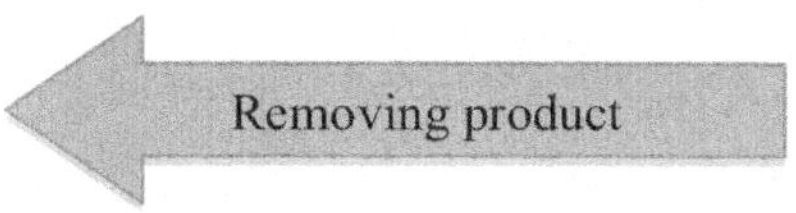

$$Hb\ (aq) + O_2\ (g) \rightleftarrows HbO_2\ (aq)$$

Another reaction that applies Le Châtelier's principle is the following.

In the example, indicate the shift in equilibrium caused by each change:

$$2\ NO_2\ (g) + \text{heat} \leftrightarrows 2\ NO\ (g) + O_2\ (g)$$

Do the following changes cause the equilibrium to shift toward reactants or products?

A. adding NO

B. lowering the temperature

C. removing O_2

D. increasing the volume

E. removing NO

Answers:

A. reactants

B. reactants

C. products

D. products

E. products

Notes

Relationship of the Equilibrium Constant and $\Delta G°$

Gibbs free energy ($\Delta G°$) is an expression for the free energy of a system. In the previous chapter, the following equation, relating Gibbs free energy to the equilibrium constant, was given:

$$\Delta G° = -RT \ln K$$

Taking the antilog (e^x) of both sides gives:

$$K = e^{-\Delta G° / RT}$$

The ΔG and K only imply the direction and extent of a reaction, not the rate.

ΔG affects the position of equilibrium in the following ways:

$$\Delta G° > 0 \text{ (positive)}$$

If $\Delta G° > 0$, equilibrium is closer to the reactants; the reverse reaction (towards reactants) is spontaneous.

$$\Delta G° < 0 \text{ (negative)}$$

If $\Delta G° < 0$, equilibrium is closer to the products; the forward reaction (towards products) is spontaneous.

$$\Delta G° = 0$$

If $\Delta G° = 0$, the reaction is at equilibrium; there is no net change to the system or surroundings.

The reaction appears spontaneous in either direction if some small change is made to some property of the system.

Consider the freezing of water at 0 °C:

$$H_2O\ (l) \rightarrow H_2O\ (s)$$

The system remains at equilibrium if no heat is added or removed. Both phases exist together indefinitely.

Below 0 °C: $\Delta G < 0$ and freezing is spontaneous

Above 0 °C: $\Delta G > 0$ and freezing is nonspontaneous

The energy profile below displays a reaction with a positive ΔG (nonspontaneous).

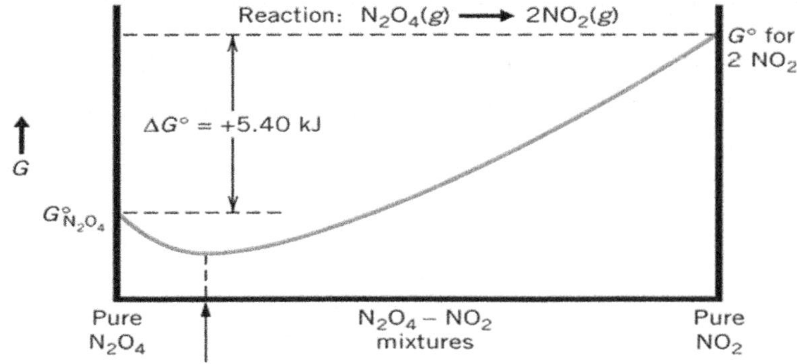

The energy profile below displays a reaction with a negative ΔG (spontaneous).

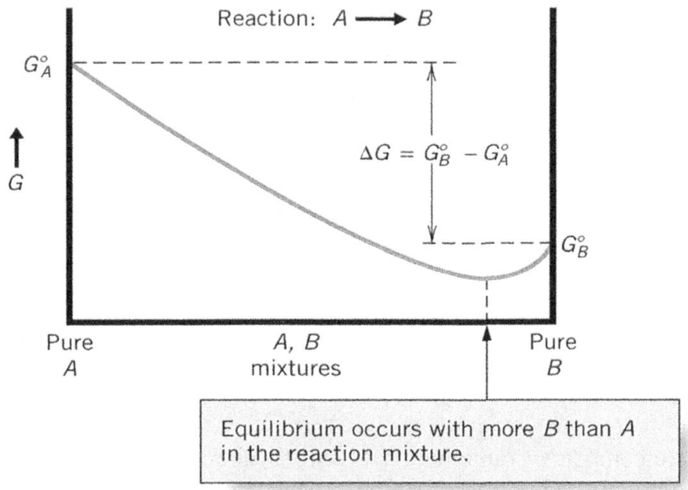

Calculate the equilibrium constant K_c at 25 °C for the decarboxylation of liquid pyruvic acid ($CH_3COCOOH$) to form gaseous acetaldehyde (CH_3COH) and carbon dioxide (CO_2).

$$CH_3COCOOH\ (l) \rightleftarrows CH_3COH\ (g) + CO_2\ (g)$$

Compound	$\Delta G°_f$ (kJ/mol)
CH_3COH	–133.30
$CH_3COCOOH$	–463.38
CO_2	–394.36

First, sum the $\Delta G°_f$ for each reaction to obtain the total $\Delta G°$.

$$\Delta G° = \Delta G°_f(CH_3COH) + \Delta G°_f(CO_2) - \Delta G°_f(CH_3COCOOH)$$

$$\Delta G° = -133.30 + (-394.36) - (-463.38)$$

$$\Delta G° = -64.28 \text{ kJ}$$

Use the $\Delta G°$ value to calculate K:

$$K = e^{-\Delta G° / RT}$$

$$\Delta G° / \text{RT} = (-64.28 \text{ kJ}) / [(8.314 \text{ J/K}) \times (298 \text{ K}) \times (1{,}000 \text{ J/kJ})]$$

$$\Delta G° / \text{RT} = -25.94$$

$$K = e^{-(-25.945)}$$

$$K = e^{25.945}$$

$$K = 1.85 \times 10^{11}$$

Notes

Relationship of the Equilibrium Constant and $\Delta H°$, $\Delta S°$

The previous section describes the relationship between the equilibrium constant and the change in Gibbs free energy ΔG.

However, enthalpy H and entropy S are also related to the equilibrium constant.

For the change in Gibbs free energy:

$$\Delta G° = \Delta H° - T\Delta S°$$

and

$$\Delta G° = -RT \ln K$$

therefore,

$$-RT \ln K = \Delta H° - T\Delta S°$$

Rearranging the equation yields the following *point-slope formula*:

$$\ln K = -(\Delta H° / R)\cdot(1 / T) + (\Delta S° / R)$$

This equation is in the form of a line equation.

On a plot of $\ln K$ vs. $1 / T$:

the slope is $-\Delta H° / RT$

the y-intercept is $\Delta S° / R$

Notes

Practice Questions

1. Which statement is NOT a correct characterization for a catalyst?

A. Catalysts are not consumed in a reaction
B. Catalysts do not actively participate in a reaction
C. Catalysts lower the activation energy for a reaction
D. Catalysts can be either solids, liquids, or gases
E. Catalysts do not alter the equilibrium of the reaction

2. What is the rate law for the following reaction that was found to be first-order in each of the two reactants and second-order overall?

$$2\ NO\ (g) + O_2\ (g) \rightarrow 2\ NO_2\ (g)$$

A. rate = $k[NO]^2 \cdot [O_2]^2$
B. rate = $k[NO_2]^2 \cdot [NO]^{-2} \cdot [O_2]^{-1/2}$
C. rate = $k[NO] \cdot [O_2]$
D. rate = $k[NO]^2$
E. rate = $k([NO] \cdot [O_2])^2$

3. Which equilibrium constant applies to a reversible reaction involving a gaseous mixture at equilibrium?

A. Ionization equilibrium constant, K_w
B. General equilibrium constant, K_p
C. Solubility product equilibrium constant, K_{sp}
D. Ionization equilibrium constant, K_i
E. None of the above

4. What is the equilibrium constant (K_{eq}) expression for the following reaction:

$$CaCO_3\ (s) \leftrightarrow CaO\ (s) + CO_2\ (g)$$

A. $K_{eq} = [CO_2]$
B. $K_{eq} = 1 / [CO_2]$
C. $K_{eq} = [CaO] \cdot [CO_2] / [CaCO_3]$
D. $K_{eq} = [CaO] \cdot [CO_2]$
E. None of the above

5. Which of the following statements about catalysts is NOT true?

A. A catalyst does not change the energy of the reactants or the products
B. A catalyst increases the rate of slow reactions
C. A catalyst is consumed in a reaction
D. A catalyst alters the rate of a chemical reaction
E. All of the above are true

6. Increasing the temperature of a chemical reaction increases the rate of reaction because:

A. both the collision frequency and collision energies of reactant molecules increase
B. the collision frequency of reactant molecules increases
C. the activation energy increases
D. the activation energy decreases
E. the stability of the products increases

7. If $K_{eq} = 6.1 \times 10^{-11}$, which statement is true?

A. Slightly more products are present
B. The number of reactants equals products
C. Mostly products are present
D. Mostly reactants are present
E. Cannot be determined

8. Which is true before a reaction reaches chemical equilibrium?

A. The number of reactants and products are equal
B. The number of reactants and products are constant
C. The number of products is decreasing
D. The number of reactants is increasing
E. The number of products is increasing

9. Why might increasing temperature alter the rate of a chemical reaction?

A. The molecules combine with other atoms at high temperature to save space
B. The density decreases as a function of temperature that increases volume and decreases the reaction rate
C. The molecules have higher kinetic energy and have more force when colliding
D. The molecules are less reactive at higher temperatures
E. None of the above

10. In the following reaction, how does increasing the pressure affect the equilibrium?

$$2\ SO_2\ (g) + O_2\ (g) \leftrightarrow 2\ SO_3\ (g) + \textit{heat}$$

A. Remains unchanged, but the reaction mixture gets warmer
B. Remains unchanged, but the reaction mixture gets cooler
C. Shifts to the right towards products
D. Shifts to the left towards reactants
E. Pressure does not affect the equilibrium

11. According to Le Châtelier's principle, which changes shifts the equilibrium to the left for the following reactions?

$N_2\ (g) + 3\ H_2\ (g) \leftrightarrow 2\ NH_3\ (g) + heat$

A. Decreasing the temperature
B. Increasing $[H_2]$
C. Increasing $[N_2]$ the pressure
D. Decreasing the pressure on the system
E. Increasing $[N_2]$ and $[H_2]$

12. What is the term for a dynamic state of a reversible reaction in which the rates of the forward and reverse reactions are equal?

A. dynamic equilibrium
B. rate equilibrium
C. reversible equilibrium
D. concentration equilibrium
E. chemical equilibrium

13. When a system is at equilibrium, the:

A. reaction rate of the forward reaction is low compared to the reverse
B. number of products and reactants is equal
C. reaction rate of the forward reaction is equal to the rate of the reverse
D. reaction rate of the reverse reaction is low compared to the forward
E. none of the above

Detailed Explanations

1. B is correct.

Enzymes (i.e., biological catalysts) bind to substrates to form an enzyme-substrate complex. While in this complex, enzymes often align reactive chemical groups and hold them closer together. Enzymes can induce structural changes that strain substrate bonds. Therefore, catalysts can actively participate in reactions.

The primary function of catalysts is lowering the reaction's activation energy (energy barrier), thus increasing its rate (k).

Although catalysts do decrease the amount of energy required to reach the rate-limiting transition state, they do *not* decrease the relative energy of the products and reactants. Therefore, a catalyst does not affect ΔG.

A catalyst provides an alternative pathway for the reaction to proceed to product formation. It lowers the energy of activation (i.e., relative energy between reactants and transition state) and therefore speeds the rate of the reaction.

Catalysts do not affect the Gibbs free energy (ΔG: stability of products vs. reactants) or the enthalpy (ΔH: bond breaking in reactants or bond making in products).

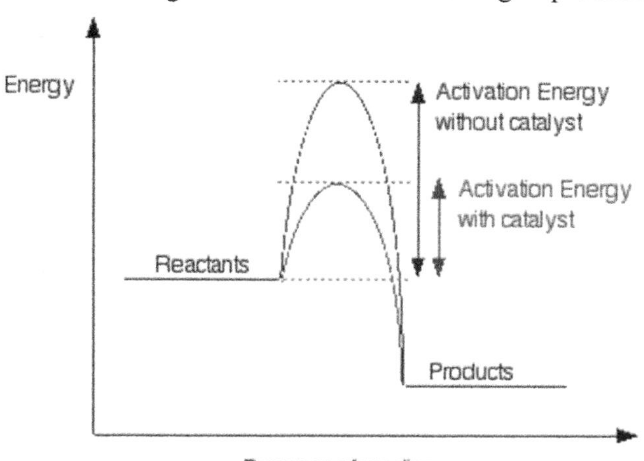

2. C is correct.

The rate law is calculated by comparing trials and determining how changes in the initial concentrations of the reactants affect the rate of the reaction.

rate = $k[A]^x \cdot [B]^y$

where k is the rate constant, and the exponents x and y are the partial reaction orders (i.e., determined experimentally). They are not equal to the stoichiometric coefficients.

To write a complete rate law, the order of each reactant is required. The order of each reactant has been provided by the problem (first order in each).

The total order is a sum of the individual orders: 1 + 1 = 2. However, the total order is usually not expressed in the rate law.

3. B is correct.

To determine the amount of each compound at equilibrium, consider the chemical reaction written in the form:

$$aA + bB \leftrightarrow cC + dD$$

The equilibrium constant is:

$$K_{eq} = ([C]^c \times [D]^d) / ([A]^a \times [B]^b)$$

K_w is the notation for the self-ionization (i.e., auto ionization) constant for water.

K_{sp} is for "solubility" and is only applicable for equilibriums for solids dissolving in liquids.

K_i is for "ionization" and is only applicable for equilibrium systems that involve liquids.

For equilibria in the gas phase, the equilibrium equation (K_p) is a function of the partial pressures (P) of the reactants and products.

Therefore:

$$K_p = ([P_C]^c \times [P_D]^d) / ([P_A]^a \times [P_B]^b)$$

where P represents partial pressure, usually in atm.

4. A is correct. General formula for the equilibrium constant of a reaction:

$$aA + bB \leftrightarrow cC + dD$$

$$K_{eq} = ([C]^c \times [D]^d) / ([A]^a \times [B]^b)$$

For equilibrium constant calculation, only include species in aqueous or gas phases:

$$K_{eq} = [CO_2]$$

5. C is correct.

The primary function of catalysts is lowering the reaction's activation energy (energy barrier), thus increasing its rate (k).

Although catalysts do decrease the amount of energy required to reach the rate-limiting transition state, they do *not* decrease the relative energy of the products and reactants.

Therefore, a catalyst does not affect ΔG.

A catalyst is never consumed in a reaction; reagents are consumed during the reaction.

A catalyst lowers the energy of the high-energy transition state (i.e., the activation energy), but it does not change the energy of the reactants or the products.

Catalysts increase the rate of chemical reactions.

6. A is correct.

Temperature is a measure of the average kinetic energy of the molecules.

Increasing the temperature increase both the collision frequency (i.e., due to an increased probability of molecules striking each other) and collision energies (kinetic energy = $\frac{1}{2}mv^2$) of reactant molecules.

7. D is correct.

For the general equation:

$$aA + bB \leftrightarrow cC + dD$$

The equilibrium constant is:

$$K_{eq} = ([C]^c \times [D]^d) / ([A]^a \times [B]^b)$$

or

$$K_{eq} = [products] / [reactants]$$

If K_{eq} is less than 1 (e.g., 6.1×10^{-11}), then the numerator (i.e., products) is smaller than the denominator (i.e., reactants) and fewer products have formed relative to the reactants.

If the reaction favors reactants compared to products, the equilibrium lies to the left.

8. E is correct.

A reaction proceeds with the formation of products.

Therefore, the rate of the forward reaction is higher than the rate of the reverse reaction until the reaction achieves equilibrium. At equilibrium, the rate (not relative concentrations of products and reactants) decreases for the forward reaction while the rate of the reverse reaction increases. At equilibrium, the rate of the forward reaction equals the rate of the reverse reaction.

An endergonic reaction (i.e., +ΔG) has fewer products formed than reactants remaining (i.e., the reaction is nonspontaneous) with the products higher in energy than the reactants.

Exergonic reactions (i.e., –ΔG) have more products formed than reactants remaining (i.e., the reaction is spontaneous) with the products lower in energy than the reactants.

The primary function of catalysts is lowering the reaction's activation energy (energy barrier), thus increasing its rate (k).

Although catalysts do decrease the amount of energy required to reach the rate-limiting transition state, they do *not* decrease the relative energy of the products and reactants. Therefore, a catalyst does not affect ΔG.

A catalyst provides an alternative pathway for the reaction to proceed to product formation. They lower the energy of activation (i.e., relative energy between reactants and transition state), and therefore speed the rate of the reaction.

Catalysts do not affect the Gibbs free energy (ΔG: stability of products vs. reactants) or the enthalpy (ΔH: bond breaking in reactants or bond making in products).

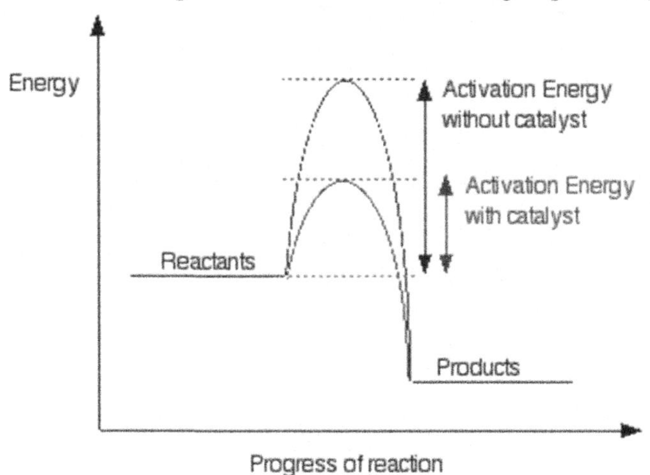

9. C is correct.

As the average kinetic energy (i.e., $KE = \frac{1}{2}mv^2$) *increases*, the particles move faster and collide more frequently per unit time and possess higher energy when they collide. This *increases* the *reaction rate*.

Hence the *reaction rate* of most *reactions increases* with *increasing temperature*.

10. C is correct.

When changing the conditions of a reaction, Le Châtelier's principle states that the position of equilibrium shifts to counteract the change. If the reaction temperature, pressure, or volume change, the position of equilibrium changes.

In general, increasing the pressure tends to favor the side of the reaction that has a lower molar coefficient sum (i.e., toward the products in this example).

11. D is correct.

When changing the conditions of a reaction, Le Châtelier's principle states that the position of equilibrium shifts to counteract the change. If the reaction temperature, pressure, or volume is changed, the position of equilibrium will change.

Removing reactants or adding products shifts the equilibrium to the left.

In this reaction, reducing the pressure of the system favors the reactants because of their respective molar concentrations of reactants (i.e., 1 + 3) and products (i.e., 2).

All other modifications listed shift the equilibrium toward products (i.e., to the right).

12. E is correct.

Equilibrium refers to the state when the *rate* of the forward reaction equals the rate of the reverse reaction.

Equilibrium does not describe the state when the relative energy of the reactants and products is the same, nor does it describe the state when the relative amounts of reactants and products are the same.

13. C is correct.

Chemical equilibrium refers to a dynamic process whereby the *rate* at which a reactant molecule is being transformed into a product is the same as the rate for a product molecule to be transformed into a reactant.

Therefore, the reaction rate of the forward reaction is equal to the rate of the reverse.

Chapter 7

Solution Chemistry

- **Introduction to Solutions**
- **Ions in Solution**
- **Solubility**

Introduction to Solutions

A *solution* is a homogeneous mixture of two or more substances in which individual components in the mixture are indistinguishable from one another. A solution is always composed of one or more solutes and a solvent.

The *solvent* is the substance that is present in the most substantial quantity.

Aqueous solutions are solutions that have water as the solvent. The formation of a solution involves disruption of the crystalline lattice structure of solute and the mixing of solute particles with solvent molecules.

Solute particles diffuse into the solution and become uniformly dispersed. For solute particles to diffuse into the solution, there must be sufficiently strong interactions between solute particles and solvent molecules (i.e., strong enough to overcome the attractive intermolecular forces between solute particles in either the liquid state or crystalline solids).

Solution processes may be divided into three main stages:

1. pure solvent → separated solvent molecules (endothermic);
   $\Delta H_1 > 0$

2. pure solute → separated solute particles (endothermic);
   $\Delta H_2 > 0$

3. separated solvent and solute molecules → solution (exothermic);
   $\Delta H_3 < 0$

Net:

Solute(s) + Solvent → Solution

$$\Delta H_{\text{soln}} = \Delta H_1 + \Delta H_2 + \Delta H_3$$

Depending on the magnitudes of ΔH_1, ΔH_2 and ΔH_3, solution processes can be:

Exothermic: if $|\Delta H_3| > |\Delta H_1 + \Delta H_2|$

Endothermic: if $|\Delta H_3| < |\Delta H_1 + \Delta H_2|$

For solutions, remember the phrase "like dissolves like."

Polar solutes dissolve in polar solvents.

Nonpolar solutes dissolve in nonpolar solvents.

Notes

Ions in Solution

Many ionic compounds dissolve in water but not in organic (nonpolar) solvents.

Water molecules are very polar and interact strongly with the ionic species (cations and anions) through ion-dipole interactions. Such interactions result in negative enthalpies, called the *solvation* or *hydration energy*.

The overall solute–solution formation process may be either exothermic or endothermic, depending on if the solute–solute or solute–solution enthalpy is larger.

Ionic compounds dissolve in water and dissociate (ionize) to produce free ions.

For example, when NaCl dissolves, it produces free Na^+ and Cl^- ions, which become hydrated with water molecules. Sodium chloride and many other ionic compounds dissolve in water because of the strong ion-dipole interactions between the charged solute particles and the polar solvent molecules.

The energy needed to overcome the lattice energy and disrupt ions in the crystalline solids (e.g., NaCl) is provided by the hydration energy produced from these ion-dipole interactions. For NaCl, the lattice energy is only slightly higher than the sum of hydration energy for Na^+ and Cl^- ions, and the solution process for NaCl is only slightly endothermic.

Strong electrolytes (solutions of ionic compounds) completely dissociate, and the dissolved ions are good conductors of electric current.

Weak electrolytes (solutions of polar covalent compounds) only partially dissociate in solution and are classified as weak electrolytes.

Nonelectrolytes do not dissociate into ions (e.g., sugar as a nonelectrolyte). Nonelectrolytes dissolve in the solution as the original molecules rather than separate into ions

Anion, cation: common names, formulas, and charges for familiar ions, such as ammonium (NH^{4+}); phosphate (PO_4^{3-}); sulfate (SO_4^{2-})

The following table lists common ions. The listing for the charged ions begins with anions (negatively charged), follows by cations (positively charged). The charge is written as a superscript. A single negative sign indicates a charge of –1, and a single positive sign indicates a charge of +1.

Common name	Formula
Anions	
Hydroxide	OH^-
Chloride	Cl^-
Hypochlorite	ClO^-
Chlorite	ClO_2^-
Chlorate	ClO_3^-
Perchlorate	ClO_4^-
Halide, hypohalite, etc.	X^-, XO^-, etc.
Carbonate	CO_3^{2-}
Hydrogen Carbonate (Bicarbonate)	HCO_3^-
Sulfate	SO_4^{2-}
Hydrogen Sulfate (Bisulfate)	HSO_4^-
Sulfite	SO_3^{2-}
Thiosulfate	$S_2O_3^{2-}$
Nitrate	NO_3^-
Nitrite	NO_2^-
Phosphate	PO_4^{3-}
Hydrogen Phosphate	HPO_4^{2-}
Dihydrogen Phosphate	$H_2PO_4^-$
Phosphite	PO_3^{3-}
Cyanide	CN^-
Common name	**Formula**
Thiocyanate	SCN^-
Peroxide	O_2^{2-}
Oxalate	$C_2O_4^{2-}$

Acetate	$C_2H_3O_2^-$
Chromate	CrO_4^{2-}
Dichromate	$Cr_2O_7^{2-}$
Permanganate	MnO_4^-
Cations	
Hydronium	H_3O^+
Ammonium	NH_4^+
Metal	M^{n+}

Qualitative analysis is a method for the separation and identification of ions in a mixture. The technique uses the differences of solubility for ionic compounds in aqueous solution, as well as the ability of certain cations to form *complex ions* with ligands. The complex ions of many transition metals are often colored, which can be used in their visual identification. The general approach in the qualitative analysis of cations is to separate them sequentially into various *ion groups*.

Ion group 1: Insoluble chlorides. Treating the mixture with 6 M HCl precipitates Ag^+, Hg_2^{2+}, and Pb^{2+} ions as chlorides, leaving other cations in solution. The formation of a *white precipitate* indicates the presence of at least one of these cations in the mixture.

Ion group 2: Acid-insoluble sulfides. The supernatant from the above treatment with HCl is adjusted to pH $\approx$ 0.5 and then treated with aqueous H_2S. The high $[H_3O^+]$ in solution keeps $[HS^-]$ low, which precipitates only the following group of cations: Cu^{2+}, Cd^{2+}, Hg^{2+}, Sn^{2+}, and Bi^{3+}. Centrifuging and decanting give the next solution.

Ion group 3: Base-insoluble sulfides. The supernatant from acidic sulfide treatment is treated with an NH_3/NH_4^+ buffer to make the solution slightly basic (pH $\approx$ 8). The excess OH^- in solution increases $[HS^-]$, which causes the precipitation of the more soluble sulfides and some hydroxides. The cations that precipitate under this condition are Zn^{2+}, Mn^{2+}, Ni^{2+}, Fe^{2+}, and Co^{2+}, as sulfides, and Al^{3+}, Cr^{3+}, and Fe^{3+} as hydroxides. The precipitate is centrifuged, and the supernatant decanted to give the next solution.

Ion group 4: Insoluble phosphates. The slightly basic supernatant separated from the group 3 ions are treated with $(NH_4)_2HPO_4$, which precipitates $Mg_3(PO_4)_2$, $Ca_3(PO_4)_2$ and $Ba_3(PO_4)_2$.

Ion group 5: Alkali metal and ammonium ions. The final solution contains any of the following ions: Na^+, K^+, and NH_4^+.

Hydration, the hydronium ion

Hydration, also known as *solvation*, is where water forms a shell around ions in solution. The oxygen atom on the water is partially negative, so it surrounds any cations. The hydrogen atoms on the water are partially positive, so they surround anions.

H^+ does not exist as a proton in water; it exists as the hydronium ion (H_3O^+). This is because the high charge density of a proton attracts it to any nearby molecule that has a partial or full negative charge.

Since water has two lone pairs of electrons on the oxygen atom, the positive proton interacts with one of the lone pairs, forming the hydronium ion.

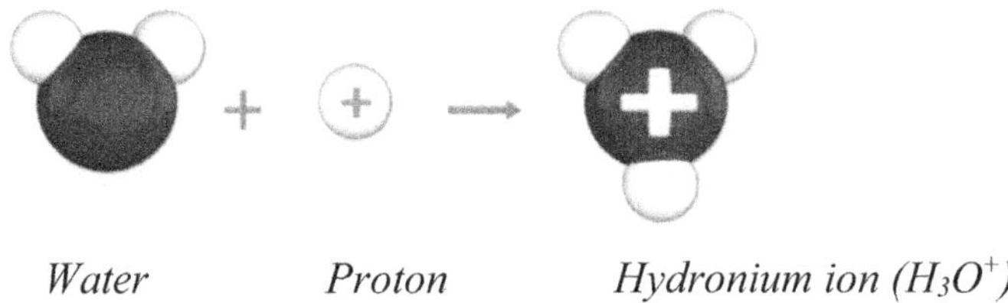

Water *Proton* *Hydronium ion (H_3O^+)*

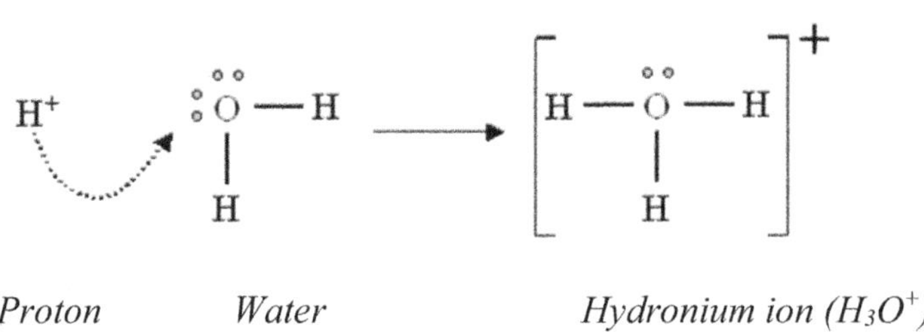

Proton *Water* *Hydronium ion (H_3O^+)*

Solubility

The *solubility* of a substance is the amount of solute (in grams) dissolves in a quantity of solvent to give a saturated solution at a specific temperature.

A *saturated* solution contains the maximum quantity of dissolved solute that is usually possible at a given temperature, and where a state of dynamic equilibrium exists between dissolution and crystallization.

Solubility is temperature dependent. For example, the solubility of KNO_3 is about 30 g per 100 g water at 20 °C and 63 g per 100 g water at 40 °C. When a solution that is almost saturated at a higher temperature is cooled slowly to a lower temperature where the solubility is less, the excess solute typically will precipitate to give a saturated solution at the lower temperature.

However, if the solution is cooled too rapidly, precipitates do not form, and the resulting solution contains more dissolved solute than it would in a typical saturated solution. This process produces a *supersaturated* solution. The supersaturated solution is unstable. Crystallization occurs readily by seeding (introducing particles that provide nuclei for precipitation).

A *concentrated solution* contains a relatively large amount of dissolved solute.

A *dilute solution* contains few dissolved solutes in a relatively large amount of solvent. A solution containing 40 g or more of dissolved solute in 100 mL of water is considered concentrated, whereas one with less than 10 g of solute per 100 mL of water is a dilute solution. A concentrated solution does not necessarily imply a saturated solution. A saturated solution does not necessarily imply a concentrated solution.

The solubility of most liquids and solids in water *increases* with temperature, but the solubility of gases *decreases* with temperature.

The pressure of a gas strongly affects its solubility, as described by Henry's Law. *Henry's Law* states that as pressure increases, solubility increases (i.e., directly proportional).

Other factors that affect solubility include the surface area of the solute (e.g., granulated sugar dissolves faster than a sugar cube) and heating or agitating a solution (e.g., stirring leads to faster dissolving).

The terms *miscible* and *immiscible* are often used to describe liquids that dissolve or do not dissolve in another liquid, respectively. Thus, ethanol and water are entirely miscible because they dissolve in each other freely when mixed, while oil and water are immiscible since they remain separate.

Units of concentration

The concentration of a solution is expressed regarding the amount of solute dissolved in the solution. The concentration may be expressed in different ways, as listed below.

1. Mass Percent, % (w/w) = $\frac{\text{mass of solute}}{\text{mass of solution}} \times 100\%$

2. Volume Percent % (v/v) = $\frac{\text{volume of solute}}{\text{volume of solution}} \times 100\%$

3. Molarity (M) = $\frac{\text{number of moles of solute}}{\text{liters of solution}}$

4. Molality (m) = $\frac{\text{number of moles of solute}}{\text{kg of solvent}}$

5. Normality (N) = $\frac{\text{number of equivalents of solute}}{\text{liters of solution}}$

where

Number of equivalents of solute = $\frac{\text{grams of solute}}{\text{equivalent weight of solute}}$

6. Mole fraction, (X_i) = $\frac{\text{moles of a component}}{\text{total moles of components in solution}}$

Normality is based on a unit of chemical mass known as the equivalent weight. The *equivalent weight* is the amount of solute needed to be the equivalent of one mole of hydrogen ions, and the equivalent weight is thus dependent on the valence of the solute.

For solutes with a valence of 1 (e.g., HCl), the molecular weight and equivalent weight are the same.

For solutes with a valence of more than 1 (e.g., H_3PO_4, the valence of 3), the equivalent weight is equal to the molecular weight divided by the valence.

Therefore, molarity is related to normality, as in the examples below:

$1\ M\ HCl = 1\ N\ HCl$

$1\ M\ H_2SO_4 = 2\ N\ H_2SO_4$

$1\ M\ H_3PO_4 = 3\ N\ H_3PO_4$

Since ionic compounds dissociate into ions, the total number of particles in a solution of an ionic compound is always greater than that in a solution of a nonionic (molecular) compound. The total concentration of ions in solution is the sum of the concentrations of cations and anions, which is dependent on the formula of the compound.

A solution of 1.0 M aluminum nitrate, $Al(NO_3)_3$, contains 1.0 M of Al^{3+} and 3.0 M NO_3^- ions. The total concentration of ions in solution is 4.0 M.

Before dissolving		After dissolving	
$Al(NO_3)_3\ (aq)$	$\rightarrow$	$Al^{3+}\ (aq)$ +	$3\ NO_3^-\ (aq)$
(1.0 M)	$\rightarrow$	(1.0 M)	(3 × 1.0 M)

For example, a solution of 1.0 M $MgCl_2$ contains 1.0 M of Mg^{2+} and 2.0 M of Cl^- ions, and the total ion concentration is 3.0 M.

The solubility product constant, the equilibrium expression (K_{sp})

Solubility product constants K_{sp} describe saturated solutions of ionic compounds of relativity low solubility.

For example, when a slightly soluble salt such as silver chloride (AgCl) is dissolved in water, a saturated solution is quickly obtained. This is because only a minimal amount of the solid dissolves, while most of the AgCl salt remains undissolved.

The equilibrium between solid AgCl and the free ions occurs in solution:

$$AgCl\ (s) \rightleftarrows Ag^+\ (aq) + Cl^-\ (aq)$$

$$K_{sp} = [Ag^+]\cdot[Cl^-]$$

A general expression of the solubility product constant K_{sp} for solubility equilibrium is:

$$M_aX_b\ (s) \rightleftarrows aM^{b+}\ (aq) + bX^{a-}\ (aq)$$

$$K_{sp} = [M^{b+}]^a\cdot[X^{a-}]^b$$

In some problems, the K_{sp} may be given, and the solubility may need to be calculated, or vice versa. How to proceed depends on the type of equilibria.

Some examples of solving solubility problems are given below.

A. For ionic equilibria of the type:

$MX\ (s) \rightleftarrows M^{n+}\ (aq) + X^{n-}\ (aq)$

$K_{sp} = [M^{n+}]\cdot[X^{n-}]$

If the solubility of compounds is S mol/L:

$K_{sp} = S^2$

$S = \sqrt{(K_{sp})}$

For example, the solubility equilibrium for $BaSO_4$ is:

$BaSO_4\ (s) \rightleftarrows Ba^{2+}\ (aq) + SO_4^{2-}\ (aq)$

$K_{sp} = [Ba^{2+}]\cdot[SO_4^{2-}]$

$K_{sp} = 1.5 \times 10^{-9}$

If the solubility of $BaSO_4$ is S mol/L, a saturated solution of $BaSO_4$ has:

$[Ba^{2+}] = [SO_4^{2-}] = S$ mol/L

$K_{sp} = S^2$

$S = \sqrt{(K_{sp})} = \sqrt{(1.5 \times 10^{-9})}$

$S = 3.9 \times 10^{-5}$ mol/L

B. For ionic equilibria of the type:

$MX_2\ (s) \rightleftarrows M^{2+}\ (aq) + 2\ X^{-}\ (aq)$

$K_{sp} = [M^{2+}]\cdot[X^{-}]^2$

For the type:

$M_2X\ (s) \rightleftarrows 2M^{+}\ (aq) + X^{2-}\ (aq)$

$K_{sp} = [M^{+}]^2\cdot[X^{2-}]$

For both types if the solubility is S mol/L:

$K_{sp} = 4S^3$

$S = (K_{sp}/4)^{1/3}$

For example:

$$CaF_2\,(s) \rightleftarrows Ca^{2+}\,(aq) + 2\,F^-\,(aq)$$

$$K_{sp} = [Ca^{2+}]\cdot[F^-]^2$$

$$K_{sp} = 4.0 \times 10^{-11}$$

The solubility of calcium fluoride is:

$$S = (K_{sp}/4)^{1/3}$$

$$S = (4.0 \times 10^{-11}/4)^{1/3}$$

$$S = 2.2 \times 10^{-4}\text{ mol/L}$$

C. For solubility equilibria of the type:

$$MX_3\,(s) \rightleftarrows M^{3+}\,(aq) + 3\,X^-\,(aq)$$

$$K_{sp} = [M^{3+}]\cdot[X^-]^3$$

Or of the type:

$$M_3X\,(s) \rightleftarrows 3\,M^+\,(aq) + X^{3-}\,(aq)$$

$$K_{sp} = [M^+]^3\cdot[X^{3-}]$$

If the solubility of the compound (MX_3 or M_3X) is S mol/L:

$$K_{sp} = 27S^4$$

$$S = \sqrt[4]{(K_{sp}/27)}$$

For example:

$$Ag_3PO_4\,(s) \rightleftarrows 3Ag^+\,(aq) + PO_4^{3-}\,(aq)$$

$$K_{sp} = [Ag^+]^3\cdot[PO_4^{3-}]$$

$$K_{sp} = 1.8 \times 10^{-18}$$

The solubility of silver phosphate is:

$$S = \sqrt[4]{(K_{sp}/27)}$$

$$S = \sqrt[4]{(1.8 \times 10^{-18}/27)}$$

$$S = 1.6 \times 10^{-5}\text{ mol/L}$$

D. For solubility equilibria:

$M_2X_3\ (s) \rightleftarrows 2\ M^{3+}\ (aq) + 3\ X^{2-}\ (aq)$

$K_{sp} = [M^{3+}]^2 \cdot [X^{2-}]^3$

Or of the type:

$M_3X_2\ (s) \rightleftarrows 3\ M^{2+}\ (aq) + 2\ X^{3-}\ (aq)$

$K_{sp} = [M^{2+}]^3 \cdot [X^{3-}]^2$

If the solubility of the compound M_3X_2 is S mol/L:

$[M^{2+}] = 3S$ and $[X^{3-}] = 2S$

$K_{sp} = (3S)^3 \cdot (2S)^2$

$K_{sp} = 108S^5$

$S = {}^5\sqrt{(K_{sp} / 108)}$

For example:

$Ca_3(PO_4)_2\ (s) \rightleftarrows 3\ Ca^{2+}\ (aq) + 2\ PO_4^{3-}\ (aq)$

$K_{sp} = [Ca^{2+}]^3 \cdot [PO_4^{3-}]^2$

$K_{sp} = 1.3 \times 10^{-32}$

The solubility of $Ca_3(PO_4)_2$ is:

$S = 5\sqrt{(1.3 \times 10^{-32}) / 108)}$

$S = 1.6 \times 10^{-7}$ mol/L

Calculating K_{sp} from solubility:

The solubility of $PbSO_4$ in water is 4.3×10^{-3} g/100 mL solution at 25 °C. What is the K_{sp} of $PbSO_4$ at 25 °C?

$$\text{Solubility of } PbSO_4 \text{ in mol/L} = \frac{4.3 \times 10^{-3}\ \text{g}}{100\ \text{mL}} \times \frac{1000\ \text{mL/L}}{303.26\ \text{g/mol}}$$

$S = 1.40 \times 10^{-4}$ mol/L

A saturated solution of $PbSO_4$, contains $[Pb^{2+}] = [SO_4^{2-}] = 1.4 \times 10^{-4}$ mol/L

For the equilibrium:

$PbSO_4\ (s) \rightleftarrows Pb^{2+}\ (aq) + SO_4^{2-}\ (aq)$

$K_{sp} = [Pb^{2+}] \cdot [SO_4^{2-}]$

$S^2 = (1.4 \times 10^{-4}\ \text{mol/L})^2$

$S = 2.0 \times 10^{-8}$

The next example displays how to determine solubility from K_{sp}.

If the K_{sp} of $Mg(OH)_2$ is 6.3×10^{-10} at 25 °C, what is its solubility in mol/L at 25 °C?

Solubility equilibrium for $Mg(OH)_2$:

$Mg(OH)_2\,(s) \rightleftarrows Mg^{2+}\,(aq) + 2\ OH^-\,(aq)$

$K_{sp} = [Mg^{2+}]\cdot[OH^-]^2$

$4S^3 = 6.3 \times 10^{-10}$

Solubility of $Mg(OH)_2$:

$S = \sqrt[3]{(K_{sp}/4)}$

$S = \sqrt[3]{(6.3 \times 10^{-10}/4)}$

$S = 5.4 \times 10^{-4}\ \text{mol/L}$

If the K_{sp} of $Mg(OH)_2$ is 6.3×10^{-10} at 25 °C, what is its solubility in a g/100 mL solution at 25 °C? Solubility in g/100 mL solution:

$S = (5.4 \times 10^{-4}\ \text{mol/L})\cdot(58.32\ \text{g/mol})\cdot(0.1\ \text{L}/100\ \text{mL})$

$S = 3.1 \times 10^{-3}\ \text{g}/100\ \text{mL}$ solution.

Common-ion effect, its use in laboratory separations

The *common-ion effect* is Le Châtelier's principle applied to K_{sp} reactions. The common-ion effect states that the presence of common ion decreases the solubility of a slightly soluble ionic compound.

A common ion is an ion in the solution that is common to the ionic compound.

For example, in the following equilibrium:

$PbCl_2\,(s) \rightleftharpoons Pb^{2+}\,(aq) + 2\ Cl^-\,(aq)$

If some NaCl is added to a saturated solution of $PbCl_2$, the $[Cl^-]$ increases, and according to Le Châtelier's principle, the equilibrium shifts in the direction that tend to reduce $[Cl^-]$. In this example, the equilibrium shifts left, to form more $PbCl_2$ solid, hence decreasing the amount of $PbCl_2$ that dissolves into solution. More $PbCl_2$ can dissolve in pure water than in water containing Cl^- ions.

In laboratory separations, the common-ion effect can be used to precipitate one component in a mixture selectively. For example, to separate AgCl from a mixture of AgCl and Ag_2SO_4, add NaCl. The addition of NaCl selectively displaces AgCl by the common-ion effect (Cl^- being the common ion).

Complex ion formation

A *complex ion* consists of a central metal ion that is covalently bonded to two or more *ligands*, which can be anions such as ^-OH, Cl^-, F^- and ^-CN, or neutral molecules such as H_2O, CO, and NH_3.

For example, in the complex ion $[Cu(NH_3)_4]^{2+}$, Cu^{2+} is the central metal ion, with four NH_3 molecules covalently bonded to it. All complex ions are Lewis adducts (i.e., the addition of a Lewis acid and a Lewis base). The metal ions act as Lewis acids (electron-pair acceptors), and the ligands are Lewis bases (electron-pair donors).

$$\text{Metal}^+ + \text{Lewis base} \rightarrow \text{Complex ion}$$

$$M^+ + L \rightarrow M - L_n^+$$

The Lewis base can be charged or uncharged.

The K_{eq} for this reaction is K_f, or the *formation constant.*

In aqueous solutions, metal ions form complex ions with water molecules as ligands. When another ligand is introduced into the solution, ligand exchanges occur, and equilibrium is established.

For example, when NH_3 is added to an aqueous solution containing Cu^{2+} ion, the following equilibrium occurs:

$$Cu(H_2O)_6^{2+}\,(aq) + 4\,NH_3\,(aq) \rightleftarrows [Cu(NH_3)_4]^{2+}\,(aq) + 6\,H_2O$$

$$K_f = [Cu(NH_3)_4^{2+}] \,/\, [Cu(H_2O)_6^{2+}]\cdot[NH_3]^4$$

At the molecular level, the ligand exchange process occurs in a stepwise manner; each water molecule is replaced with an NH_3 molecule, one at a time, to give a series of intermediate species, each with its formation constant.

For convenience, the water molecules can be omitted from the equation.

1. $Cu^{2+}\,(aq) + NH_3\,(aq) \rightleftarrows Cu(NH_3)^{2+}\,(aq)$

   $K_{f1} = [Cu(NH_3)^{2+}] \,/\, [Cu^{2+}]\cdot[NH_3]$

2. $Cu(NH_3)^{2+}\,(aq) + NH_3\,(aq) \rightleftarrows Cu(NH_3)_2^{2+}\,(aq)$

   $K_{f2} = [Cu(NH_3)_2^{2+}] \,/\, [Cu(NH_3)^{2+}]\cdot[NH_3]$

3. $Cu(NH_3)_2^{2+}$ (*aq*) + NH_3 (*aq*) ⇄ $Cu(NH_3)_3^{2+}$ (*aq*)

   $K_{f3} = [Cu(NH_3)_3^{2+}] / [Cu(NH_3)_2^{2+}]\cdot[NH_3]$

4. $Cu(NH_3)_3^{2+}$ (*aq*) + NH_3 (*aq*) ⇄ $Cu(NH_3)_4^{2+}$ (*aq*)

   $K_{f4} = [Cu(NH_3)_4^{2+}] / [Cu(NH_3)_3^{2+}]\cdot[NH_3]$

The formation constant K_f is the product of the intermediate formation constants:

$$K_f = K_{f1} \times K_{f2} \times K_{f3} \times K_{f4}$$

$$K_f = [Cu(NH_3)_4^{2+}] / [Cu^{2+}]\cdot[NH_3]^4$$

Complex ions and solubility

The *complex-ion effect* is the opposite of the common ion effect. A ligand increases the solubility of slightly soluble ionic compounds if complex ions form with the metal ions.

For example, silver chloride, AgCl, is more soluble in ammonia solution because silver ions form complex ions with NH_3:

AgCl (*s*) ⇄ Ag^+ (*aq*) + Cl^- (*aq*)

$$K_{sp} = 1.6 \times 10^{-10}$$

Ag^+ (*aq*) + 2 NH_3 (*aq*) ⇄ $Ag(NH_3)_2^+$ (*aq*)

$$K_f = 1.7 \times 10^7$$

AgCl (*s*) + 2 NH_3 (*aq*) ⇄ $Ag(NH_3)_2^+$ (*aq*) + Cl^- (*aq*)

$$K_{net} = K_{sp} \times K_f$$

$$K_{net} = (1.6 \times 10^{-10})\cdot(1.7 \times 10^7)$$

$$K_{net} = 2.7 \times 10^{-3}$$

When a complex ion forms, the Cl^- ion is reduced, so more of AgCl dissolves.

Alternatively:

AgCl (*s*) ↔ Ag^+ (*aq*) + Cl^- (*aq*)

NH_3 + Ag^+ ↔ Ag-$(NH_3)_n$ complex ion

The complex ion formation reduces Ag^+, causing more AgCl to dissolve.

Solubility and pH

The pH affects the solubility of slightly soluble compounds containing anions that are conjugate bases of weak acids, such as F^-, NO_2^-, OH^-, SO_3^{2-} and PO_4^{3-}.

The pH *does not* affect the solubility of slightly soluble compounds containing anions that are conjugate bases of strong acids, such as SO_4^{2-}, Cl^- and Br^-.

In a saturated solution of calcium fluoride, CaF_2, an equilibrium exists:

$$CaF_2\ (s) \rightleftarrows Ca^{2+}\ (aq) + 2\ F^-\ (aq)$$

If a strong acid is added to the saturated solution, the following reaction occurs:

$$H^+\ (aq) + F^-\ (aq)\ \rightarrow\ HF\ (aq)$$

This reaction has the net effect of reducing the concentration of F^- ions, which causes the equilibrium to shift to the right, and more of CaF_2 to dissolve.

Acids are more soluble in bases.

$$HA \rightarrow H^+ + A^-$$

Placing the above reactions in a base reduces the H^+.

Thus, more HA dissolves, according to Le Châtelier's principle.

Bases are more soluble in acids.

$$B + H^+ \rightarrow BH^+$$

Putting the above reactions in an acid adds more H^+, and thus, more B dissolves, according to Le Châtelier's principle.

Practice Questions

1. Which of the following statements describing solutions is NOT correct?

A. Solutions are colorless
B. The particles in a solution are atomic or molecular
C. Making a solution involves a physical change in size
D. Solutions are homogeneous
E. Solutions are transparent

2. Which of the following describes a saturated solution?

A. When the ratio of solute to solvent is small
B. When it contains less solute than it can hold at 25 °C
C. When it contains as much solute as it can hold at a given temperature
D. When it contains 1 g of solute in 100 mL of water
E. When it is equivalent to a supersaturated solution

3. An ionic compound that strongly attracts atmospheric water is said to be:

A. immiscible
B. miscible
C. diluted
D. hygroscopic
E. soluble

4. What happens when the molecule-to-molecule attractions in the solute are less than those in the solvent?

A. The material has only limited solubility in the solvent
B. The solution will become saturated
C. The solute can have infinite solubility in the solvent
D. The solute does not dissolve in the solvent
E. None of the above

5. What does the negative heat of solution indicate about solute-solvent bonds as compared to solute-solute bonds and solvent-solvent bonds?

A. Solute-solute and solute-solvent bond strengths are greater than solvent-solvent bond strength
B. The heat of solution does not support a conclusion about bond strength
C. Solute-solute and solvent-solvent bonds are stronger than solute-solvent bonds
D. Solute-solute and solvent-solvent bond strengths are equal to solute-solvent bond strength
E. Solute-solute and solvent-solvent bonds are weaker than solute-solvent bonds

6. Which statement supports the fact that calcium fluoride is much less soluble in water than sodium fluoride?

I. calcium fluoride is not used in toothpaste
II. calcium fluoride is not used to fluoridate city water supplies
III. sodium fluoride is not used to fluoridate city water supplies

A. I only
B. II only
C. I and II only
D. I and III only
E. I, II and III

7. The principle *like dissolve like* is NOT applicable for predicting solubility when the solute is a/an:

A. nonpolar liquid
B. polar gas
C. nonpolar gas
D. ionic compound
E. covalent compound

8. Which of the following is the most soluble in benzene (C_6H_6)?

A. glucose ($C_6H_{12}O_6$)
B. sodium benzoate
C. octane (C_8H_{18})
D. hydrobromic acid
E. CH_2Cl_2

9. Apply the *like dissolves like* rule to predict which of the liquids is immiscible in water.

A. ethanol, CH_3CH_2OH
B. acetone, C_3H_6O
C. acetic acid, CH_3COOH
D. formaldehyde, H_2CO
E. none of the above

10. What is the equilibrium constant expression (K_{sp}) for slightly soluble silver sulfate in an aqueous solution?

$$Ag_2SO_4\,(s) \leftrightarrow 2\,Ag^+\,(aq) + SO_4^{2-}\,(aq)$$

A. $K_{sp} = [Ag^+]\cdot[SO_4^{2-}]^2$
B. $K_{sp} = [Ag^+]^2\cdot[SO_4^{2-}]\,/\,[Ag_2SO_4]$
C. $K_{sp} = [Ag^+]\cdot[SO_4^{2-}]$
D. $K_{sp} = [Ag^+]\cdot[SO_4^{2-}]^2\,/\,[Ag_2SO_4]$
E. $K_{sp} = [Ag^+]^2\,[SO_4^{2-}]$

11. In the reaction KHS (*aq*) + HCl (*aq*) → KCl (*aq*) + H_2S (*g*), which ions are the spectator ions?

A. H^+ and HS^-
B. K^+ and HS^-
C. K^+ and Cl^-
D. K^+ and H^+
E. HS^- and Cl^-

12. What is the mass of a 10.0% blood plasma sample that contains 2.50 g of dissolved solute?

A. 21.5 g **B.** 25.0 g **C.** 0.215 g **D.** 0.430 g **E.** 12.5 g

13. What is the molarity of KCl in seawater, if KCl is 12.5% (m/m), and the density of seawater is 1.06 g/mL?

A. 1.78 M
B. 17.8 M
C. 2.78 M
D. 0.845 M
E. 27.8 M

14. If 25.0 mL of urine has a mass of 25.5 g and contains 1.8 g of solute, what is the mass/mass percent concentration of solute in the urine sample?

A. 12.48% **B.** 17.42% **C.** 7.06% **D.** 3.82% **E.** 47.4%

15. With increasing temperature, many solvents expand to occupy greater volumes. What happens to the concentration of a solution made with such a solvent as temperature increases?

A. The concentration of a solution decreases because the solution has a greater ability to dissolve more solute at a higher temperature
B. The concentration of a solution increases because the solution has a greater ability to dissolve more solute at a higher temperature
C. The concentration of a solution decreases as volume increases because concentration depends on how much mass is dissolved in a given volume
D. The concentration of a solution increases as the solute fits into the new spaces between the molecules
E. The concentration of a solution increases as volume increases because concentration depends on how much mass is dissolved in a given volume

Detailed Explanations

1. A is correct. Some solutions are colored (e.g., Kool-Aid powder dissolved in water).

The color of chemicals is a physical property of chemicals from (most common) the excitation of electrons due to the absorption of energy by the chemical. The observer sees not the absorbed color, but the wavelength that is reflected.

Most simple inorganic (e.g., sodium chloride) and organic compounds (e.g., ethanol) are colorless.

Transition metal compounds are often colored because of transitions of electrons between *d*-orbitals of different energy.

Organic compounds tend to be colored when there is extensive conjugation (i.e., alternating double and single bonds). Conjugation causes the energy gap between the HOMO (i.e., highest occupied molecular orbital) and LUMO (i.e., lowest occupied molecular orbital) to decrease, bringing the absorption band from the UV to the visible region. Color is due to energy absorbed by the compound when an electron transitions from the HOMO to the LUMO.

2. C is correct.

A saturated solution contains the maximum amount of dissolved material in the solvent under normal conditions. Increased heat allows for a solution to become supersaturated. A supersaturated solution refers to the vapor of a compound that has a higher partial pressure than the vapor pressure of that compound.

A saturated solution forms a precipitate more solute is added to the solution.

3. D is correct.

A hygroscopic substance (e.g., honey, glycerin) is one that readily attracts water from its surroundings, through either adsorption or absorption.

Adsorption is the process in which atoms, ions, or molecules from a substance (e.g., gas, liquid, or dissolved solid) adhere to a surface of the *adsorbent.*

4. C is correct.

If the attraction between solute molecules is less than the attraction between solvent molecules, a solute has (a virtually) infinite solubility because solvent-solute attraction is stronger than solute-solute attraction.

5. E is correct.

Bond formation releases energy: if the heat of the solution is negative, then energy is released.

The heat of solution is the net of enthalpy changes for making and breaking bonds.

During solution formation, solvent-solvent bonds and solute-solute bonds break while solute-solvent bonds form.

The breaking of bonds absorbs energy while the formation of bonds releases energy.

6. C is correct.

Fluorides (e.g., BaF_2, MgF_2 PbF_2) are frequently insoluble.

7. D is correct.

Like dissolves like means that polar substances tend to dissolve in polar solvents and nonpolar substances in nonpolar solvents.

Molecules that can form hydrogen bonds with water are soluble.

Ionic compounds form anions and cations that bond with the polar water molecule and therefore are soluble in water.

8. C is correct.

Like dissolves like means that polar substances tend to dissolve in polar solvents and nonpolar substances in nonpolar solvents.

Benzene is a nonpolar molecule, and octane is also nonpolar.

Therefore, nonpolar octane is soluble in nonpolar benzene.

9. E is correct.

The "like dissolves like" rule applies when a solvent is miscible (soluble) with a solute that has similar properties.

A polar solute is miscible with a polar solvent.

All the molecules are polar and therefore are miscible in water.

10. E is correct.

The calculation of solubility constant (K_{sp}) is similar to the equilibrium constant.

For K_{sp}, only aqueous species are included in the calculation.

The concentration of each species is raised to the power of their coefficients and multiplied with each other.

Therefore,

$$K_{sp} = [Ag^+]^2 \cdot [SO_4^{2-}]$$

11. C is correct.

Spectator ions appear on both sides of the net ionic equation.

Ionic equation of the reaction:

$$K^+ (aq) + HS^- (aq) + H^+ (aq) + Cl^- (aq) \rightarrow BaSO_4 (s) + 2\ K^+ (aq) + 2\ NO_3^- (aq)$$

12. B is correct.

Mass % = mass of solute / mass of solution

Rearrange that equation to solve for mass of solution:

mass of solution = mass of solute / mass %

mass of solution = 2.50 g / 10.0%

mass of solution = 2.50 g / 0.1

mass of solution = 25.0 g

13. A is correct.

Assume 1 L of seawater.

Molarity is the number of moles in 1 L of solution, so starting with 1 L makes the calculation easier.

Mass of seawater = volume × density

Mass of seawater = (1 L × 1,000 L/mL) × 1.06 g/mL

Mass of seawater = 1,060 g

Use the mass of the seawater to determine the mass of KCl:

Mass of KCl = mass % of KCl × mass of seawater

Mass of KCl = 12.5% × 1,060 g

Mass of KCl = 132.5 g

Calculate the number of moles:

Moles of KCl = mass of KCl / molar mass of KCl

Moles of KCl = 132.5 g / (39.1 g/mol + 35.45 g/mol)

Moles of KCl = 1.78 moles

Divide moles by volume to calculate molarity:

Molarity of KCl = moles of KCl / volume of KCl

Molarity of KCl = 1.78 moles / 1 L

Molarity of KCl = 1.78 M

14. C is correct.

Mass % of solute = (mass of solute / total urine mass) × 100%

Mass % of solute = (1.8 g / 25.5 g) × 100%

Mass % of solute = 7.06%

15. C is correct.

As the volume of a solution increases, the concentration of a solution decreases because concentration depends on how much mass is dissolved in a given volume.

Notes

Chapter 8

Acids and Bases

- **Acid–Base Equilibria**
- **Titration**

Acid–Base Equilibria

Before the twentieth century, acids, bases, and salts were characterized by properties such as taste and their ability to change the color of *litmus* (a water-soluble mixture of organic dyes).

Acids taste sour (e.g., lemon juice), bases taste bitter (e.g., mustard), and salts, as their name suggests, taste salty (e.g., sodium chloride, or table salt).

Acids cause blue litmus paper to turn red, bases turn red litmus paper blue, while neutral compounds do not affect the color of litmus paper.

Bases are recognized by their slippery feel (e.g., soap).

The position on the pH scale categorically classifies acids and bases.

The chemistry of acids and bases has a significant role in processes within nature, as well as industry. Some complex metabolic processes in the human body are controlled by physiological pH ($\approx$ 7.4); even a small change in this pH may lead to severe illness and death. Another example is the acidity of the soil, which is essential to plant growth.

Acids and bases are particularly essential in manufacturing industries. Sulfuric acid (H_2SO_4) is the most widely produced chemical. It is needed in the production of fertilizers, polymers, steel, and many other materials. The extensive use of sulfuric acid has also led to environmental problems, such as the phenomenon of acid rain.

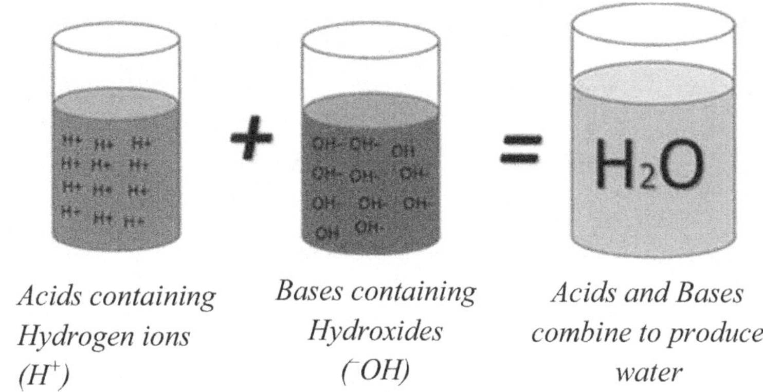

Acids containing Hydrogen ions (H^+) | *Bases containing Hydroxides (^-OH)* | *Acids and Bases combine to produce water*

Acids have names that are slightly different from the ionic naming rules discussed earlier, while bases follow the ionic naming rules.

There are two types of acids:

binary acids (acids that do not contain oxygen atoms)

oxoacids (those containing oxygen atoms).

Binary acids' names start with *hydro-*, followed by the first syllable of the anion's name, and end with *~ic*:

IIF - *hydro*fluor*ic* acid HCl - *hydro*chlor*ic* acid HBr - *hydro*brom*ic* acid

HI - *hydro*iod*ic* acid H_2S - *hydro*sulfur*ic* acid HCN - *hydro*cyan*ic* acid

Oxoacids' names are derived from the name of the oxyanion of the acid.

If the anion's name ends with *~ate*, the acid's name starts with the first syllable of the anion name and ends with *~ic*.

If the anion's name ends with *~ite*, the name of acid starts with the first syllable of the anion's name and ends with *-ous*.

Anion	Name of Anion	Acid	Name of Acid
NO_3^-	nitrate ion	HNO_3	nitric acid
NO_2^-	nitrite ion	HNO_2	nitrous acid
SO_4^{2-}	sulfate ion	H_2SO_4	sulfuric acid
SO_3^{2-}	sulfite ion	H_2SO_3	sulfurous acid
PO_4^{3-}	phosphate ion	H3PO4	phosphoric acid
$C_2H_3O_2^-$	acetate ion	$HC_2H_3O_2$	acetic acid
ClO^-	hypochlorite	HClO	hypochlorous acid
ClO_2^-	chlorite	$HClO_2$	chlorous acid
ClO_3^-	chlorate	$HClO_3$	chloric acid
ClO_4^-	perchlorate	$HClO_4$	perchloric acid

Arrhenius definition of acids and bases

In 1884, Swedish chemist Svante Arrhenius defined for acids and bases:

acids are substances that dissociate in water to produce *hydrogen ions* (H^+)

bases are substances that dissociate to produce *hydroxide ions* (OH^-)

HCl is an acid: $HCl\ (aq) \rightarrow H^+\ (aq) + Cl^-\ (aq)$

NaOH is a base: $NaOH\ (aq) \rightarrow Na^+\ (aq) + OH^-\ (aq)$

Arrhenius acids and bases

Arrhenius acids increase the hydronium ion concentration $[H_3O^+]$ in aqueous solution.

Arrhenius bases increase the hydroxide ion concentration $[OH^-]$ in aqueous solution.

Acids according to the Arrhenius concept:

1. $HCl\ (aq) + H_2O \rightarrow H_3O^+\ (aq) + Cl^-\ (aq)$
2. $HNO_3\ (aq) + H_2O \rightarrow H_3O^+\ (aq) + NO_3^-\ (aq)$
3. $CH_3COOH\ (aq) + H_2O \rightleftarrows H_3O^+\ (aq) + CH_3COO^-\ (aq)$

Bases according to the Arrhenius concept:

1. $NaOH\ (aq) \rightarrow Na^+\ (aq) + OH^-\ (aq)$
2. $Ba(OH)_2\ (aq) \rightarrow Ba^{2+}\ (aq) + 2\ OH^-\ (aq)$
3. $NH_3\ (aq) + H_2O \rightleftarrows NH_4^+\ (aq) + OH^-\ (aq)$

Lewis definition of acids and bases

The *Lewis definition of acids* is a reactant that has an empty orbital and *shares a pair of electrons* from another reactant to form a covalent bond.

The *Lewis definition of bases* is a reactant that *provides a lone pair of electrons* from another reactant to form a covalent bond.

According to the Lewis definition proposed in 1923, the hydrogen ion (H^+) is a Lewis acid and water and ammonia are the Lewis bases in the following reactions:

H^+	$+\ H_2O$	$\rightarrow H_3O^+$		H^+	$+\ NH_3$	$\rightarrow NH_4^+$
Lewis acid	Lewis base			Lewis acid	Lewis base	

In reactions that involve the formation of new covalent bonds, the species with an incomplete octet (i.e., an electron-deficient molecule) may act as Lewis acids, and those with a lone pair of electrons may act as Lewis bases.

In the following reactions,

BF_3, $AlCl_3$, and $FeBr_3$ are Lewis acids, while

NH_3, Cl^- and Br^- are Lewis bases.

$BF_3 + NH_3 \rightarrow F_3B{:}NH_3$ $\quad$ $AlCl_3 + Cl^- \rightarrow AlCl_4^-$ $\quad$ $FeBr_3 + Br^- \rightarrow FeBr_4^-$

In the formation of complex ions, the positive ions act as Lewis acids, and the ligands (anions or small molecules) act as Lewis bases:

$Cu^{2+}(aq)$ + $4\,NH_3(aq) \rightleftarrows Cu(NH_3)_4^{2+}(aq)$
Lewis acid Lewis base

$Al^{3+}(aq)$ + $6\,H_2O \rightleftarrows [Al(H_2O)_6]^{3+}(aq)$
Lewis acid Lewis base

Note that the ionizable hydrogen in oxoacids is bonded to the oxygen in the molecule.

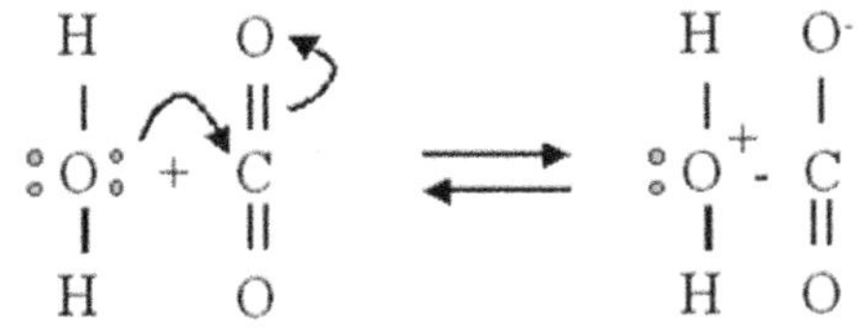

Lewis base *Lewis acid*

Brønsted–Lowry definition of acids and bases

The *Brønsted-Lowry theory* was proposed, in 1923, independently by Danish chemist Johannes Nicolaus Brønsted (1879-1947) and British chemist Martin Lowry (1874-1936). The Brønsted-Lowry theory states that:

an acid is a substance that acts as a proton donor, and

a base is a substance that acts as a proton acceptor in a chemical reaction.

From this exchange of protons,

acids form its *conjugate base*

bases form its *conjugate acid*

The Brønsted-Lowry acid-base reaction can be represented as follows:

$HA + B \rightleftarrows BH^+ + A^-$

HA	B	BH⁺	A⁻
acid	base	conjugate acid	conjugate base

Brønsted-Lowry acids, bases, conjugate acids, and conjugate bases examples:

$HCl + H_2O$	$\rightarrow$	$H_3O^+ (aq) + Cl^- (aq)$
acid base		conjugate acid, conjugate base

$HC_2H_3O_2 + H_2O$	$\rightleftarrows$	$H_3O^+ (aq) + C_2H_3O_2^- (aq)$
acid base		conjugate acid, conjugate base

$NH_3 + H_2O$	$\rightleftarrows$	$NH_4^+ (aq) + OH^- (aq)$
base acid		conjugate acid, conjugate base

The transfer of protons in a Brønsted-Lowery acid-base reaction

$HF (aq)$	+	NH_3	$\rightleftarrows$	F^-	+	$^+NH_4 (aq)$
Acid donates H^+ to NH_3		Base accepts H^+ from HF		Conjugate base accepts H^+ / $^+NH_4$		Conjugate acid donates H^+ to F^-

Ionization of water

Water is an *amphoteric* substance; it can act as an acid or a base.

A water molecule acts as a base by accepting a hydrogen ion, a second water molecule acts as an acid.

Therefore, ions form in pure water. As ions are formed, the ions react with to produce water again, and thus the following equilibrium occurs:

H_2O + H_2O	$\rightleftarrows$	$H_3O^+ (aq)$ + $OH^- (aq)$
acid base		conjugate acid, conjugate base

This process of exchange is the *auto-ionization of water*.

Ions formed in the auto-ionization of water (diagram below)

$$H_2O\ (l) + H_2O\ (l) \rightleftarrows H_3O^+ (aq) + OH^- (aq)$$

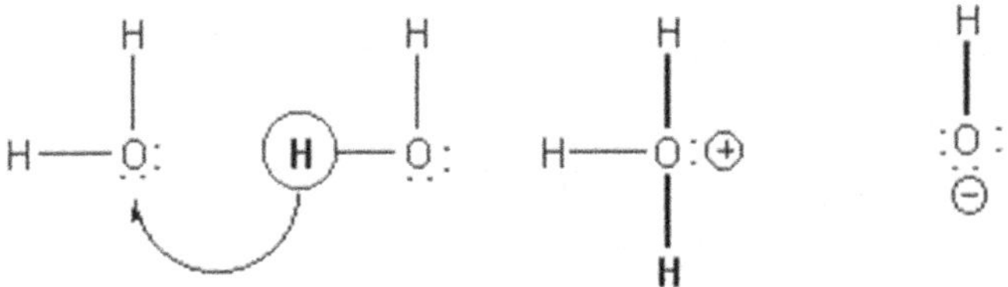

H^+ acceptor base *H^+ donor acid* *Conjugate acid* *Conjugate base*

K_w, its approximate value ($K_w = [H^+]\cdot[^-OH] = 10^{-14}$ at 25 °C, 1 atm)

The equilibrium constant expression for the above equilibrium is:

$$K_W = [H_3O^+]\cdot[^-OH] / [H_2O]^2$$

At standard temperature and pressure (25 °C, 1 atm), the equilibrium constant for water K_W (the *ion-product constant* for water:

$$K_W = [H_3O^+]\cdot[^-OH]$$

$$K_W = 1.0 \times 10^{-14}$$

If $[H_3O^+]$ increases ($> 1.0 \times 10^{-7}$ *M*), $[^-OH]$ decreases ($< 1.0 \times 10^{-7}$ M), and vice versa.

If $[H_3O^+] = [^-OH] = 1.0 \times 10^{-7}$ M, neutral solution (e.g., pure water)

If $[H_3O^+] > 1.0 \times 10^{-7}$ M, $[^-OH] < 1.0 \times 10^{-7}$ M, solution is acid ($[H^+] > [^-OH]$)

If $[H_3O^+] < 1.0 \times 10^{-7}$ M, $[^-OH] > 1.0 \times 10^{-7}$ M, solution is basic ($[H^+] < [^-OH]$)

pH definition, pH of pure water

The pH scale measures the acidity or basicity of a solution, especially when the hydrogen ion concentration is extremely low.

$$pH = -\log[H^+]$$

$$pOH = -\log[^-OH]$$

Neutral solutions

$$[H^+] = 1.0 \times 10^{-7} \text{ M}$$

$$pH = -\log(1.0 \times 10^{-7})$$

$$pH = 7.00$$

Neutral solutions also contain,

$[^-OH] = 1.0 \times 10^{-7}$ M

$pOH = -\log(1.0 \times 10^{-7})$

$pOH = 7.00$

Acidic solutions,

$[H^+] > 1.0 \times 10^{-7}$ M

$pH < 7.00$

Basic solutions,

$[H^+] < 1.0 \times 10^{-7}$ M

$pH > 7.00$

thus

$pH = 7 \rightarrow$ a neutral solution

$pH < 7 \rightarrow$ an acidic solution

$pH > 7 \rightarrow$ a basic solution

Note

$K_w = [H^+]\cdot[^-OH]$

$K_w = 1.0 \times 10^{-14}$

$pK_w = -\log(K_w)$

$pK_w = -\log(1.0 \times 10^{-14})$

$pK_w = 14.00$

but

$pK_w = pH + pOH = 14.00$

$pOH = 14.00 - pH$

$pH = 14.00 - pOH$

if

$pH = 7$

$pOH = 7$

if

$pH < 7$

$pOH > 7$

if

$pH > 7$

$pOH < 7$

Thus

$pH < 7$

$[H^+] > [^-OH]$

and

$pH > 7$

$[^-OH] > [H^+]$

For example. if

$[H^+] = 1.0 \times 10^{-4}$ M

$pH = -\log(1.0 \times 10^{-4})$

$pH = 4.00$

when

$[^-OH] = 1.0 \times 10^{-4}$ M

$[H^+] = 1.0 \times 10^{-14} / 1.0 \times 10^{-4}$ M

$[H^+] = 1.0 \times 10^{-10}$ M

$pH = -\log(1.0 \times 10^{-10})$

$pH = 10.00$

Alternatively,

$[^-OH] = 1.0 \times 10^{-4}$ M

$pOH = -\log[^-OH]$

$pOH = -\log(1.0 \times 10^{-4}\text{ M})$

$pOH = 4.00$

$pH = 14.00 - 4.00$

$pH = 10.00$

Reference values for $[H^+]$ and pH

$[H^+]$, M	pH	$[H^+]$, M	pH
1.0×10^{-1}	1.00	1.0×10^{-8}	8.00
1.0×10^{-2}	2.00	1.0×10^{-9}	9.00
1.0×10^{-3}	3.00	1.0×10^{-10}	10.00
1.0×10^{-4}	4.00	1.0×10^{-11}	11.00
1.0×10^{-5}	5.00	1.0×10^{-12}	12.00
1.0×10^{-6}	6.00	1.0×10^{-13}	13.00
1.0×10^{-7}	7.00	1.0×10^{-14}	
14.00			

To calculate the pH of an acidic solution, use the expression:

$$pH = -\log[H_3O^+]$$

For example, if

$$[H_3O^+] = 1.0 \times 10^{-2}\ M$$

$$pH = -\log(1.0 \times 10^{-2})$$

$$pH = -\ (-2.00)$$

$$pH = 2.00\ (\rightarrow \text{acidic})$$

To calculate the pH of a basic solution, use the expression:

$$pOH = -\log[^-OH]$$

If a solution has

$$[OH^-] = 1.0 \times 10^{-2}\ M$$

$$pOH = -\log(1.0 \times 10^{-2})$$

$$pOH = -\ (2.00)$$

$$pOH = 2.00\ (\rightarrow \text{basic})$$

Since at 25 °C

$K_w = [H_3O^+]\cdot[OH^-]$

$K_w = 1.0 \times 10^{-14}$

$pK_w = -\log(K_w)$

$pK_w = -\log[H_3O^+] + (-\log[^-OH])$

$pK_w = -\log(1.0 \times 10^{-14})$

$pK_w = -(-14.00)$

$pK_w = 14.00$

$pK_w = pH + pOH$

$pK_w = 14.00$

$pOH = 14.00 - pH$

Thus, in aqueous solutions

$pH = 2$

$pOH = 12$

and

$pOH = 2$

$pH = 12$

Equilibrium constants K_a and K_b (pK_a and pK_b)

Equilibrium constants K_a and K_b measure the extent to which an acid or base dissociates (dissociation constants).

The strength of an acid is defined by its dissociation (ionization) in *aq* solution.

$HX\ (aq) + H_2O \rightleftarrows H_3O^+\ (aq) + X^-\ (aq)$

The equilibrium constant, K_a, for the acid ionization:

$K_a = [H_3O^+]\cdot[X^-] / [HX]$

For K_a, the products (conjugate acid H_3O^+ and conjugate base X^-) of the dissociation are the numerator, while the parent acid (HX) is the denominator.

The strength of a base is defined by its dissociation (ionization) in aqueous solution.

$X^-\ (aq) + H_2O \rightleftarrows OH^-\ (aq) + HX\ (aq)$

$K_b = [HX^+]\cdot[^-OH] / [X]$

For K_b, the products (conjugate acid HX^+ and conjugate base OH^-) of the dissociation are the numerator, while the parent base (X) is the denominator.

The value of K_a measures the extent of acid dissociation, hence the relative strength of the acid. Stronger acids have larger K_a values.

For strong acids (e.g., $HClO_4$, HCl, H_2SO_4, HNO_3), their K_a is exceptionally large (not in the tables of reported K_a values).

For weak acids, $K_a << 10^{-1}$.

Strong bases have larger K_b values; weak bases have exceedingly small K_b values.

If the K_a value for a conjugate acid-base pair is known, the K_b can be calculated (and vice versa), by using the following relationship:

$$K_aK_b = K_w$$

When sodium acetate dissolves in water, it dissociates into sodium and acetate ions:

$$NaC_2H_3O_2\,(aq) \rightarrow Na^+\,(aq) + C_2H_3O_2^-\,(aq)$$

If the K_a is 1.8×10^{-5} (provided in a reference table), what is the K_b?

The acetate ion reacts with water, and the following equilibrium is established:

$$C_2H_3O_2^-\,(aq) + H_2O \rightleftarrows HC_2H_3O_2\,(aq) + OH^-\,(aq)$$

$$K_b = [HC_2H_3O_2]\cdot[OH^-] / [C_2H_3O_2^-]$$

For the dissociation of acetic acid:

$$HC_2H_3O_2\,(aq) + H_2O \rightleftarrows H_3O^+\,(aq) + C_2H_3O_2^-\,(aq)$$

$$K_a = [H_3O^+]\cdot[C_2H_3O_2^-] / [CH_3COOH]$$

$$K_a \times K_b = [H_3O^+]\cdot[C_2H_3O_2^-] / [CH_3COOH] \times [HC_2H_3O_2]\cdot[\,OH^-] / [C_2H_3O_2^-]$$

$$K_w = [H_3O^+]\cdot[OH^-]$$

$$K_w = 1.0 \times 10^{-14}$$

Thus, for acetate ion $C_2H_3O_2^-$ in solution

$$K_b = K_w / K_a \text{ (for } HC_2H_3O_2)$$

$$K_b = (1.0 \times 10^{-14}) / (1.8 \times 10^{-5})$$

$$K_b = 5.6 \times 10^{-10}\ (> K_w)$$

An aqueous solution of 0.10 M $NaC_2H_3O_2$ has $[OH^-] \sim 7.5 \times 10^{-6}$ *M* and pH ~ 8.9.

pK_a and pK_b are measures of acidity and basicity.

The operator "p" means "take the negative logarithm of."

Therefore,

$$pK_a = -\log K_a$$

As K_a gets larger, pK_a gets smaller.

The smaller the pK_a, the stronger the acid (e.g., pK_a for H_2SO_4 = 1.92).

$$pK_b = -\log K_b$$

As K_b gets larger, pK_b gets smaller.

The smaller the value of pK_b, the stronger the base.

Conjugate acids and bases

A *conjugate base* is the species that remains from the acid after it loses a proton.

A *conjugate acid* is the species that remains from the base after it gains a proton.

The *conjugate acid-base pairs* (acid$_1$–conjugate base$_1$ and acid$_2$–conjugate base$_2$) are related substances by the loss or gain of a single proton (H^+).

Thus, H_2O and H_3O^+, and H_2O and OH^- are conjugate acid-base pairs, but H_3O^+ and OH^- are not conjugate acid-base pairs.

Strong acids have weak conjugate bases.

Weak acids have strong conjugate bases.

The weaker the acid, the stronger its conjugate base.

Weak bases have strong conjugate acids.

The weaker the base, the stronger the conjugate acid it produces.

HCl is a strong acid, and Cl^- is a very weak base.

HF is a weak acid, and F^- is a strong conjugate.

A Brønsted-Lowry acid-base reaction involves a competition between two bases for a proton, in which the stronger base is the most protonated at equilibrium.

$$HCl + H_2O \rightarrow H_3O^+ (aq) + Cl^- (aq)$$

H_2O is a much stronger base than Cl^-. At equilibrium, the HCl solution contains mostly H_3O^+ and Cl^- ions.

$$HC_2H_3O_2\,(aq) + H_2O \rightleftarrows H_3O^+\,(aq) + C_2H_3O_2^-\,(aq)$$

$C_2H_3O_2^-$ is the stronger base. At equilibrium, acetic acid contains mostly $HC_2H_3O_2$ and a small amount of H_3O^+ and $C_2H_3O_2^-$ ions.

$$NH_3\,(aq) + H_2O \rightleftarrows NH_4^+\,(aq) + {}^-OH\,(aq)$$

H_2O is an acid. Competition for protons occurs between NH_3 and ^-OH, in which ^-OH is the stronger base. The equilibrium favors the reactants. An aqueous ammonia solution contains mostly NH_3 molecules and smaller amounts of NH_4OH, NH_4^+, and ^-OH.

According to Brønsted-Lowry, the net acid-base reactions are favored in the direction from strong acid-strong base combinations to weak acid-weak base combinations.

The following acid-base reactions proceed in the forward direction:

$HCl\,(aq) + NH_3\,(aq) \rightarrow NH_4^+\,(aq) + Cl^-\,(aq)$

(HCl is a strong acid)

$HSO_4^-\,(aq) + CN^-\,(aq) \rightarrow HCN\,(aq) + SO_4^{2-}\,(aq)$

HSO_4^- (pK_a = 6.91) is a stronger acid than HCN (pK_a = 9.21)

Many acid-base reactions reach a state of equilibrium.

For the following acid-base reactions, the equilibrium may favor the products or the reactants, depending on the relative strength of the acid:

$H_2PO_4^-\,(aq) + C_2H_3O_2^-\,(aq) \rightleftarrows HC_2H_3O_2\,(aq) + HPO_4^{2-}\,(aq)$

Equilibrium shifts to the left (reactants):

$HC_2H_3O_2$ (pK_a = 4.75) is the stronger acid

HPO_4^{2-} (pK_a = 7.21) is the stronger base

$HNO_2\,(aq) + C_2H_3O_2^-\,(aq) \leftrightarrows HC_2H_3O_2\,(aq) + NO_2^-\,(aq)$

Equilibrium shifts to the right (products).

HNO_2 (pK_a = 3.39) is the stronger acid

$C_2H_3O_2^-$ (pK_a = 4.75) is the stronger base

Common ions are ions produced by more than one solute in the same solution.

For example, in a solution containing sodium acetate and acetic acid, the acetate ion ($C_2H_3O_2^-$) is the common ion.

According to Le Châtelier's principle, the following equilibrium for acetic acid,

$$HC_2H_3O_2\,(aq) + H_2O\,(l) \leftrightharpoons H_3O^+\,(aq) + C_2H_3O_2^-\,(aq)$$

The reaction shifts to the left if $C_2H_3O_2^-$, from another source, is introduced. This effect reduces the degree of dissociation of the acid, decreases $[H_3O^+]$, and increases the pH of the solution.

The addition of ammonium chloride, NH_4Cl, causes the reaction to shift to the left in the following equilibrium of ammonia in aqueous solution:

$$NH_3\,(aq) + H_2O\,(l) \leftrightharpoons NH_4^+\,(aq) + OH^-\,(aq)$$

In solution, NH_4Cl dissociates to produce NH_4^+ and Cl^- ions, where NH_4^+ is a common ion in the equilibrium of ammonia.

The equilibrium shift caused by NH_4^+ ion reduces the extent of ionization of ammonia, which decreases $[OH^-]$ in the system and lowers the pH of the solution

Salts

Salts are the products of acid-base reactions.

The general reaction to produce salt is:

$$\text{Acid} + \text{Base} \rightarrow \text{Salt} + \text{Water}$$

The acid and base are neutralized, and H^+ and OH^- combine to form water.

The nonmetallic ions of the acid and the metal ions of the base form the salt.

For example, NaCl is a product of the following acid-base reaction:

$$HCl\,(aq) + NaOH\,(aq) \rightarrow NaCl\,(aq) + H_2O$$

In the chemical formula of salt, the cation is contributed by the base (e.g., Na^+), while the acid contributes the anion (e.g., Cl^-).

There are four types of salts:

- Salts of strong acid-strong base reactions (e.g., NaCl, KNO_3, $NaClO_4$, etc.)

- Salts of weak acid-strong base reactions (e.g., $NaC_2H_3O_3$, K_2CO_3, KCN, $NaCHO_2$, etc.)

- Salts of strong acid-weak base reactions (e.g., NH_4Cl, NH_4NO_3, $HONH_3Cl$, etc.)

- Salts of weak acid-weak base reactions (e.g., $NH_4C_2H_3O_2$, NH_4CN, NH_4HS, etc.)

These salts, when dissolved in water, produce acidic, basic, or neutral solutions; see diagram below.

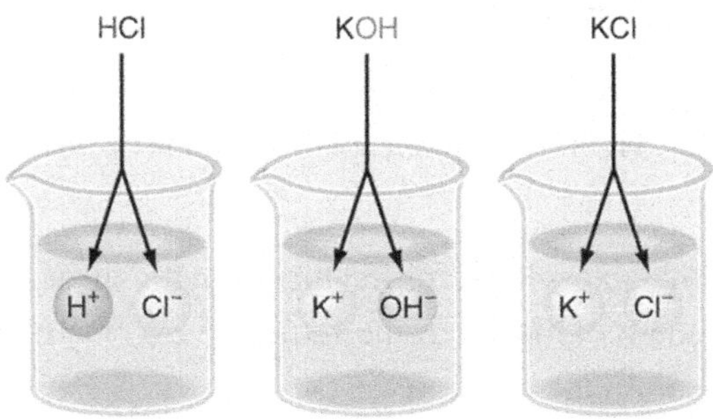

Strength of acids and bases

A *strong acid* ionizes entirely in aqueous solutions.

Strong acids include $HClO_4$, HCl, H_2SO_4, HNO_3, HBr, and HI. For these strong acids, the equilibrium lies far to the right.

Weak acids only partially ionize, and their ionization equilibriums lie far to the left.

Weak acids include $HC_2H_3O_2$, HNO_2, H_2SO_3, H_3PO_4 HClO, and others.

Consider the reversible process when an acid dissolves in water:

$$HA\ (aq) + H_2O \leftrightarrow H_3O^+\ (aq) + A^-\ (aq)$$

In the forward reaction above, the acid HA donates a proton to the water molecule to form hydronium ion (H_3O^+) and conjugate base (A^-). Water acts as a Brønsted-Lowry base. The acid strength is measured by their degree of ionization (or dissociation) in water. If an acid ionizes completely, it is a strong acid.

A strong acid has a weak conjugate base (i.e., the conjugate base loses its proton to water quite readily). Cl^-, Br^-, I^-, ClO_4^-, HSO_4^-, and NO_3^- are weak conjugate bases.

A strong acid is less able to compete with water for a proton. The ionization constant K_a for a strong acid is exceptionally large, and the equilibrium shifts far to the right (forming the dissociated proton and stable anion).

A weak acid does not readily give up its proton to water, and it has a strong conjugate base. Strong conjugate bases include $C_2H_3O_2^-$, F^-, CN^-, NO_2^-, HSO_3^-, SO_3^{2-}, $H_2PO_4^-$, HPO_4^{2-} and PO_4^{3-}.

The weaker the acid, the stronger its conjugate base. The ionization equilibrium for weak acids shifts far to the left.

An aqueous solution of a strong acid contains only hydronium ions (H_3O^+) and the acid's conjugate base. For example, in an aqueous HCl solution, there are H_3O^+ and Cl^- ions, but virtually no HCl molecules.

Conversely, an aqueous solution of a weak acid, such as acetic acid, contains mainly the undissociated molecules, $HC_2H_3O_2$, with a small fraction (~1%) of H_3O^+ and $C_2H_3O^-$ ions.

A combination of factors determines the strength of acids. These include the strength and polarity of the X–H bond in the molecule and the hydration energy of the ionic species in aqueous solution.

For inorganic binary acids (e.g., HF, HCl, HBr, HI), H– X bond strength decreases down the group. The weaker the bond, the easier it ionizes in aqueous solution. The stronger bond is the acid; their strength increases down the group: HF < HCl < HBr < HI.

Among the hydrohalic acids, HF is the only weak acid; the others are strong acids. The relative strength of HCl, HBr, and HI cannot be differentiated in aqueous solution, because each of them dissociates almost completely. Less polar solvents are used to determine their relative strength. For example, HCl, HBr, and HI ionize only partially in acetone or methanol, which have a weaker ionizing strength than water.

The ionization of HCl in acetone is represented by:

$$(CH_3)_2CO\ (l)\ (\textit{acetone}) + HCl \rightleftarrows (CH_3)_2COH^+ + Cl^-$$

The degree of ionization in acetone increases in the order of HCl < HBr < HI. The acidity of hydrogen halides increases down the group, as stated previously.

A similar trend of relative acidity is observed for the hydrides of Group 15:

$$H_2O < H_2S < H_2Se < H_2Te.$$

For the same period hydrides, the relative acidity increases from left to right, such:

$$CH_4 < NH_3 < H_2O < HF$$

$$PH_3 < H_2S << HCl$$

Water is a stronger acid than ammonia, and in an acid-base reaction, H_2O acts as a Brønsted-Lowry acid, which donates a proton to NH_3:

$$H_2O + NH_3\,(aq) \rightleftarrows NH_4^+\,(aq) + OH^-\,(aq)$$

In reaction with HF, water acts as a Brønsted-Lowry base, which accepts a proton:

$$HF\,(aq) + H_2O \rightleftarrows H_3O^+\,(aq) + F^-\,(aq)$$

Oxoacids contain one or more ~OH groups covalently bonded to a central atom, which can be a metal or a nonmetal. The ~OH group ionizes completely or partially in an aqueous solution, producing hydrogen ions.

Examples of oxoacids:

H_2CO_3, HNO_3, H_3PO_4, H_2SO_4, $HClO_4$, $HC_2H_3O_2$

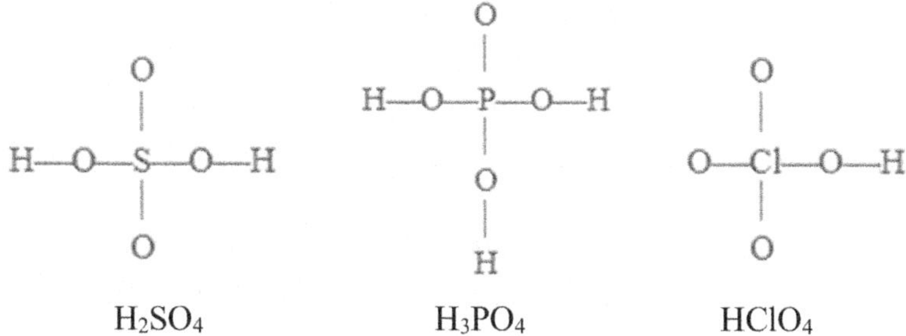

H_2SO_4 H_3PO_4 $HClO_4$

For these types of acids, their relative strengths depend on the *electronegativity* of the central atoms. The more electronegative the central atom, the more polarized the O–H bond, and the more readily it ionizes in aqueous solution to release the H^+ ion.

For example, N, S, and Cl are more electronegative than P; and HNO_3, H_2SO_4, and $HClO_4$ are stronger acids, whereas H_3PO_4 is a weak acid.

The order of electronegativity of the relative acid strength are as follows:

Electronegativity: $Cl \approx N > S > P$

Acid strength: $HClO_4 > HNO_3 > H_2SO_4 > H_3PO_4$

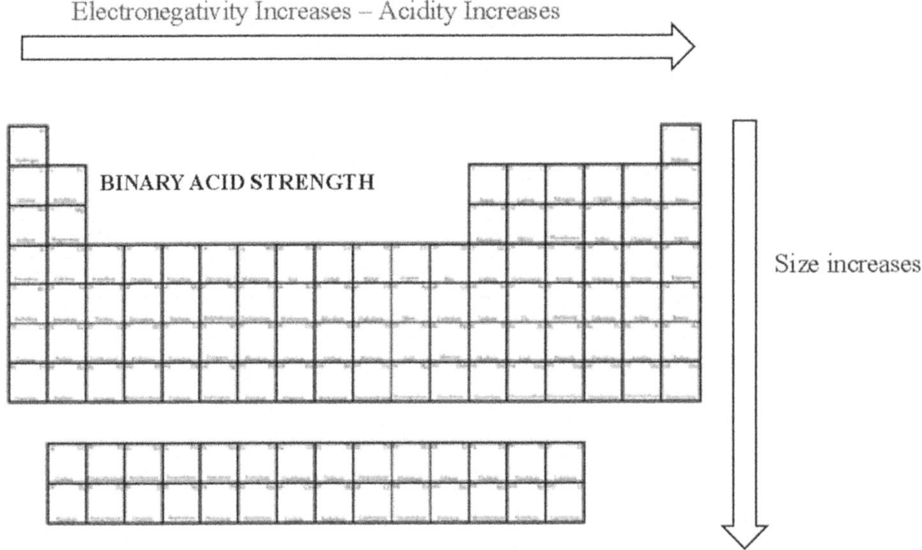

Acidity increases with increases in electronegativity and with increasing atomic size

For oxoacids that have the central atoms with elements of the same group in the periodic table, their relative strength decreases from top to bottom (as the electronegativity of the central atom decreases):

$HOCl > HOBr > HOI$ $HClO_2 > HBrO_2 > HIO_2$ $HClO_4 > HBrO_4 > HIO_4$

For oxoacids containing identical central atoms, their acidity increases as more oxygen atoms bond. Acidity increases as follows:

$HOCl < HClO_2 < HClO_3 < HClO_4$

$H_2SO_3 < H_2SO_4$

$HNO_2 < HNO_3$

When more oxygen atoms bond to the central atom, the O–H bond in the molecule becomes highly polarized (due to the inductive electronegative effect). The O–H bond ionizes more readily to release an H^+ ion.

Acetic acid (CH_3COOH) is an organic acid, which contains the carboxyl (~COOH) group. In an aqueous solution, ionization of an acetic acid only involves the breaking of O–H bond of the carboxyl group, but not the C–H bonds in the methyl group (CH_3).

However, if one or more of the hydrogen atoms in the methyl group is substituted with a more electronegative atom, the inductive effect causes the electron cloud to be drawn away from the carbonyl group. The O–H bond becomes more polarized and ionizes more readily, increasing the acidity.

The following K_a values illustrate the effect on the acidity of acetic acid and its derivatives when methyl hydrogens are substituted with electronegative atoms.

The stronger the acid, the higher the K_a value:

CH_3COOH (*aq*) (*acetic acid*) + $H_2O \rightleftarrows CH_3COO^-$ (*aq*) + H_3O^+ (*aq*)

$K_a = 1.8 \times 10^{-5}$

$ClCH_2COOH$ (*aq*) (*chloroacetic acid*) + $H_2O \rightleftarrows ClCH_2COO^-$ (*aq*) + H_3O^+ (*aq*)

$K_a = 1.4 \times 10^{-3}$

FCH_2COOH (*aq*) (fluoroacetic acid) + $H_2O \rightleftarrows FCH_2COO^-$ (*aq*) + H_3O^+ (*aq*)

$K_a = 2.6 \times 10^{-3}$

CCl_3COOH (*aq*) (trichloroacetic acid) + $H_2O \rightleftarrows CCl_3COO^-$ (*aq*) + H_3O^+ (*aq*)
$K_a = 3.0 \times 10^{-1}$

The *monoprotic acids* include HCl, HF, HOCl, HNO_2 and $HC_2H_3O_2$ because each molecule contains a single ionizable hydrogen ion.

The *polyprotic acids* contain more than one ionizable hydrogen, and examples include H_2SO_4, H_2SO_3, $H_2C_2O_4$ and H_3PO_4. The hydrogen ionizes in stages and with different ionization constants, such as the following example with H_3PO_4.

H_3PO_4 (*aq*) $\rightleftarrows H^+$ (*aq*) + $H_2PO_4^-$ (*aq*); $K_{a1} = 7.5 \times 10^{-3}$

$H_2PO_4^-$ (*aq*) $\rightleftarrows H^+$ (*aq*) + HPO_4^{2-} (*aq*); $K_{a2} = 6.2 \times 10^{-8}$

HPO_4^{2-} (*aq*) $\rightleftarrows H^+$ (*aq*) + PO_4^{3-} (*aq*); $K_{a3} = 4.8 \times 10^{-13}$

From above, acid strength decreases in the order: $H_3PO_4 >> H_2PO_4^- >> HPO_4^{2-}$.

Sulfuric acid (H_2SO_4) is a strong acid, but only the first hydrogen ionizes completely:

H_2SO_4 (*aq*) $\rightarrow H^+$ (*aq*) + HSO_4^- (*aq*) K_{a1} = exceptionally large

The second hydrogen does not dissociate complete and HSO_4^- is a weak acid:

HSO_4^- (*aq*) $\rightleftarrows H^+$ (*aq*) + SO_4^{2-} (*aq*) $K_{a2} = 1.2 \times 10^{-2}$

Strong bases, like strong acids (e.g., NaOH, KOH, and $Ba(OH)_2$) ionize completely when dissolved in water. A strong base tends to accept a proton.

Weak bases do not ionize completely when dissolved in water and show minimal tendency to accept a proton.

Examples of weak bases are NH_3 (or NH_4OH), NH_2OH, $Mg(OH)_2$ and all hydroxides and oxides that are only slightly soluble in water.

Hydroxides of Group I metals (LiOH, NaOH, KOH, RbOH, and CsOH) are strong bases, but only NaOH and KOH are commercially important and commonly used laboratory bases. These bases are soluble in water, and they dissociate entirely in aqueous solution, producing a high concentration of hydroxide ions.

A moderately dilute solution of NaOH contains [^-OH], [Na+], and [NaOH].

Among the hydroxides of the alkaline earth metals, $Ba(OH)_2$ is a relatively strong base. The other metal hydroxides are only sparingly soluble in water, which limits their basicity. A saturated solution of these hydroxides contains a very low concentration of OH^-.

These hydroxides can react with strong acids:

$$Na_2O\ (s) + H_2O \rightarrow 2\ NaOH\ (aq)$$

$$BaO\ (s) + H_2O \rightarrow Ba(OH)_2\ (aq)$$

$$MgO\ (s) + 2\ HCl\ (aq) \rightarrow MgCl_2\ (aq) + H_2O$$

Hydroxides of some metals (e.g., $Al(OH)_3$, $Cr(OH)_3$, $Zn(OH)_2$, $Sn(OH)_2$ and $Pb(OH)_2$) exhibit amphoteric properties (i.e., can act as an acid or a base). For example:

$$Al(OH)_3\ (s) + OH^-\ (aq) \rightleftarrows Al(OH)_4^-\ (aq)$$

$$Al(OH)_3\ (s) + 3\ H_3O^+\ (aq) \rightleftarrows [Al(H_2O)_6]^{3+}\ (aq)$$

Hydrides of reactive metals (e.g., NaH, MgH_2, CaH_2) form strongly basic solutions when dissolved in water.

The hydride ion reacts with water to produce hydroxide ions and hydrogen gas:

$$H^-\ (aq) + H_2O \rightarrow H_2\ (g) + OH^-\ (aq)$$

The oxide ion O^{2-} has an extraordinarily strong affinity for protons and reacts with water to produce hydroxide ions.

$$O^{2-}\ (aq) + H_2O \rightarrow 2\ OH^-\ (aq)$$

Oxides of nonmetals are acidic. They form acidic solutions when dissolved in water.

$$CO_2\ (g) + H_2O \rightleftarrows H_2CO_3\ (aq) \rightleftarrows H^+\ (aq) + HCO_3^-\ (aq)$$

$$SO_2\ (g) + H_2O \rightleftarrows H_2SO_3\ (aq) \rightleftarrows H^+\ (aq) + HSO_3^-\ (aq)$$

Ammonia is the only weak base that is of commercial importance. It does not contain hydroxide ions, but it reacts with water and ionizes as follows:

$$NH_3\,(aq) + H_2O \rightleftarrows NH_4^+\,(aq) + OH^-\,(aq)$$

The base dissociation constant K_b is given by the expression:

$$K_b = [NH_4^+]\cdot[^-OH] / [NH_3]$$

$$K_b = 1.8 \times 10^{-5}$$

Dissociation of acids and bases without added salt

The *percent dissociation* (or *degree of ionization*) of a weak acid is:

Percent dissociation = [acid ionized] / [initial acid] × 100%

For strong acids, the percent dissociation at equilibrium is almost 100%.

For weak acids, the percent dissociation depends on the K_a of the acid and its initial concentration.

For example, the percent dissociation of acetic acid ($HC_2H_3O_2$, $K_a = 1.8 \times 10^{-5}$) at 0.10 M concentration is:

$$(1.3 \times 10^{-3}\text{ M} / 0.10\text{ M}) \times 100\% = 1.3\ \%$$

The stronger the acid, the larger is the K_a and the greater is the percent ionization.

The percent ionization of a weak acid depends on the K_a and the extent of dilution.

The more an acid solution is diluted, the higher is the percentage ionization.

Consider a solution of 0.010 M acetic acid and its ionization products.

Use an ICE table (**I**nitial, **C**hange, **E**quilibrium), which is a method to simplify the calculations in reversible equilibrium reactions.

Once the equilibrium row is completed (by summing the initial and change rows), its contents can be substituted into the equilibrium constant expression to solve for K_a.

$$CH_3COOH\,(aq) \rightleftarrows H^+\,(aq) + CH_3COO^-\,(aq)$$

Initial [ ], M:	0.010	0.00	0.00
Change, Δ[ ], M:	$-x$	$+x$	$+x$
Equilibrium [ ], M :	$(0.010 - x)$	x	x

The acid ionization constant, K_a, is given by the expression:

$$K_a = [H_3O^+]\cdot[C_2H_3O_2^-] / [CH_3COOH]$$

$$K_a = x^2 / (0.010 - x)$$

$$K_a = 1.8 \times 10^{-5}$$

Since $K_a \ll 0.010$, approximate that $x \ll 0.010$, and $(0.010 - x) \sim 0.010$

Then

$$K_a = x^2 / (0.010 - x)$$

$$K_a \approx x^2 / 0.010$$

$$K_a = 1.8 \times 10^{-5}$$

$$x^2 = 1.8 \times 10^{-7}$$

and

$$x = \sqrt{(1.8 \times 10^{-7})}$$

$$x = 4.2 \times 10^{-4}$$

$$x = [H_3O^+]$$

$$x = 4.2 \times 10^{-4}\ M$$

The degree of ionization of acetic acid at this concentration is

$$(4.2 \times 10^{-4}\ M / 0.010\ M) \times 100\% = 4.2\%$$

The degree of ionization of the acid increases as the solution is more diluted. For 0.1 M acetic acid, the degree of ionization is only 0.42%, which is 10-fold lower than in 0.010 M acid solution.

The percent dissociation of weak bases depends on the K_b value and dilution of the base solution; larger K_b and greater extent of dilution results in higher percent dissociation.

Hydrolysis of salts

The presence of salt affects the dissociation of acids and bases due to the *hydrolysis* of salts. For example, CH_3COOH dissociates less in a solution containing CH_3COONa salt, while NH_4OH dissociates less in a solution containing NH_4Cl salt.

When salts (ionic compounds) dissolve in water, it is generally assumed that they dissociate entirely into separate ions. Some of these ions can react with water and behave as acids or bases.

The acidic or basic nature of a salt solution depends on whether it is a product of a 1) strong acid-strong base reaction, 2) a weak acid-strong base reaction, 3) a strong acid-weak base reaction, 4) or a weak acid-weak base reaction.

Salts of Strong Acid-Strong Base Reactions: (e.g., NaCl, $NaNO_3$, KBr)

- Salts of this type form neutral solution, because neither the cation nor the anion reacts with water and offsets the equilibrium concentrations of H_3O^+ and OH^- in the solution.

Salts of Weak Acid-Strong Base Reactions: (e.g., NaF, $NaNO_2$, $NaC_2H_3O_2$)

- Salts that are products of reactions between weak acids and strong bases form basic solutions when dissolved in water.

  The anions of such salts react with water that increases [$^-$OH].

- Sodium acetate ($NaC_2H_3O_2$) is a product of the reaction between acetic acid ($HC_2H_3O_2$), which is a weak acid, and a strong base (NaOH).

$$HC_2H_3O_2\ (aq) + NaOH\ (aq) \rightarrow NaC_2H_3O_2\ (aq) + H_2O$$

Salts of Strong Acid-Weak Base Reactions: (e.g., NH_4Cl, NH_4NO_3, $(CH_3)_2NH_2Cl$, C_5H_5NHCl)

- Aqueous solutions of salts that are products of strong acid-weak base reactions are acidic.

- The cations react with water and increase [H_3O^+] in solutions.

- Example: NH_4Cl is produced when HCl (strong acid) reacts with NH_3 (a weak base):

$$HCl\ (aq) + NH_3\ (aq) \rightarrow NH_4Cl\ (aq) \rightarrow NH_4^+\ (aq) + Cl^-\ (aq)$$

In an aqueous solution, NH_4^+ establishes the following equilibrium that increases [H_3O^+], and creates an acidic solution:

$$NH_4^+\ (aq) + H_2O \rightleftarrows H_3O^+\ (aq) + NH_3\ (aq)$$

$$K_a = [H_3O^+]\cdot[NH_3]\ /\ [NH_4^+]$$

While in a solution of NH_3, the following equilibrium occurs:

$NH_3\ (aq) + H_2O \rightleftarrows NH_4^+\ (aq) + OH^-\ (aq)$

$K_b = [NH_4^+]\cdot[^-OH] / [NH_3]$

$K_a \times K_b = \{[H_3O^+]\cdot[NH_3] / [NH_4^+]\} \times \{[NH_4^+]\cdot[^-OH] / [NH_3]\}$

$K_a \times K_b = K_w = [H_3O^+]\cdot[^-OH]$

$K_w = 1.0 \times 10^{-14}$

For NH_4^+

$K_a = K_w / K_b$ (for NH_3)

$K_a = (1.0\ x\ 10^{-14}) / (1.8\ x\ 10^{-5})$

$K_a = 5.6 \times 10^{-10}$

An aqueous solution of NH_4^+ has a $K_a = 5.6 \times 10^{-10}$ at 25 °C (which is > K_w).

A 0.10 *M* solution of NH_4Cl or NH_4NO_3 has $[H_3O^+] \approx 7.5 \times 10^{-6}$ M and pH ≈ 5.1

Salts of Weak Acid-Weak Base Reactions: (e.g., $NH_4C_2H_3O_2$, NH_4CN, NH_4NO_2)

- Solutions of salts that are products of weak acid-weak base reactions can be neutral, acidic, or basic, depending on the relative magnitude of the K_a of the weak acid and the K_b of the weak base.

  If $K_a \approx K_b$, the salt forms an approximately neutral solution.

Example:

K_a of $HC_2H_3O_2 = 1.8 \times 10^{-5}$ and K_b of $NH_3 = 1.8 \times 10^{-5}$

When $NH_4C_2H_3O_2$ dissolves in water and dissociates, the following equilibria exist:

$NH_4C_2H_3O_2\ (aq) \rightarrow NH_4^+\ (aq) + C_2H_3CO_2^-\ (aq)$

$NH_4^+\ (aq) + H_2O \rightleftarrows H_3O^+\ (aq) + NH_3\ (aq)$

$K_a = 5.6 \times 10^{-10}$

$C_2H_3O_2^-\ (aq) + H_2O \rightleftarrows HC_2H_3O_2\ (aq) + OH^-\ (aq)$

$K_b = 5.6 \times 10^{-10}$

Since K_a(for NH_4^+) = K_b(for $C_2H_3O_2^-$), at equilibrium $[H_3O^+] = [^-OH]$, $NH_4C_2H_3O_2$ solution is neutral.

- If $K_a > K_b$, the salt solution is acidic.

For NH_4NO_2:

$K_a(HNO_2) = 4.0 \times 10^{-4}$

$K_b(NH_3) = 1.8 \times 10^{-5}$

$NH_4NO_2\,(aq) \rightarrow NH_4^+\,(aq) + NO_2^-\,(aq)$

$NH_4^+\,(aq) + H_2O \rightleftarrows H_3O^+\,(aq) + NH_3\,(aq)$

$K_a = 5.6 \times 10^{-10}$

$NO_2^-\,(aq) + H_2O \rightleftarrows HNO_2\,(aq) + OH^-\,(aq)$

$K_b = 2.5 \times 10^{-11}$

$K_a > K_b \rightarrow$ acidic solution, because the hydrolysis results in a solution with $[H_3O^+] > [^-OH]$.

- If $K_a < K_b$, the salt solution is basic. For NH_4CN in solution,

  $K_a(HCN) = 6.2 \times 10^{-10}$,

  $K_b(NH_3) = 1.8 \times 10^{-5}$

  the following equilibria exist:

  $NH_4CN\,(aq) \rightarrow NH_4^+\,(aq) + CN^-\,(aq)$

  $NH_4^+\,(aq) + H_2O \rightleftarrows H_3O^+\,(aq) + NH_3\,(aq)$

  $K_a = 5.6 \times 10^{-10}$

  $CN^-\,(aq) + H_2O \rightleftarrows HCN\,(aq) + OH^-\,(aq)$

  $K_b = 1.6 \times 10^{-5}$

Since $K_b\,(CN^-) > K_b\,(NH_4^+)$, at equilibrium $[^-OH] > [H_3O^+]$ and an aqueous solution of NH_4CN is basic.

Calculation of pH

Calculating the pH of Strong Acid and Strong Base Solutions

Strong acids are assumed to ionize completely in aqueous solution.

For monoprotic acids (i.e., acids with a single ionizable hydrogen) such as HCl and HNO_3, the [hydronium ion] in solution is the same as the molar concentration of the acid:

$$[H_3O^+] = [HX]$$

Consider 0.10 M HCl (*aq*)

$[H_3O^+] = 0.10$ M

$pH = -\log(0.10)$

pH − 1.00

A strong base such as NaOH has $[^-OH]$ equal to the molar concentration of dissolved NaOH.

A solution of 0.10 M NaOH (*aq*) has:

$[^-OH] = 0.10$ M

$pOH = -\log[^-OH] = -\log(0.10)$

pOH = 1.00

pH = 14.00 – 1.00

pH = 13.00

A strong base such as $Ba(OH)_2$ produces twice the concentration of ^-OH as the molar concentration of $Ba(OH)_2$ in solution:

$Ba(OH)_2\,(aq) \rightarrow Ba^{2+}\,(aq) + 2\,OH^-\,(aq)$

$[^-OH] = 2 \times [Ba(OH)_2]$

In a solution of 0.010 M $Ba(OH)_2$,

$[^-OH] = 0.020$ M

pOH = 1.70, and pH = 12.30

Calculating the pH of Weak Acid Solutions

Unlike strong acids, weak acids do not ionize completely.

At equilibrium, $[H^+]$ is much less than the concentration of the acid.

The concentration of H^+ in a weak acid solution depends on the initial acid concentration and the K_a of the acid.

To determine $[H^+]$ of a weak acid. the "ICE" table is set as follows:

Consider a solution of 0.10 M acetic acid and its ionization products.

$HC_3H_3O_2\,(aq) \rightleftarrows H^+\,(aq) + C_2H_3O_2^-\,(aq)$

Initial [ ], M:	0.10	0.00	0.00
Change, Δ[ ], M:	$-x$	$+x$	$+x$
Equilibrium [ ], M :	$(0.10 - x)$	x	x

The acid ionization constant, K_a, is given by the expression:

$K_a = [H_3O^+]\cdot[C_2H_3O_2^-] / [CH_3COOH]$

$K_a = x^2 / (0.010 - x)$

$K_a = 1.8 \times 10^{-5}$

Since $K_a << 0.10$, approximate that $x << 0.10$, and $(0.10 - x) \sim 0.10$

$K_a = x^2 / (0.10 - x)$

$K_a \approx x^2 / 0.10$

$K_a = 1.8 \times 10^{-5}$

$x^2 = 1.8\ x\ 10^{-6}$

$x = \sqrt{}(1.8\ x\ 10^{-6})$

$x = 1.3 \times 10^{-3}$

Note that

$x = [H_3O^+]$

$[H_3O^+] = 1.3 \times 10^{-3}$ M

$pH = -\log(1.3 \times 10^{-3})$

$pH = 2.89$

Calculating the pH of Weak Base Solutions

The concentration of ^-OH in a weak base, such as NH_3 (*aq*), depends on its K_b value and the initial concentration of the base.

To determine [^-OH] and pH of 0.10 M NH_3 (*aq*), set the following "ICE" table:

Concentration:

$$NH_3\,(aq) + H_2O \rightleftarrows NH_4^+\,(aq) + {}^-OH\,(aq)$$

Initial [ ], *M*:	0.10	0.00	0.00
Change, Δ[ ], *M*:	$-x$	$+x$	$+x$
Equilibrium [ ], *M*:	$(0.10 - x)$	x	x

$K_b = [NH_4^+]\cdot[^-OH] / [NH_3]$

$K_b = x^2 / (0.10 - x)$

$K_b = 1.8 \times 10^{-5}$

Using the approximation method

$K_b = x^2 / (0.10 - x)$

$K_b \approx x^2 / 0.10$

$K_b = 1.8 \times 10^{-5}$

$x^2 = 1.8 \times 10^{-6}$

$x = \sqrt{(1.8 \times 10^{-6})}$

$x = 1.3 \times 10^{-3}$

where

$x = [^-OH]$

$x = 1.3 \times 10^{-3}\ M$

$pOH = -\log(1.3 \times 10^{-3})$

$pOH = 2.87$

$pH = 11.13$

Calculating the pH of basic or acidic salt solutions

1. Consider a solution of 0.050 *M* sodium acetate, which dissociates completely and establishes the following equilibrium:

   $NaC_2H_3O_2\,(aq) \rightarrow Na^+\,(aq) + C_2H_3O_2^-\,(aq)$

   The acetate ion establishes the equilibrium in aqueous solution:

   $C_2H_3O_2^-\,(aq) + H_2O \rightleftarrows HC_2H_3O_2\,(aq) + {}^-OH\,(aq)$

   $K_b = [HC_2H_3O_2]\cdot[^-OH] / [C_2H_3O_2^-]$

   $K_b = 5.6 \times 10^{-10}$

   By approximation

   $[^-OH] = \sqrt{(K_b[C_2H_3O_2^-])}$

   $[^-OH] = \sqrt{\{(5.6 \times 10^{-10})\cdot(0.050)\}}$

   $[^-OH] = 5.3 \times 10^{-6}$ M

   $pOH = -\log(5.3 \times 10^{-6})$

   $pOH = 5.28$

   $pH = 8.72$ (solution is basic)

2 Consider a solution of 0.050 M NH_4Cl, which dissociates and establishes the following equilibrium:

   $NH_4Cl\,(aq) \rightarrow NH_4^+\,(aq) + Cl^-\,(aq)$

   $NH_4^+\,(aq) + H_2O \rightleftarrows H_3O^+\,(aq) + NH_3\,(aq)$

   $K_a = [H_3O^+]\cdot[NH_3] / [NH_4^+]$

   $K_a = 5.6 \times 10^{-10}$

   By approximation

   $[H_3O^+] = \sqrt{(K_a[NH_4^+])}$

   $[H_3O^+] = \sqrt{\{(5.6 \times 10^{-10})\cdot(0.050)\}}$

   $[H_3O^+] = 5.3 \times 10^{-6}$ M

   $pH = -\log(5.3 \times 10^{-6})$

   $pH = 5.28$, (solution is acidic)

Buffers

A *buffer* is a solution that maintains its pH (with little change) even when a small amount of strong acid or strong base is added. A buffer solution contains a weak acid and the "salt" of its conjugate base or a weak base and the "salt" of its conjugate acid.

A given buffer is effective within a range of pH that is typically within approximately ±1 of the pK_a of its acid component.

Below are some examples of common buffer systems:

Buffer	pK_a	pH Range
$HCHO_2$ – $NaCHO_2$	3.74	2.75–4.75
$HC_2H_3O_2$ – $NaC_2H_3O_2$	4.74	3.75–5.75
KH_2PO_4 – K_2HPO_4	7.21	6.20–8.20 (a buffer system in the blood)
CO_2/H_2O – $NaHCO_3$	6.37	5.40–7.40 (a buffer system in the blood)
NH_3 – NH_4Cl	9.25	8.25–10.25

Buffered solutions are vital to living organisms. All metabolic reactions are controlled or accelerated by biological catalysts called *enzymes*, which are often proteins that function only within a narrow pH range.

The fluid of the human body must be maintained at a specific (narrow) pH range. Human blood is maintained at the pH range of 7.30 - 7.40. A drop below pH 7 or rise above pH 7.5 can be fatal.

When 0.01 mol of HCl is added to 1 L of pure water, $[H^+]$ increases from 10^{-7} to 10^{-2} M and the pH changes from about 7 to 2. This change in pH indicates that water is not a buffer.

When the same amount of HCl is added to a solution containing a mixture of 1 M acetic acid ($HC_2H_3O_2$) and 1 M sodium acetate ($NaC_2H_3O_2$), the pH of the solution changes little – it goes from 4.74 to 4.66. A solution that is composed of acetic acid and sodium acetate is a buffer solution.

Consider a buffered solution composed of KH_2PO_4 and K_2HPO_4. The species present in solution are primarily K^+, $H_2PO_4^-$ and HPO_4^{2-}. (K^+ is a spectator ion and not involved in the buffering reaction.)

If a small amount of strong acid is added to this solution, the H^+ ions from the acid react with the base component of the buffer (HPO_4^{2-}):

$$H^+ (aq) + HPO_4^{2-} (aq) \rightarrow H_2PO_4^- (aq) \ldots \text{(buffering reaction 1)}$$

When a strong base such as NaOH is added, the ^-OH reacts with the acid component of the base ($H_2PO_4^-$).

$$^-OH (aq) + HC_2H_3O_2 (aq) \rightarrow H_2O + C_2H_3O_2^- (aq) \ldots \text{(buffering reaction 2)}$$

Reactions (1) and (2) are critical buffering reactions that maintain the pH of the solution. Two buffer systems – the phosphate buffer ($H_2PO_4^-$-HPO_4^{2-}) and carbonic acid-bicarbonate buffer (H_2CO_3-HCO_3^-) are essential to maintaining the normal blood pH. The buffering reactions of bicarbonate buffer are:

$$H^+ (aq) + HCO_3^- (aq) \rightarrow H_2O + CO_2 (aq)$$

$$^-OH (aq) + CO_2 (aq) + H_2O \rightarrow 2\ HCO_3^- (aq)$$

Consider a buffer made up of acetic acid and sodium acetate, in which the significant species present in solution are $HC_2H_3O_2$ and $C_2H_3O_2^-$. If a small amount of HCl (*aq*) is added to this solution, most of H^+ (from HCl) is absorbed by the conjugate base, $C_2H_3O_2^-$, in the following reaction:

$$H^+ (aq) + C_2H_3O_2^- (aq) \rightarrow HC_2H_3O_2^- (aq)$$

Since $C_2H_3O_2^-$ is present in a much larger quantity than the added H^+, the reaction shifts almost entirely to the right. This buffering reaction prevents a significant increase in $[H^+]$ and minimizes the change in its pH.

If a strong base, such as NaOH (*aq*), is added, most of the ^-OH ions (from NaOH) react with the acidic component of the buffer.

$$^-OH (aq) + HC_2H_3O_2 (aq) \rightarrow H_2O + C_2H_3O_2^- (aq)$$

Because of the larger concentration of $HC_2H_3O_2$ compared to ^-OH, this reaction also goes almost to completion. This buffering reaction prevents a significant increase in the $[^-OH]$ and minimizes a change in the pH of the solution.

For a buffer containing the weak acid HB and the salt NaB, such that B^- is the conjugate base to the acid, the concentration $[H^+]$ and pH of the buffer depending on the dissociation constant, K_a, of the acid component and the concentration ratio $[B^-] / [HB]$ in the buffer solution.

Consider the equilibrium:

$$HB (aq) \leftrightharpoons H^+ (aq) + B^- (aq)$$

$$K_a = [H^+]\cdot[B^-] / [HB]$$

Rearranging the expression:

$[H^+] = K_a \times ([HB] / [B^-])$

$pH = pK_a + \log([B^-] / [HB])$ *Henderson-Hasselbalch equation*

The last expression is the *Henderson-Hasselbalch equation*, which is useful for calculating the pH of solutions when both the K_a and the ratio $[B^-] / [HB]$ are known.

The *buffering capacity* of a buffered solution represents the amount of H^+ ion or OH^- ion the buffer can absorb without significantly altering its pH. A buffer that contains large concentrations of buffering components and can absorb significant quantities of strong acid or strong base, with little change in its pH, has a large buffering capacity.

The pH of a buffered solution is determined by the ratio $[B^-] / [HB]$.

The capacity of a buffer is determined by the sizes of [AB] and $[B^-]$.

The following example illustrates how to calculate the change in pH and the buffering capacity of a buffered solution after a strong acid is added.

Calculate the change in pH when 0.010 mol of HCl is added to 1.0 L of each of the following buffers:

Buffer A: 1.0 M $HC_2H_3O_2$ + 1.0 M $NaC_2H_3O_2$

Buffer B: 0.020 M $HC_2H_3O_2$ + 0.020 M $NaC_2H_3O_2$

For both buffers, the pH is calculated by the Henderson-Hasselbalch equation:

$pH = pK_a + \log([B^-] / [HB])$

$pH = pK_a + \log([C_2H_3O_2^-] / [HC_2H_3O_2])$

$pH = -\log(1.8 \times 10^{-5}) + \log(1)$

$pH = 4.74$

When 0.010 mol HCl is added to Buffer A, the following reaction takes place:

$H^+ (aq) + C_2H_3O_2^- (aq) \rightarrow HC_2H_3O_2 (aq)$

[ ] before reaction:	0.010 M	1.0 M	1.0 M
[ ] after reaction:	0	0.99 M	1.01 M

The new pH:

$pH = 4.74 + \log (0.99 / 1.01)$

$pH = 4.74 - 0.010$

$pH = 4.73$ (pH is changed $\approx 0.21\%$)

When 0.010 mol HCl is added to Buffer B, the following reaction takes place:

	$H^+ (aq)$ +	$C_2H_3O_2^- (aq)$ →	$HC_2H_3O_2 (aq)$
[ ] before rxn:	0.010 M	0.020 M	0.020 M
[ ] after rxn:	0	0.010 M	0.030 M

The expression to calculate the pH:

$$pH = 4.74 + \log(0.010 / 0.030)$$

$$pH = 4.74 - 0.48$$

$$pH = 4.26 \text{ (pH decreases by 10\%)}$$

From above, Buffer A, which contains larger quantities of buffering components, has a much higher buffering capacity than Buffer B.

For Buffer A to decrease its pH by 0.48 unit (or 10%), it must absorb the equivalent of 0.50 mol of HCl.

In summary, buffer solutions have the following characteristics:

1. The solution contains a weak acid HX and its conjugate base X^-, or a weak base B and its conjugate acid BH^+ in appreciable amounts.

2. A buffer solution maintains its pH by absorbing H^+ or OH^- produced by a strong acid or strong base, so these ions do not accumulate.

3. The buffering reaction involves the reaction of H^+ with the conjugate base $\mathbf{X}^-$ in the buffer, or the reaction of ^-OH with the acid component (HX) of the buffer:

   $$H^+ (aq) + \mathbf{X}^- (aq) \rightarrow HX (aq)$$

   $$^-OH (aq) + \mathbf{HX} (aq) \rightarrow H_2O + X^-$$

   These two reactions prevent a significant increase in $[H^+]$ or $[^-OH]$ in the solution.

4. The *buffering capacity* of a solution implies the amount of H^+ or ^-OH it can absorb without significantly changing its pH. This depends on the concentration of the weak acid and its conjugate base in the solution.

5. The *buffering range* of a solution depends on the pK_a of the acid component of the buffer. A given buffer is most effective when the pH range = pK_a ± 1.

Neutralization

As discussed previously, the products of acid-base reactions are salt and water.

Neutralization is a chemical reaction where an acid and a base react quantitatively. This results in no excess of hydrogen or hydroxide ions in the aqueous solution.

The following are examples of acid-base reactions:

$$HCl\ (aq) + NaOH\ (aq) \rightarrow H_2O\ (l) + NaCl\ (aq)$$

$$HclO_4\ (aq) + KOH\ (aq) \rightarrow H_2O\ (l) + KclO_4\ (aq)$$

$$HC_2H_3O_2\ (aq) + NaOH\ (aq) \rightarrow H_2O\ (l) + NaC_2H_3O_2\ (aq)$$

Note that the substances involved in these reactions are subject to dissociation.

Listed below are examples of molecular, complete (or total) ionic and net ionic equations for strong acid-strong base reactions:

Molecular:

$$HCl\ (aq) + NaOH\ (aq) \rightarrow H_2O\ (l) + NaCl\ (aq)$$

Complete ionic:

$$H^+\ (aq) + Cl^-\ (aq) + Na^+\ (aq) + OH^-\ (aq) \rightarrow H_2O\ (l) + Na^+\ (aq) + Cl^-\ (aq)$$

Net ionic:

$$H^+\ (aq) + OH^-\ (aq) \rightarrow H_2O\ (l)$$

Spectator ions:

Na^+ and Cl^-

Weak acids and weak bases only ionize partially. Most weak acids and bases remain in the molecular form in solution. Therefore, weak acids and bases should NOT be written in the ionized forms, even when writing the ionic equations.

The three equations for the reaction between acetic acid (a weak acid) and sodium hydroxide (a strong base) are as follows:

Molecular:

$HC_2H_3O_2\,(aq) + NaOH\,(aq) \rightarrow H_2O\,(l) + NaC_2H_3O_2\,(aq)$

Complete ionic:

$HC_2H_3O_2\,(aq) + Na^+\,(aq) + OH^-\,(aq) \rightarrow H_2O\,(l) + Na^+\,(aq) + C_2H_3O_2^-\,(aq)$

Net ionic:

$HC_2H_3O_2\,(aq) + OH^-\,(aq) \rightarrow H_2O\,(l) + C_2H_3O_2^-\,(aq)$

Spectator ion: Na^+

The following illustrates the stoichiometry of acid-base neutralization reactions.

1. How many mL of 0.1725 M NaOH (*aq*) are needed to neutralize 25.00 mL of 0.2040 M HCl (*aq*)?

   According to the equation:

   $HCl\,(aq) + NaOH\,(aq) \rightarrow H_2O\,(l) + NaCl\,(aq)$

   Moles of NaOH needed = Moles of HCl present

   (Liters NaOH × Molarity NaOH) = (Liters HCl × Molarity HCl)

   (Liters of NaOH × 0.1725 mol/L) = (0.02500 L × 0.2040 mol/L)

   Divide both side by 0.1725 mol/L:

   $$\text{Liters of NaOH} = \frac{0.02500\text{ L} \times 0.2040\text{ mol/L}}{0.1725\text{ mol/L}}$$

   Liters of NaOH = 0.02957 L = 29.57 mL

2. If 10.00 mL of acetic acid $HC_2H_3O_2$ of an unknown concentration requires 38.64 mL of 0.2250 M KOH to neutralize, what is the molarity of the acetic acid?

From the reaction:

$HC_2H_3O_2\,(aq) + NaOH\,(aq) \rightarrow H_2O\,(l) + NaC_2H_3O_2\,(aq)$

Moles of acetic acid = Moles of NaOH

(Liter of acid × Molarity of acid) = (Liter of base × Molarity of base)

(0.01000 L × Molarity $HC_2H_3O_2$) = (0.03864 L × 0.2250 M NaOH)

Divide both side by 0.01000 L:

$$\text{Molarity of } HC_2H_3O_2 = \frac{0.03864 \text{ L} \times 0.2250 \text{ M}}{0.01000 \text{ L}}$$

$HC_2H_3O_2 = 0.8694$ M

Titration

Titration (*volumetric analysis*) is an essential application of neutralization reactions. It involves adding an exact amount of one reactant (*titrant*) from a buret to another reactant (*analyte*) in a flask or beaker. The primary objective of titration is to determine the molar concentration of one solution (the analyte) using the volume and concentration of another solution (titrant). In the process, the titrant is carefully added from a buret until the *equivalence point* is reached (i.e., the point of neutralization).

The *endpoint* of a titration is marked by the change in the color of the *indicator*. The indicator changes color when a solution transitions from being acidic to slightly basic. In a useful titration, the equivalence point should match the *endpoint*. If the volume and concentration of the titrant are known, its number of moles can be calculated. From the reaction stoichiometry (the balanced equation), the number of moles of the analyte and its concentration can be determined.

A successful titration experiment depends on the following factors:

1. The reaction between titrant and analyte occurs rapidly.

2. The balanced equation is known.

3. The endpoint occurs precisely at or close to the equivalence point.

4. The volume of titrant to reach the equivalence point is accurately measurable.

When an acid (e.g., HCl (*aq*)) is titrated with aqueous NaOH as follows:

$$HCl\ (aq) + NaOH\ (aq) \rightarrow NaCl\ (aq) + H_2O$$

The stoichiometric ratio of HCl to NaOH is 1 mole HCl to 1 mole NaOH. In a titration, the volume and concentration of the standard solution (acid or base) are known. However, only the volume of the other solution (acid or base), whose concentration to be determined, is known.

The above stoichiometry enables the calculation of the unknown concentration.

Suppose that 25.00 mL of 0.2250 M HCl (*aq*) is required to neutralize 27.45 mL of aqueous NaOH solution, whose concentration is not known. The moles of each reactant and the concentration of NaOH can be calculated as follows:

No. of mol of HCl reacted:

$$25.0\ mL \times (1\ L/1{,}000\ mL) \times (0.2250\ mol/L) = 0.005625\ mol$$

Since HCl and NaOH react in a 1:1 ratio,

No. of mol of NaOH = No. of mol of HCl

No. of mol of NaOH = 0.005625 mol

Molarity of NaOH = (0.005625 mol / 0.02745 L)

Molarity of NaOH = 0.2049 M

Sometimes, the stoichiometric ratio is not 1 to 1, as in the example below.

20.0 mL H_2SO_4 of unknown concentration requires 32.0 mL of 0.205 M NaOH. Calculate the concentration of the H_2SO_4 solution.

$$H_2SO_4\,(aq) + 2\,NaOH\,(aq) \rightarrow Na_2SO_4\,(aq) + 2\,H_2O$$

The stoichiometric ratio above is 1 mole of H_2SO_4 to 2 moles of NaOH.

$$\text{No. of mol of NaOH reacted} = \frac{0.205\text{ mol NaOH}}{1\text{ L solution}} \times 32.0\text{ mL} \times \frac{1\text{ L}}{1{,}000\text{ mL}}$$

No. of mol of NaOH reacted = 0.00656 mol

No. of mol of H_2SO_4 = Mol NaOH × (1 mol H_2SO_4 / 2 mol NaOH)

No. of mol of H_2SO_4 = 0.00656 mol NaOH × (1 mol H_2SO_4 / 2 mol NaOH)

No. of mol of H_2SO_4 = 0.00328 mol

$$\text{Molarity of } H_2SO_4 = \frac{0.00328\text{ mol } H_2SO_4}{0.0200\text{ L}}$$

Molarity of H_2SO_4 = 0.164 M

Consider the strong acid-strong base titration of 20.0 mL of 0.100 M HCl (*aq*) with 0.100 M NaOH (*aq*) solution. Calculate the pH of the acid solution: (a) before any of the NaOH is added, (b) after 15.0 mL of NaOH is added, (c) after 19.5 mL of NaOH is added, (d) after 20.0 mL of NaOH is added, (e) after 21.0 mL of NaOH is added and (f) after 25.0 mL of NaOH is added.

(a) Before titration:

$[H^+]$ = 0.100 M

pH = 1.000

(b) When 15.0 mL of NaOH is added, the following reaction occurs:

$H^+ (aq) + OH^- (aq) \rightarrow H_2O$

[ ] before mixing:

0.100 M 0.100 M

[ ] after mixing, but before reaction:

0.0571 M 0.0429 M

[ ] after reaction:

0.0142 M 0

$[H^+] = 0.0142$ M

$pH = -\log(0.0142)$

$pH = 1.848$

(c) After 19.5 mL of NaOH is added, calculation of $[H^+]$ is as follows:

$H^+ (aq) + {}^-OH (aq) \rightarrow H_2O$

[ ] before mixing:

0.100 M 0.100 M

[ ] after mixing, but before reaction:

0.0506 M 0.0494 M

[ ] after reaction:

0.0012 M 0

$[H^+] = 0.0012$ M

$pH = -\log(0.0012)$

$pH = 2.92$

Before the equivalent point, $[H^+]$ can be calculated:

$$[H^+] = \frac{(\text{initial mol of } H^+ - \text{mol of } {}^-OH \text{ added})}{(\text{L of HCl titrated} + \text{L of NaOH added})}$$

$$[H^+] = \frac{(0.00200 \text{ mol } H^+ - 0.00195 \text{ mol } {}^-OH)}{(0.0200 \text{ L of HCl} + 0.0195 \text{ L NaOH})}$$

$[H^+] = (0.000050 \text{ mol} / 0.0395 \text{ L}) = 0.0013$ M

$pH = 2.90$

(d) When 20.0 mL of 0.100 M NaOH has been added,

$$H^+ (aq) + {}^-OH (aq) \rightarrow H_2O$$

[ ] after mixing, but before reaction:

0.0500 M 0.0500 M

[ ] after reaction:

0 0

This point of the titration is the equivalence point, whereby only Na^+ and Cl^- occur in the solution. Since neither of them reacts with water, the solution has a pH = 7.00.

(e) When 21.0 mL of 0.100 M NaOH has been added, there is excess OH^-:

$$H^+ (aq) + {}^-OH (aq) \rightarrow H_2O$$

[ ] after mixing, but before reaction:

0.0488 M 0.0512 M

[ ] after reaction:

0 0.0024 M

$[{}^-OH] = 0.0024$ M

pOH = 2.62

pH = 11.38

(f) When 25.0 mL of NaOH is added,

$$H^+ (aq) + {}^-OH (aq) \rightarrow H_2O$$

[ ] after mixing, but before reaction:

0.0444 M 0.0556 M

[ ] after reaction:

0 0.0112 M

$[{}^-OH] = 0.0112$ M

pOH = 1.953

pH = 12.047

Note that in strong acid-strong base titrations, an abrupt change from about pH 3 to 11 occurs within ±0.5 mL of NaOH, added near the equivalent point.

An example of a weak acid-strong base titration follows.

General reaction:

$$HA\ (aq) + {}^{-}OH\ (aq) \rightarrow H_2O + A^{-}\ (aq)$$

When acetic acid (weak) is titrated with sodium hydroxide (strong), the net reaction is:

$$HC_2H_3O_2\ (aq) + {}^{-}OH\ (aq) \rightarrow C_2H_3O_2^{-}\ (aq) + H_2O$$

Consider the titration of 20.0 mL of 0.100 M $HC_2H_3O_2$ (*aq*) with 0.100 M NaOH (*aq*) solution. Calculate the pH of the solution: (a) before any of the NaOH is added; (b) after 10.0 mL of NaOH is added; (c) after 15.0 mL of NaOH is added; (d) after 20.0 mL of NaOH is added; (e) after 25.0 mL of NaOH is added.

(a) Before titration:

$$[H^+] = \sqrt{(0.100\ M)\cdot(1.8 \times 10^{-5})}$$

$$[H^+] = 1.34 \times 10^{-3}\ M$$

$$pH = -\log(1.34 \times 10^{-3})$$

$$pH = 2.873$$

(b) After adding 10.0 mL of 0.100 M NaOH, $[H^+]$ is calculated as follows:

$$HC_2H_3O_2\ (aq) + \ OH^{-}\ (aq) \rightarrow C_2H_3O_2^{-}\ (aq) + H_2O$$

Before mixing:

0.100 M	0.1000 M	0.0000 M

After mixing (before rxn):

0.0667 M	0.0333 M	0.0000 M

After reaction:

0.0333 M	0.0000 M	0.0333 M

Using the Henderson-Hasselbalch equation,

$$pH = pK_a + \log([B^{-}] / [HB])$$

$$[H^+] = K_a \times \frac{[HC_2H_3O_2]}{[C_2H_3O_2^{-}]}$$

$[H^+] = 1.8 \times 10^{-5} \times (0.0333\ M / 0.0333\ M)$

$[H^+] = 1.8 \times 10^{-5}\ M$

$pH = pK_a + \log([C_2H_3O_2^-] / [HC_2H_3O_2])$

$pH = 4.74 + \log(1)$

$pH = 4.74$

Therefore, when a weak acid is half-neutralized, 50% of the acid is converted to its conjugate base. That is, at halfway to the equivalence point of the titration, $[C_2H_3O_2^-] = [HC_2H_3O_2]$. Under this condition, $[H^+] = K_a$, and $pH = pK_a$.

(c) After adding 15.0 mL of 0.100 M NaOH, $[H^+]$ is calculated as follows:

$$HC_2H_3O_2\,(aq) + {}^-OH\,(aq) \rightarrow C_2H_3O_2^-\,(aq) + H_2O$$

Before mixing:	0.100 M	0.100 M	0.000 M
After mixing (before rxn):	0.0571 M	0.0429 M	0.000 M
After reaction:	0.0142 M	0.000 M	0.0429 M

Using the Henderson-Hasselbalch equation:

$pH = pK_a + \log([B^-] / [HB])$

$[H^+] = K_a \times [HC_2H_3O_2] / [C_2H_3O_2^-]$

$[H^+] = 1.8 \times 10^{-5} \times (0.0142\ M / 0.0429\ M)$

$[H^+] = 6.0 \times 10^{-6}\ M$

$pH = pK_a + \log([C_2H_3O_2^-] / [HC_2H_3O_2])$

$pH = 4.74 + \log(3.02)$

$pH = 4.74 + 0.48$

$pH = 5.22$

(d) At the equivalent point, when 20.0 mL of 0.100 M NaOH has been added, all the acid has been reacted and converted to its conjugate base. The latter undergoes hydrolysis (reacts with water) as follows:

$$C_2H_3O_2^-\,(aq) + H_2O\,(l) \leftrightarrows HC_2H_3O_2\,(aq) + {}^-OH\,(aq)$$

Initial [ ], M:

0.0500	0.000	0.000

Change, Δ[ ], M:

$-x$ $+x$ $+x$

Equilibrium [ ], M:

$(0.0500 - x)$ x x

By approximation:

$[OH^-] = x = \sqrt{(K_b \times [C_2H_3O_2^-]_0)}$

$[^-OH] = \sqrt{(5.6 \times 10^{-10}) \cdot (0.0500)}$

$[^-OH] = 5.3 \times 10^{-6}$ M

$pOH = -\log(5.3 \times 10^{-6})$

$pOH = 5.28$

$pH = 8.72$

Since the conjugate base of a weak acid undergoes hydrolysis (reacts with water), the pH of the solution at the equivalence point is greater than 7.00. In the case of acetic-NaOH titration, the pH at the equivalence point is about 8.72.

For weak acids with larger K_as, the pH at the equivalence point is closer to neutral pH; for those with smaller K_as, the pH at the equivalence point is much higher than neutral pH.

Indicators

An *indicator* is a substance that changes color to mark the *endpoint* of a titration. Most indicators used in an acid-base titration are weak organic acids. Indicators exhibit one color in the acid or protonated form (HIn) and another color in the base or deprotonated form (In^-).

Each indicator has a range of pH = pK_a ± 1, where the change of colors occurs. A suitable indicator gives the *endpoint* that corresponds to the equivalence point of the titration; this is one with a range of pH that falls within the sharp increase (or decrease) of pH in the titration curves.

Like a weak acid, an indicator has the following equilibrium in aqueous solution:

$HIn\ (aq) \leftrightarrows H^+\ (aq) + In^-\ (aq)$

$$K_a = \frac{[H^+] \cdot [In^-]}{[HIn]}$$

Rearranging obtains:

$[H^+] = K_a \times ([HIn] / [In^-])$

$pH = pK_a + \log([In^-] / [HIn]$

when

$[HIn] = 10 \times [In^-]$

$pH = pK_a + \log([In^-] / 10[In^-])$

$pH = pK_a - 1.0$; the indicator assumes the color of acid form.

when

$[In^-] = 10 \times [HIn]$

$pH = pK_a + \log(10[HIn] / [HIn])$

$pH = pKa + 1.0$; the indicator assumes the color of base form.

Phenolphthalein, which is the most common acid-base indicator, has $K_a \sim 10^{-9}$. Its acid form (HIn) is colorless, and the conjugate base form (In^-) is pink. It is colorless when the solution's $pH \leq 8$ when 90% or more of the species are in the acid form (HIn) and pink at $pH \geq 10$, when 90% or more of the species are in the conjugate base form (In^-).

The pH range at which an indicator change depends on the K_a. For phenolphthalein, which has $K_a \sim 10^{-9}$, its color changes in the pH range 8–10. It is a suitable indicator for strong acid-strong base titrations and weak acid-strong base titrations.

The pH ranges for other common acid-base indicators

Indicators	Acid color	Base color	pH Range	Type of Titrations
Methyl orange	orange	yellow	3.2–4.5	strong acid-strong base strong acid-weak base
Bromocresol green	yellow	blue	3.8–5.4	strong acid-strong strong acid-weak base
Methyl red	red	yellow	4.5–6.0	strong acid-strong base strong acid-weak base
Bromothymol blue	yellow	blue	6.0–7.6	strong acid-strong base
Phenol Red	orange	red	6.8–8.2	strong acid-strong base weak acid-strong base

Interpretation of titration curves

A *pH curve* is a graph of the pH of the solution *vs*. the volume of titrant in an acid-base titration.

The data used to plot a pH curve may be obtained either by computation or by measuring the pH directly with a pH meter during titration. The acid is incrementally added to the alkali (how titrations usually are performed).

Four types of pH curves for the different types of acid-base titrations

1. Strong acid-strong base titration

$$NaOH\ (aq) + HCl\ (aq) \rightarrow NaCl\ (aq) + H_2O\ (l)$$

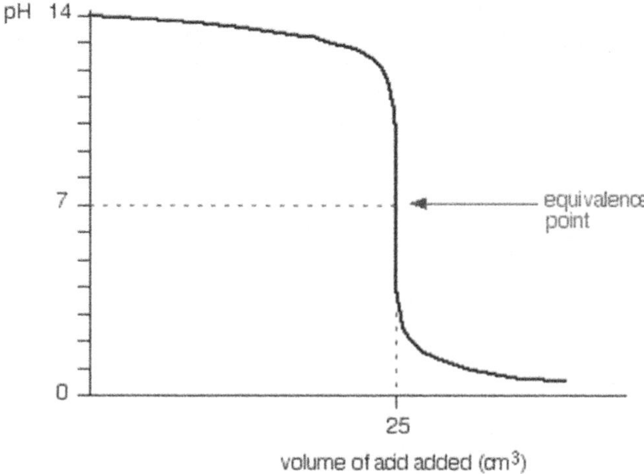

The pH falls only a minimal amount until near the equivalence point; then, there is a very steep drop.

2. Strong acid-weak base titration

$NH_3\,(aq) + HCl\,(aq) \rightarrow NH_4Cl\,(aq)$

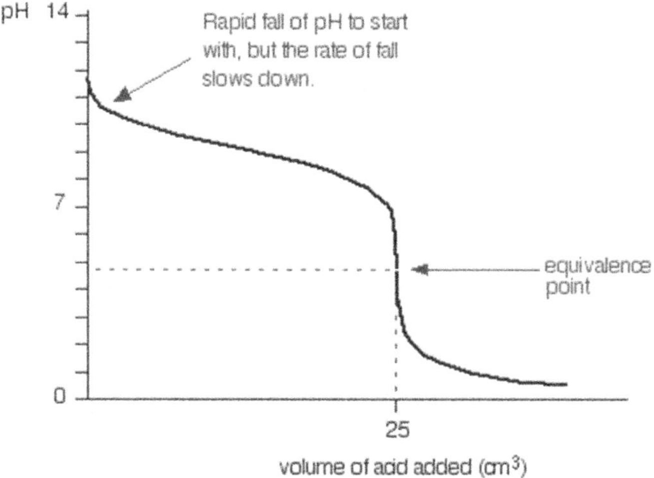

Since it is a weak base, rather than a strong base, the beginning of the titration curve is vastly different. The pH decreases steeply as the acid is added, but the curve soon becomes less steep. This is because a buffer solution, composed of excess ammonia and formed ammonium chloride, is being created. (The pH of the buffering region is usually $14 - pK_b \pm 1$).

The equivalence point is now acidic (pH ~ 5) but is still on the steepest part of the curve.

3. Weak acid-strong base titration

$$CH_3COOH\ (aq) + NaOH\ (aq) \rightarrow CH_3COONa\ (aq) + H_2O\ (aq)$$

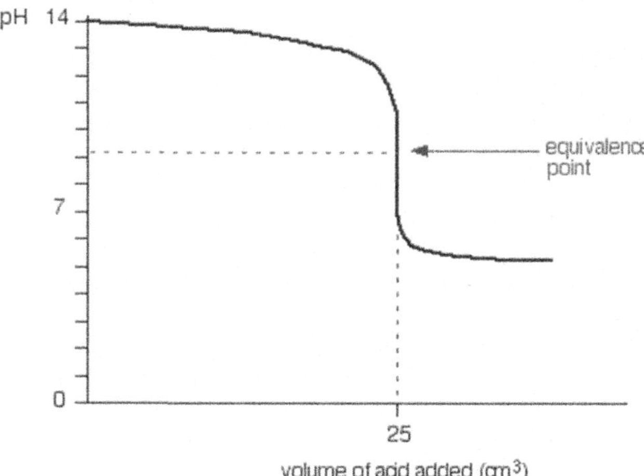

In the first part of the graph, there is an excess of sodium hydroxide, and this part of the curve is like the strong acid-strong base titration.

However, once the acid is in excess, there is a difference, due to the formation of a buffer solution that contains sodium ethanoate and ethanoic acid.

The pH of the buffering region is usually $pK_a \pm 1$, which resists changes in pH before the smooth (horizontal) portion of the curve is reached.

4. Weak acid-weak base titration.

The following titration curve is for the reaction:

$$CH_3COOH\ (aq) + NH_3\ (aq) \rightarrow CH_3COONH_4\ (aq)$$

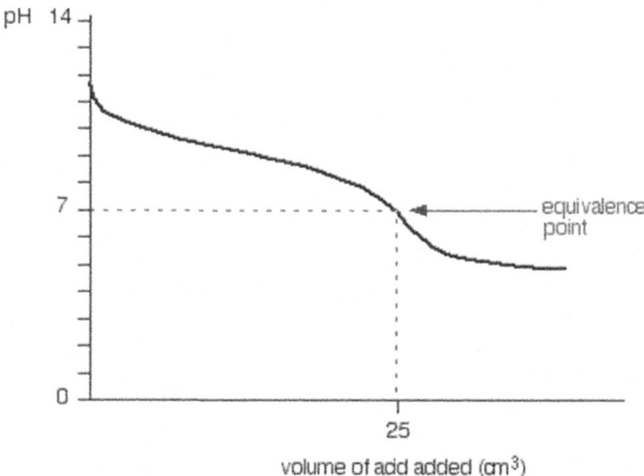

The acid and the base are equally weak, so the equivalence point is around pH 7. This titration curve is essentially a combination of the two previous graphs.

Before the equivalence point, it is like a strong acid-weak base case.

After the equivalence point, it is like the end of the weak acid-strong base case. There is no steep portion of this graph; the lack of a steep portion is an important identifying factor of a weak acid-weak base titration curve.

Four titration curves for comparison

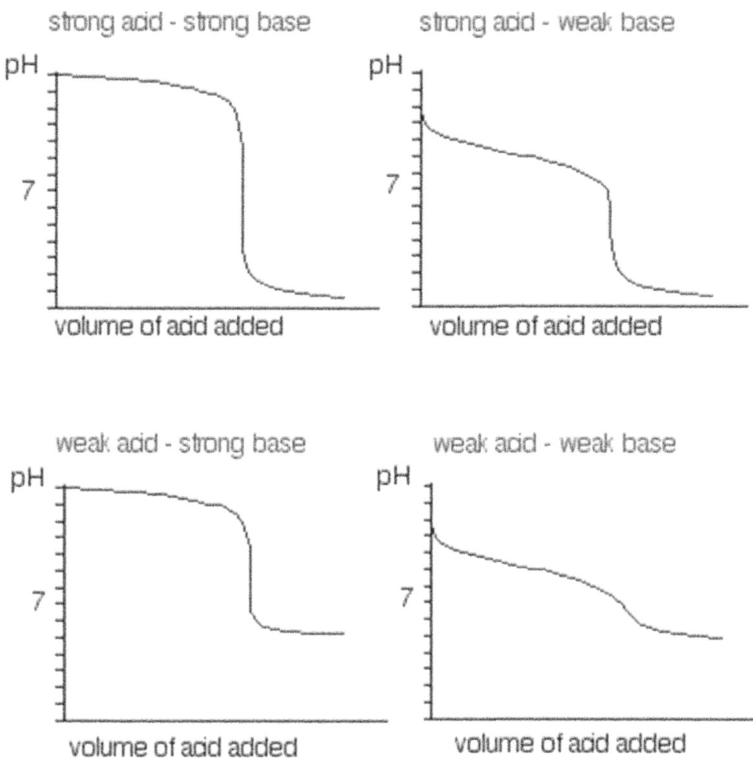

Redox titration

A redox titration is a type of titration based on the transfer of electrons. In redox titration, standard solutions of oxidizing agents are usually used because solutions of reducing agents may react with oxygen in the air. The species being analyzed must be in a single oxidation state before titration.

For example, when iron ore is dissolved in hydrochloric acid solution, both Fe^{2+} and Fe^{3+}, known as iron(II) and iron(III), are present in solution. However, since it must be in a single oxidation state before titration with a standard $KMnO_4$ solution, iron(III) is reduced to iron(II) by reaction with excess zinc:

$$2\ Fe^{3+}\ (aq) + Zn\ (s) \rightarrow 2\ Fe^{2+}\ (aq) + Zn^{2+}\ (aq)$$

$$MnO_4^-\ (aq) + 5\ Fe^{2+}\ (aq) + 8\ H^+\ (aq) \rightarrow Mn^{2+}\ (aq) + 5\ Fe^{3+}\ (aq) + 4\ H_2O\ (l)$$

$KMnO_4$ and $K_2Cr_2O_7$ are the most used oxidizing agents in redox titration because the solutions change color during reduction; this color change serves as the titration indicator. $KMnO_4$ is purple but becomes colorless when MnO_4^- is reduced to Mn^{2+}. The bright orange color of $K_2Cr_2O_7$ changes to purplish-blue when $Cr_2O_7^{2-}$ is reduced to Cr^{3+}.

An example problem of redox titration.

A solution contains both iron(II) and iron(III) ions. A 50.0 mL sample of the solution is titrated with 35.0 mL of 0.00280 M $KMnO_4$, which oxidizes Fe^{2+} to Fe^{3+}. The permanganate ion is reduced to manganese(II) ion. Another 50.0 mL sample of the solution is treated with zinc metal, which reduces all the Fe^{3+} to Fe^{2+}. The resulting solution is again titrated with 0.00280 M $KMnO_4$. Therefore, 48.0 mL is required. What are the concentrations of Fe^{2+} and Fe^{3+} in the solution?

For this problem, several steps are required.

1) The stoichiometric relationship of Fe(II) to permanganate is five to one:

   $$MnO_4^-\ (aq) + 5\ Fe^{2+}\ (aq) + 8\ H^+\ (aq) \rightarrow Mn^{2+}\ (aq) + 5\ Fe^{3+}\ (aq) + 4\ H_2O\ (l)$$

2) Calculate moles of Fe(II) reacted:

   $(0.00280\ \text{mol/L})\ (0.0350\ \text{L}) = 0.000098\ \text{mol of}\ MnO_4^-$

   $(0.0000980\ \text{mol Mn})\cdot(5\ \text{mol Fe}\ /\ 1\ \text{mol Mn}) = 0.000490\ \text{mol Fe(II)}$

3) Determine the total iron content:

   $(0.00280\ \text{mol/L})\cdot(0.0480\ \text{L}) = 0.0001344\ \text{mol of}\ MnO_4^-$

   $(0.0000980\ \text{mol Mn})\cdot(5\ \text{mol Fe}\ /\ 1\ \text{mol Mn}) = 0.000672\ \text{mol total Fe}$

4) Determine Fe(III) in solution and its molarity:

   $0.000672\ \text{mol} - 0.000490\ \text{mol} = 0.000182\ \text{mol}$

   $(0.000182\ \text{mol})\ /\ (0.050\ \text{L}) = 0.00364\ \text{M Fe(III)}$

5) Determine molarity of Fe(II):

   $(0.000490\ \text{mol})\ /\ (0.050\ \text{L}) = 0.0098\ \text{M Fe(II)}$

Notes

Practice Questions

1. Which of the following indicators is green at pH 7?

I. phenolphthalein II. bromothymol blue III. methyl red

A. I only
B. II only
C. III only
D. I and II only
E. I, II and III

2. Which of the following is NOT a strong acid?

A. $HClO_3$ **B.** HF **C.** HBr **D.** HCl **E.** HI

3. Which of the following compounds is the strongest base?

A. ClO_3^- **B.** NH_3 **C.** ClO^- **D.** ClO_2^- **E.** ClO_4^-

4. A Brønsted-Lowry acid is defined as a substance that:

A. acts as a proton donor in any system
B. acts as a proton acceptor in any system
C. increases $[H^+]$ when placed in water
D. decreases $[H^+]$ when placed in water
E. increases $[H^+]$ in any system

5. Which of the following acts as the best buffer solution?

A. Strong acids or bases
B. Strong acids and their salts
C. Salts
D. Weak acids or bases and their salts
E. Strong bases and their salts

6. Which of the following is true if a drain cleaner solution is a strong electrolyte?

I. Slightly reactive II. Highly reactive III. Highly ionized

A. I only
B. II only
C. III only
D. I and III only
E. II and III only

7. Which of the following statements describes a basic solution?

A. $[H_3O^+] \times [^-OH] \neq 1 \times 10^{-14}$
B. $[H_3O^+] / [^-OH] = 1 \times 10^{-14}$
C. $[H_3O^+] > [^-OH]$
D. $[H_3O^+] < [^-OH]$
E. $[H_3O^+] / [^-OH] = 1$

8. Which of the following is a general property of an acidic solution?

I. Neutralizes bases II. Tastes sour III. Turns litmus paper red

A. I only
B. II only
C. III only
D. I and III only
E. I, II and III

9. Compared to a solution with a higher pH, a solution with a lower pH has a(n):

A. decreased K_a
B. increased $[^-OH]$
C. increased pK_a
D. increased $[H^+]$
E. decreased $[H^+]$

10. Which of the following statements is a correct definition for an Arrhenius acid?

A. Decreases $[H^+]$ when placed in aqueous solutions
B. Increases $[H^+]$ when placed in aqueous solutions
C. Acts as a proton acceptor in any system
D. Acts as a proton donor in any system
E. Acts as a proton acceptor in aqueous solutions

11. A Brønsted-Lowry base is a(n):

A. electron acceptor
B. proton acceptor
C. electron donor
D. proton donor
E. both proton donor and electron acceptor

12. Which of the following is the conjugate base of water?

A. ^-OH (*aq*)
B. H^+ (*aq*)
C. H_2O (*l*)
D. H_3O^+ (*aq*)
E. O^{2-} (*aq*)

13. Which of the following explains why distilled H_2O is neutral?

A. $[H^+] = [OH^-]$
B. Distilled H_2O has no OH^-
C. Distilled H_2O has no H^+
D. Distilled H_2O has no ions
E. None of the above

14. In the reaction below, what does the symbol ⇌ indicate?

$^{-}OH + NH_4^+ \rightleftharpoons H_2O + NH_3$

A. The rate of the reverse reaction is the same as the forward reaction
B. The forward reaction does not proceed
C. The reaction does not produce an equilibrium
D. The reverse reaction does not proceed
E. The equilibrium depends on the concentrations of the reactants

15. What is the $[H_3O^+]$ of a solution that has a pH = 2.34?

A. 1.3×10^1 M
B. 2.3×10^{-10} M
C. 4.6×10^{-3} M
D. 2.4×10^{-3} M
E. 3.6×10^{-8} M

Detailed Explanations

1. B is correct.

Bromothymol blue is a pH indicator that is often used for solutions with neutral pH near 7 (e.g., managing the pH of pools and fish tanks).

Bromothymol blue acts as a weak acid in a solution that can be protonated or deprotonated.

It appears yellow when protonated (lower pH), blue when deprotonated (higher pH), and bluish-green in neutral solution.

Phenolphthalein is used as an indicator of acid-base titrations. It is a weak acid, which can dissociate protons (H^+ ions) in solution.

The phenolphthalein molecule is colorless, and the phenolphthalein ion is pink. It turns colorless in acidic solutions and pink in basic solutions.

With basic conditions, the phenolphthalein (neutral) ⇌ ions (pink) equilibrium shifts to the right, leading to more ionization as H^+ ions are removed.

Methyl red has a pK_a of 5.1 and is a pH indicator dye that changes color in acidic solutions: turns red in pH under 4.4, orange in pH 4.4-6.2, and yellow in pH over 6.2.

2. B is correct.

Strong acids dissociate a proton to produce the weakest conjugate base (i.e., most stable anion).

Weak acids dissociate a proton to produce the strongest conjugate base (i.e., least stable anion).

Hydrofluoric (HF) acid has a pK_a of about 3.8 and is a weak acid because it does not dissociate completely.

The F^- anion is the least stable of the halogen anions (due to its small valence shell).

3. C is correct.

Perchloric acid ($HClO_4$) is the strongest acid listed and therefore is the weakest conjugate base (i.e., most stable anion).

Hypochlorite (ClO^-) is the strongest base (i.e., least stable anion).

In oxyacids, the more oxygen present, the greater is the acid strength.

The weakest conjugate bases of oxyacids (containing more than one oxygen) are stabilized by resonance.

4. A is correct.

The Brønsted-Lowry acid-base theory focuses on the ability to accept and donate protons (H^+).

A Brønsted-Lowry acid is a term for a substance that donates a proton in an acid-base reaction, while a Brønsted-Lowry base is a term for the substance that accepts the proton.

5. D is correct.

A buffer is an aqueous solution that consists of a weak acid and its conjugate base, or vice versa.

Buffered solutions resist changes in pH and are often used to keep the pH at a nearly constant value in many chemical applications. It does this by readily absorbing or releasing protons (H^+) and ^-OH.

When acid is added to the solution, the buffer releases ^-OH and accepts H^+ ions from the acid.

Weak acids or bases and their salts are the best buffer systems.

6. C is correct.

An electrolyte is a substance that dissociates into cations (i.e., positive ions) and anions (i.e., negative ions) when placed in solution.

7. D is correct.

In a basic solution, the concentration of ^-OH is higher than H^+ or H_3O^+.

8. E is correct.

An acid can neutralize a base (i.e., a substance with a pH above 7) to form a salt.

Acids have a sour taste (e.g., lemon juice has a pH of about 2).

An acid is a chemical substance with a pH of less than 7.

Litmus paper is red under acidic conditions and blue under basic conditions.

Acids are known to have a sour taste (e.g., lemon juice) because the sour taste receptors on the tongue detect the dissolved hydrogen (H^+) ions.

A pH greater than 7 and feels slippery are all qualities of bases, not acids. An acid is a chemical substance with a pH of less than 7, which produces H^+ ions in water. An acid can be neutralized by a base (i.e., a substance with a pH above 7) to form a salt.

Litmus paper is red under acidic conditions and blue under basic conditions.

However, acids are not known to have a slippery feel; this is a characteristic of bases. Bases feel slippery because they dissolve the fatty acids and oils from the skin and therefore reduce the friction between the skin cells.

9. D is correct.

Acidic solutions contain hydronium ions (H_3O^+). These ions are in the aqueous form because they are dissolved in water.

Although chemists often write H^+ (*aq*), referring to a single hydrogen nucleus (a proton), it exists as the hydrogen atom.

10. B is correct.

An Arrhenius acid increases the concentration of H^+ ions in an aqueous solution.

Since pH is the negative log of the activity of H^+ ions in an aqueous solution, therefore acids have a low pH.

An acid can act as a proton donor, but this is the Brønsted-Lowry definition of an acid, not Arrhenius.

11. B is correct.

Brønsted-Lowry acids are proton donors (e.g., HCl, H_2SO_4).

Brønsted-Lowry bases are proton acceptors (e.g., HSO_4 , NO_3).

Lewis acids are electron-pair acceptors, whereas Lewis bases are electron-pair donors.

12. A is correct.

By the Brønsted-Lowry acid-base theory:

An acid (reactant) dissociates a proton to become the conjugate base (product).

A base (reactant) gains a proton to become the conjugate acid (product).

The definition is expressed in terms of an equilibrium expression

acid + base ↔ conjugate base + conjugate acid.

13. A is correct.

If the $[H_3O^+] = [^-OH]$, it has a pH of 7, and the solution is neutral.

14. A is correct.

The symbol ⇌ indicates that the reaction is in equilibrium, whereby the rate of the forward reaction is equal to the rate of the reverse reaction.

It refers to the rate and not the relative magnitude of the remaining reactants nor the formed products during the reaction.

15. C is correct.

The formula for pH is:

$$pH = -\log[H_3O^+]$$

Rearrange to solve for $[H_3O^+]$:

$$[H_3O^+] = 10^{-pH}$$

$$[H_3O^+] = 10^{-2.34}$$

$$[H_3O^+] = 4.6 \times 10^{-3}\ M$$

Chapter 9

Electrochemistry

Electrochemistry specifies the interconversion of electrical energy and chemical energy. Electrochemistry utilizes spontaneous oxidation-reduction reactions to determine the electrical energy to drive nonspontaneous reactions. Electrochemistry always involves an oxidation-reduction process. Many critical applications, such as batteries, control of corrosion, metallurgy, and electrolysis use electrochemistry.

Electrolytic Cell

- **Anode, Cathode**
- **Electrolysis**
- **Electrolytes**
- **Faraday's Law Relating Amount of Elements Deposited (or Gas Liberated) at an Electrode to Current**
- **Electron Flow; Oxidation and Reduction at the Electrodes**

Galvanic (Voltaic) Cell

- **Half-Reactions**
- **Reduction Potentials; Cell Potential**
- **Direction of Electron Flow**
- **Concentration Cell**
- **Batteries**

Electrolytic Cell

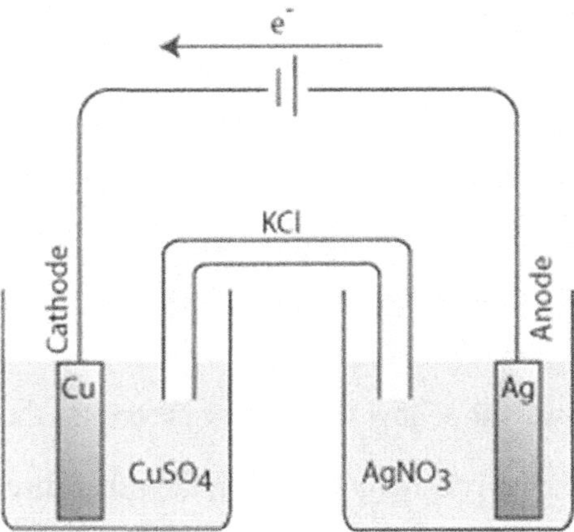

Electrolytic Cell

Anode, cathode

For an electrochemical cell, oxidation always occurs at the *anode,* and reduction occurs at the *cathode*.

Designation of Anode and Cathode

Write a balanced net ionic equation for the spontaneous cell reaction.

- The oxidizing agent (one with the more positive or less negative reduction potential E°) is the cathode, and the other is the anode.

- Oxidation occurs in the anode half-cell and reduction in the cathode half-cell.

- The anode is negative (-), and the cathode is positive (+).

Mnemonic: An Ox = ANode Oxidation Red Cat = REDuction CAThode

When drawing a galvanic cell, by convention, the anode is on the left, and the cathode is on the right. Use the mnemonic **ABC** to remember this convention (**A**node / **B**ridge / **C**athode).

Electrolysis

Electrolysis requires potential (voltage) input. On the diagram, this is represented by a battery in the circuit.

A galvanic (voltaic) cell has either a resistor or a voltmeter in place of a battery.

The potential (voltage) input + the cell potential must be > 0 for reactions to occur.

For electrolytic cells, the cell potential is negative, so a potential input greater than the magnitude of the cell potential must be present for electrolysis to occur.

Galvanic (voltaic) cells already have a positive cell potential. Thus, no input is required for galvanic (voltaic) cells.

In the diagram, arrows show how the battery is forcing the flow of electrons.

Without energy, the electrons would flow in the other direction (or not at all).

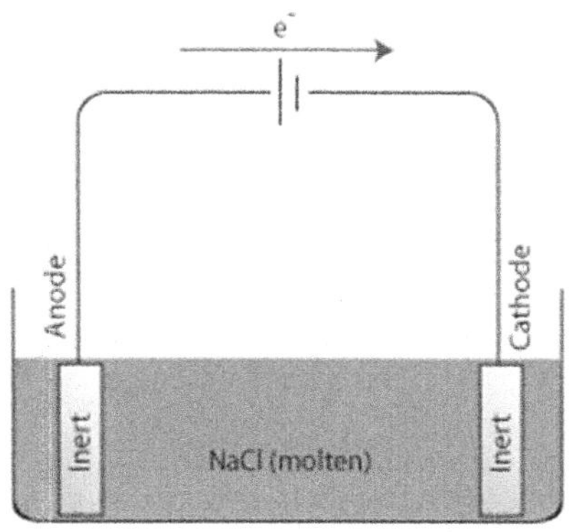

Electrolysis of Salt

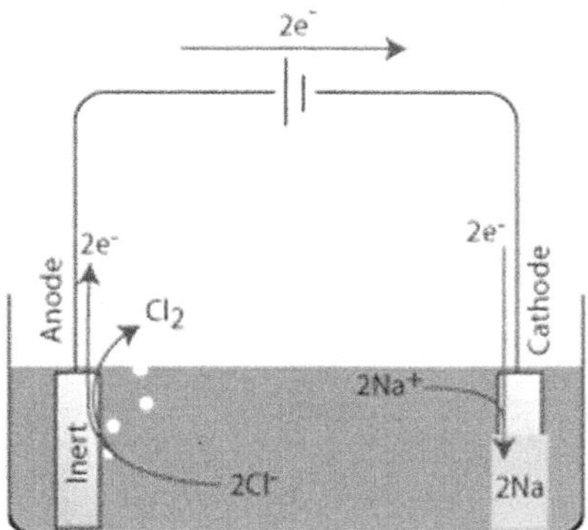

Salt Electrolysis Charge Flow

The electrolysis process is used to extract metals from their molten salt form. A current is applied to the molten salt for some time.

Current is measured in amperes (A).

1 ampere = 1 coulomb / 1 second

A galvanic (voltaic) cell produces current from a spontaneous oxidation-reaction.

An *electrolytic cell* uses electrical energy to drive nonspontaneous oxidation-reduction reactions.

Electrolysis involves forcing a current through a cell to produce a chemical change, which has a negative cell potential.

For example, the following spontaneous reaction occurs in a galvanic (voltaic) cell that consists of $Zn|Zn^{2+}$ (*aq*) and $Cu|Cu^{2+}$ (*aq*) half-cells:

$$Zn\ (s) + Cu^{2+}\ (aq) \rightarrow Zn^{2+}\ (aq) + Cu\ (s); \qquad E^{\circ}_{cell} = 1.10\ V$$

In an electrolytic cell, the reverse reaction occurs if a voltage greater than the E° cell is applied. A higher voltage than E° is needed to cause the reverse reaction. The excess voltage is referred to as *overpotential* or overvoltage.

Three types of reactions occur in electrolytic cells if sufficient voltage is applied:

- Solute ions or molecules may be oxidized or reduced.
- The solvent can be oxidized or reduced.
- Metal electrode that forms the anode can be oxidized.

Which of these reactions will occur depends both on the thermodynamic and kinetic properties of each reaction. In general, the half-reactions with the most positive (or the least negative) reduction potential occurs before others. The kinetic factor drives the reaction if the difference in the standard reduction potentials between two half-cell reactions is small, the concentration may determine the outcome of electrolysis.

The following reactions are possible for the electrolysis of 1 M NaCl (*aq*):

At anode:	$2\ Cl^-(aq) \rightarrow Cl_2(g) + 2\ e^-$	E° = −1.36 V
	$2\ H_2O \rightarrow O_2(g) + 4\ H^+(aq) + 4\ e^-$	E° = −1.23 V
At cathode:	$2\ H_2O + 2\ e^- \rightarrow H_2(g) + 2\ OH^-(aq)$	E° = −0.83 V
	$Na^+(aq) + e^- \rightarrow Na(s)$	E° = −2.71 V

In electrolysis of aqueous sodium chloride, Cl_2 is formed at the anode instead of O_2, although E° for the formation of O_2 from H_2O is less negative than for the formation of Cl_2 from Cl^-. This is because the formation of O_2 involves higher activation energy (overvoltage). Thus it is kinetically less favorable. At the cathode, H_2 is formed since it requires less voltage than the reduction of $Na^+(aq)$.

If a solution containing metal cations that require lower potential (or one that has a positive reduction potential) is electrolyzed, the metal is deposited on the cathode, and no hydrogen is produced.

The electrolysis of aqueous copper(II) chloride solution yields Cl_2 gas at the anode and copper metal deposits at the cathode. If the anode is made of copper, it is oxidized because of the lower voltage requirement relative to water.

Anode reaction:	$Cu(s) \rightarrow Cu^{2+}(aq) + 2\ e^-$	E° = −0.34 V
Cathode reaction:	$Cu^{2+}(aq) + 2\ e^- \rightarrow Cu(s)$	E° = +0.34 V

Electrolysis of Water

The decomposition of water is a nonspontaneous process that has an amount of energy of about 400 kJ/mol of water. However, the decomposition of water can be affected by electrolysis, where oxygen is formed at the anode and hydrogen at the cathode:

Anode reaction: $2\ H_2O \rightarrow O_2 + 4\ H^+(aq) + 4\ e^-$ $E° = -1.23$ V

Cathode reaction: $4\ H_2O + 4\ e^- \rightarrow 2\ H_2 + 4\ OH^-(aq)$ $E° = -0.83$ V

Net reaction: $6\ H_2O \rightarrow 2\ H_2 + O_2 + 4\ (H^+ + OH^-)$ $E° = -2.06$ V

$2\ H_2O \rightarrow 2\ H_2 + O_2$

Assumes an electrolytic cell with $[H^+] = [OH^-] = 1$ M and $P_{H2} = P_{O2} = 1$ atm.

For the electrolysis of pure water, where $[H^+] = [OH^-] = 10^{-7}$ M, the potential for the overall process is –1.23 V. A voltage of 1.23 V is enough to affect the electrolysis of water. An extra voltage of about 1 V (as overpotential or overvoltage) is needed to force the electrolysis.

Electrolysis of Solution Containing Mixtures of Ions

Consider a solution containing Cu^{2+}, Zn^{2+} and Ag^+ that is electrolyzed using a current with sufficient voltage to reduce all three cations.

In such electrolysis, the metal with the smallest potential is the first to be formed.

The reduction potentials of these elements are as follows:

$Ag^+(aq) + e^- \rightarrow Ag\ (s)$ $E° = 0.80$ V

$Cu^{2+}(aq) + 2\ e^- \rightarrow Cu\ (s)$ $E° = 0.34$ V

$Zn^{2+}(aq) + 2\ e^- \rightarrow Zn\ (s)$ $E° = -0.76$ V

Since the reduction of Ag^+ to Ag has the most positive potential, Ag deposits at the cathode before the other metals, which is then followed by Cu and Zn, respectively.

If the voltage supply is appropriately controlled, starting with the lowest voltage possible, it is possible to separate the three metals using electrolysis.

The Stoichiometry of Electrolysis – Total Charge and Theoretical Yield

An electric current that flows through a cell is measured in *ampere* (A). Ampere is the amount of charge in *coulomb* (C) that flows through the circuit per second.

Ampere = Coulomb/second (1 A = 1 C/s)

Coulomb = Ampere × time in seconds (1 C = 1 A·s) = total charge

Joule = Coulomb × Volt (1 J = 1 C.V)

The number of substances formed at the anode or cathode is calculated from the magnitude of the current (in amperes) and time (in seconds) of electrolysis.

If a current of 1.50 A flows through an aqueous solution of $CuSO_4$ for 25.0 minutes, the amount of charge passing through the solution is:

(1.50 C/s)·(25.0 minutes)·(60 seconds/minutes) = 2,250 C

The number of moles of electrons passing through the solution:

$(2{,}250\ C) \times (1\ mol\ e^-) = 0.0233\ mol$

Since the reduction of Cu^{2+} requires 2 mol e^- per mole of Cu:

$Cu^{2+} + 2\ e^- \rightarrow Cu$, the amount of copper formed at the cathode is:

$$(0.0233\ mol\ e^-) \times \frac{(1\ mol\ Cu)}{(2\ mol\ e^-)} \times \frac{(63.55\ g\ Cu)}{(1\ mol\ Cu)} = 0.741\ g\ Cu$$

If the above process uses a cell with a voltage of 3.0 V, the energy consumed is:

3.0 V × 2250 C = 6,750 C.V = 6,800 J = 6.8 kJ

Electrolytes

Ions = electrolyte.

Electrolytes conduct electricity by the motion of ions.

Without electrolytes, there is no circuit when electricity is not able to propagate.

Faraday's Law relating the number of elements deposited (or gas liberated) at an electrode to current

Current I = coulombs of charge per second

$$I = q / t$$

Faraday's constant F = coulombs of charge per mol of electron

$$F = \text{total charge / the total moles of electrons}$$

$$F = q / n$$

$$q = It \text{ and } q = \text{nF}$$

thus

$$It = nF$$

Current × time = moles of e^- × Faraday's constant.

Using this equation, solve for moles of electrons (n). Then, using the half-equation stoichiometry, determine how many moles of the element is made for every e^- transferred.

For example, 1 mol of Cu is deposited for every 2 moles of electrons for the following half-reaction, use Faraday's constant = 96,485 C/mo.

$Cu^{2+} + 2\ e^- \rightarrow Cu$

Electron flow; oxidation and reduction at the electrodes

Electrons originate from the anode because oxidation (loss of electrons) occurs.

$$M \rightarrow M^+ + e^-$$

Electrons travel into the cathode, where they join the cations on the surface of the cathode. Reduction (gain of electrons) occurs at the cathode.

$$M^+ + e^- \rightarrow M$$

Mnemonic: OIL RIG: **O**xidation **I**s **L**osing e^-, **R**eduction **I**s **G**aining e^-.

Oxidation is an increase in charge (more positive) due to the loss of e^-.

Reduction is a decrease in charge (more negative) due to the gain of e^-.

Galvanic (Voltaic) Cell

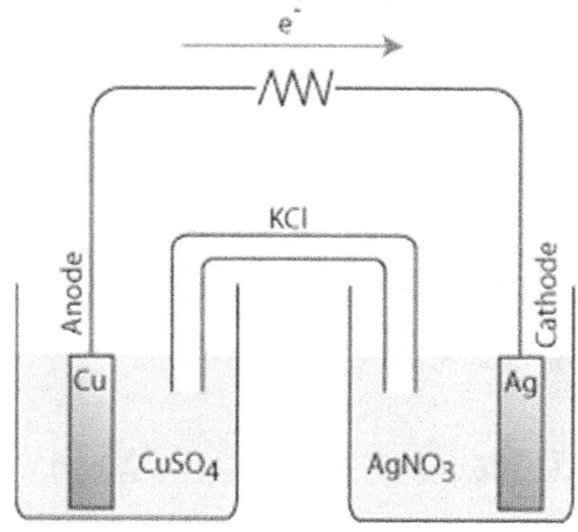

Galvanic (Voltaic) Cell

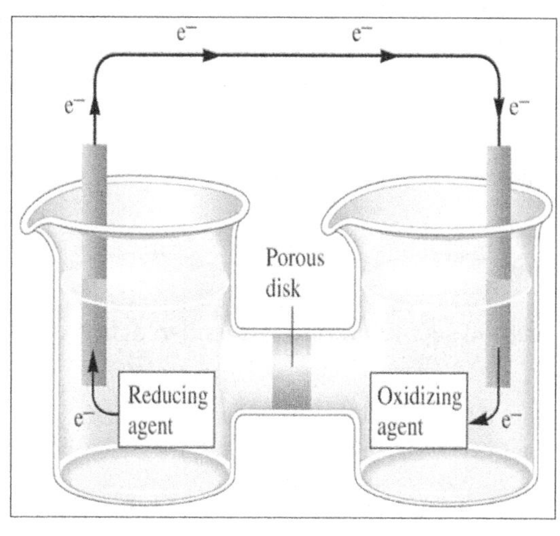

anode *cathode*

A *galvanic cell* is a common application of electrochemistry as for batteries. A galvanic cell has the capability of producing a voltage and was invented by two Italian physicists Luigi Aloisio Galvani (1737-1798) and Alessandro Volta (1745-1827).

A galvanic cell (voltaic cell) uses a spontaneous oxidation-reduction reaction to produce an electric current.

In Galvanic Cell:

- Oxidation occurs at the anode.
- Reduction occurs at the cathode.

- Salt bridges or porous disks allows ions to flow without extensive mixing of the solutions.
  - *Salt bridge* contains a strong electrolyte held in a gel-like matrix.
  - *Porous disk* contains tiny passages for the hindered flow of ions.

The following are some examples of spontaneous oxidation-reduction reactions:

1. $Zn\ (s) + CuSO_4\ (aq) \rightarrow Cu\ (s) + ZnSO_4\ (aq)$

   $Zn\ (s) + Cu\cancel{SO_4}\ (aq) \rightarrow Cu\ (s) + Zn\cancel{SO_4}\ (aq)$

   $Zn\ (s) + Cu^{2+}\ (aq) \rightarrow Cu\ (s) + Zn^{2+}\ (aq)$

2. $Cr\ (s) + 3\ AgNO_3\ (aq) \rightarrow Cr(NO_3)_3\ (aq) + 3\ Ag\ (s)$

   $Cr\ (s) + 3\ Ag\cancel{NO_3}\ (aq) \rightarrow Cr\cancel{(NO_3)_3}\ (aq) + 3\ Ag\ (s)$

   $Cr\ (s) + 3\ Ag^{+}\ (aq) \rightarrow Cr^{3+}\ (aq) + 3\ Ag\ (s)$

3. $MnO_4^{-}\ (aq) + 5\ Fe^{2+}\ (aq) + 8\ H^{+}\ (aq) \rightarrow Mn^{2+}\ (aq) + 5\ Fe^{3+}\ (aq) + 4\ H_2O\ (l)$

4. $Zn\ (s) + 2\ MnO_2\ (s) \rightarrow ZnO\ (s) + Mn_2O_3\ (s)$

Half-reactions

Redox reactions can be separated into two half-reactions:

Example 1:

Oxidation half-reaction: $Zn\ (s) \rightarrow Zn^{2+}\ (aq) + 2\ e^{-}$

Reduction half-reaction: $Cu^{2+}\ (aq) + 2\ e^{-} \rightarrow Cu\ (s)$

Overall reaction: $Zn\ (s) + Cu^{2+}\ (aq) \rightarrow Cu\ (s) + Zn^{2+}\ (aq)$

Example 2:

Oxidation half-reaction: $5\ Fe^{2+}\ (aq) \rightarrow 5\ Fe^{3+}\ (aq) + 5\ e^{-}$

Reduction half-reaction: $MnO_4^{-}\ (aq) + 8\ H^{+}\ (aq) + 5\ e^{-} \rightarrow Mn^{2+}\ (aq) + 4\ H_2O\ (l)$

Overall reaction:

$$MnO_4^{-}\ (aq) + 5\ Fe^{2+}\ (aq) + 8\ H^{+}\ (aq) \rightarrow Mn^{2+}\ (aq) + 5\ Fe^{3+}\ (aq) + 4\ H_2O\ (l)$$

Oxidation half-reaction describes the species that loses electrons (increases charge).

For example, $Cu \rightarrow Cu^{2+} + 2\ e^-$

Reduction half-reaction describes the species that gains electrons (decreases charge).

For example, $2\ Ag^+ + 2\ e^- \rightarrow 2\ Ag$

In galvanic cells, redox reactions are split into two half-reactions, each occurring in two separate compartments as *half-cells*. The chemical energy drives electrons through the external circuit connecting the two half-cells, thus producing an electric current.

Consider the following reaction:

$$Zn\ (s) + CuSO_4\ (aq) \rightarrow Cu\ (s) + ZnSO_4\ (aq)$$

If zinc metal is placed in the $CuSO_4$ solution, an exothermic reaction occurs, producing heat. In a galvanic cell, zinc metal is placed in a $ZnSO_4$ solution in one container, a copper metal in $CuSO_4$ solution in another container. Two metals are connected by a wire to complete the circuit; the two solutions are connected by a salt-bridge containing strong electrolytes such as KCl (*aq*) or K_2SO_4 (*aq*) that allow ions to flow between the two half-cells. The reactants—zinc metal and copper ions—are not allowed direct contact.

Due to the potential difference between the two half-cells, electrons are forced to flow from the zinc electrode (the anode) to the copper electrode (the cathode). At the interface between the zinc electrode and $ZnSO_4$ solution, Zn atoms are oxidized to Zn^{2+} ions; at the interface between the copper electrode and $CuSO_4$ solution, Cu^{2+} ions combine with incoming electrons and are reduced to Cu atoms.

These oxidation and reduction processes are shown in half-reaction equations:

Anode-reaction (oxidation):

$$Zn\ (s) \rightarrow Zn^{2+}\ (aq) + 2\ e^-$$

Cathode-reaction (reduction):

$$Cu^{2+}\ (aq) + 2\ e^- \rightarrow Cu\ (s)$$

The two half-cells are represented by notations $Zn|Zn^{2+}$ and $Cu^{2+}|Cu$.

In a galvanic cell, the metal that is more readily oxidized serves as an anode and the other as a cathode. Determine the tendency of a species to be oxidized from the table of standard reduction potentials, shown below.

Zinc is more easily oxidized than copper, so it serves as the anode, while copper forms the cathode. Oxidation half-reaction occurs at anode, and reduction half-reaction at the cathode. This galvanic can be represented using the following cell notation:

$Zn|ZnSO_4\,(aq)||CuSO_4\,(aq)|Cu$

A cell notation is written with the anode on the left side and cathode on the right.

At the anode half-cell, Zn^{2+} ions are continuously formed, creating an excess of positive ions.

At the cathode half-cell, Cu^{2+} ions are continuously reduced to Cu, causing a decrease in cation concentration.

To maintain electrically neutral solutions in both half-cells, anions flow into the anode half-cell, and cations flow into the cathode half-cell. Thus, an electric current is the flow of charged particles. Electrons flow from anode to cathode through the wire; cations and anions flow in the opposite direction through the salt-bridge.

Reduction potentials; cell potential

Reduction potential = potential of the reduction half-reaction

Oxidation potential = potential of the oxidation half-reaction
= reverse the sign of the reduction potential

Cell potential = reduction potential + oxidation potential

Standard Reduction Potential

In a galvanic cell, electrons flow from one electrode to the other because there is an electrical potential difference between the two half-cells, called the *cell potential.*

In the zinc-copper cell, electrons flow from $Zn|Zn^{2+}$ half-cell to $Cu|Cu^{2+}$ half-cell. The $Zn|Z^{2+}$ half-cell has a higher electrical potential than $Cu|Cu^{2+}$.

The magnitude of cell potential depends on the nature of the two half-cells, the concentration of the electrolyte in each half-cell, and temperature.

The *standard cell potential*, $E°_{cell}$, is the cell potential measured under standard conditions (1 atm pressure for gas, 1 M of electrolytes, and at 25 °C). The parameters for a standard cell require the species in solution to have a concentration of 1 M, the partial pressure of gases to be 1 atm, and the temperature to be 25 °C (298 K). Using these conditions makes it possible to describe a standard electrode potential (E°) in volts and calculate the electrical potential of the cell. The superscript ° denotes standard conditions.

The *standard half-cell potential* or *the reduction potential* of a substance is determined by connecting the half-cell of the substance (under standard conditions) to the *standard hydrogen electrode* (SHE) as reference half-cell. This reference half-cell consists of a Pt-electrode in a solution containing 1 M H^+, into which H_2 gas is purged at a constant pressure of 1 atm. Since the reference half-cell is assigned zero potential, the cell potential measured under these conditions is the standard half-cell potential of the substance.

When a $Zn|Zn^{2+}$(*aq*, 1 M) half-cell is connected to the reference half-cell (SHE), the voltage measured at 25 °C is found to be 0.763 V. Electron flows from $Zn|Zn^{2+}$ to SHE; relative to SHE, the reduction potential for $Zn|Zn^{2+}$ half-cell is the negative value of the measured voltage. (where E° is the reduction potential)

$Zn^{2+}(aq) + 2\ e^- \rightarrow Zn\ (s)$ E° = –076 V

$2\ H^+(aq) + 2\ e^- \rightarrow H_2\ (s)$ E° = 0.000 V

$Cu^{2+}(aq) + 2\ e^- \rightarrow Cu\ (s)$ E° = 0.34 V

Since values are obtained against the standard hydrogen potential, species with positive reduction potentials are easier to reduce compared to H^+ ions. In contrast, those with negative reduction potentials are more difficult to reduce.

Therefore, relative to H^+, Cu^{2+} is easier to reduce, whereas Zn^{2+} is more challenging to reduce. When $Zn|Zn^{2+}$ is connected to $Cu|Cu^{2+}$ half-cells, the spontaneous process is the flow of electrons from $Zn|Zn^{2+}$ to $Cu|Cu^{2+}$ half-cell.

In galvanic cells, species with more positive reduction potential serves as a cathode, and those with less positive or more negative reduction potential serve as the anode half-cell.

The *net cell potential* is the sum of the two half-cell potentials.

For the Zn-copper cell, the two half-cell potentials:

Anode half-cell reaction: $Zn\ (s) \rightarrow Zn^{2+}(aq) + 2\ e^-$ $E°_{Zn \rightarrow Zn2+} = 0.76$ V

Cathode half-cell reaction: $Cu^{2+}(aq) + 2\ e^- \rightarrow Cu\ (s)$ $E°_{Cu2+ \rightarrow Cu} = 0.34$ V

Overall cell reaction: $Zn\ (s) + Cu^{2+}(aq) \rightarrow Zn^{2+}(aq) + Cu\ (s)$

$E°_{cell} = E°_{Zn \rightarrow Zn2+} + E°_{Cu2+ \rightarrow Cu}$

$E°_{cell} = 0.76\ V + 0.34\ V$

$E°_{cell} = 1.10\ V$

For a cell consisting of $Cr|Cr^{3+}$ and $Ag|Ag^+$ half-cells, cell notation and potential:

$Cr|Cr^{3+}(aq, 1\ M)||Ag^+(aq, 1\ M)|Ag$
(anode) (cathode)

Anode half-cell reaction: $Cr\ (s) \rightarrow Cr^{3+}(aq) + 3\ e^-$ $E°_{Cr \rightarrow Cr3+} = 0.73$ V

Cathode half-cell reaction: $Ag^+ (aq) + e^- \rightarrow Ag (s)$ $E°_{Ag+ \rightarrow Ag} = 0.80$ V

Overall cell reaction:

$$Cr (s) + 3 Ag^+ (aq) \rightarrow Cr^{3+} (aq) + 3 Ag (s)$$

$$E°_{cell} = E°_{Cr \rightarrow Cr3+} + E°_{Ag+ \rightarrow Ag}$$

$$E°_{cell} = 0.73 \text{ V} + 0.80 \text{ V}$$

$$E°_{cell} = 1.53 \text{ V}$$

The standard electrode potentials are usually listed as reduction potentials and are relative to a standard hydrogen electrode (SHE) with 0 volt potential as a reference.

The cell potential for all galvanic (voltaic) cells is positive because the voltaic cell generates potential.

The cell potential for all electrolytic cells is negative because the electrolytic cell requires potential input.

The *Nernst* equation is used when the cell is not under standard conditions:

$$E_{cell} = E_{cell}° - (RT / nF) \ln Q$$

where E_{cell} = nonstandard cell potential in volts; $E°_{cell}$ = standard cell potential in volts; R = constant, 8.314 J/mol·K, T = temperature in Kelvin; n = moles of electrons transferred; F = Faraday's constant = 96,485 C/mol e^- (C represents coulombs) and Q = the reaction quotient expression.

The Nernst Equation may be used if the temperature is the standard temperature (298 K) and use either the natural logarithm (ln) or common logarithm, base 10 (log).

$$E_{cell} = E°_{cell} - [(0.0257 \text{ V}) / n] \ln Q$$

or

$$E_{cell} = E°_{cell} - [(0.0592 \text{ V}) / n] \log Q$$

Modifying one of the previous cells, as follows:

$Zn (s) | Zn^{2+} (aq)$, 3.00M; SO_4^{2-} (3.00M) || $Cu^{2+} (aq)$, 0.0015M; SO_4^{2-} (0.0015M) | Cu (s)

$$E°_{cell} = +1.10\text{V}$$

Assuming standard temp. (298K), the Nernst equation expression becomes:

$$E_{cell} = E°_{cell} - [(0.0257 \text{ V}) / n] \ln [Zn^{2+}] / [Cu^{2+}]$$

$$E_{cell} = +1.10 \text{ V} - (0.0257 \text{ V} / 2 \text{ mol } e^-) \ln (3.00 \text{ M} / 0.0015 \text{ M})$$

Evaluating the above expression gives:

$E_{cell} = +1.10\ V - 0.098\ V$

$E_{cell} = 1.00\ V$

Cell Potential and Free Energy

The cell potential indicates if the cell reaction is spontaneous or nonspontaneous. Spontaneity for a process is often associated with its Gibbs free energy change (ΔG).

The mathematical relationship between cell potential and ΔG is:

$\Delta G° = -\ nFE°$ under standard conditions

$\Delta G = -\ nFE$ under nonstandard conditions

where ΔG = change in Gibbs free energy in Joules, n = the moles of electrons transferred, the Faraday's constant F = 96,485 C/mol e^-, E = cell potential

Using the units of Faraday's constant, coulombs/mol e^- and volts for cell potential, the Gibbs free energy units are coulomb-volts. Coulomb-volts are equivalent to Joules.

The cell potential measures the potential difference between the two half-cells.

A potential difference of 1 V is equivalent to 1 Joule of work done per coulomb of charge that flows between two points in the circuit. (1 V = 1 J/C or 1 J = 1 C·V)

Maximum work produced = charge × maximum potential

$w_{max} = -qE_{max} = \Delta G$

Electrical charge,

$q = nF$

$\Delta G = -nFE_{cell}$ or $\Delta G° = -nFE°_{cell}$

where n = mole of electrons transferred or that flow through circuit, Faraday's constant F = 96,485 C/mole e^-.

For example:

$Zn\ (s) + Cu^{2+}\ (aq) \rightarrow Zn^{2+}\ (aq) + Cu\ (s)$ $E°_{cell} = 1.10\ V$

$\Delta G° = -2\ mol\ e^- \times (96,485\ C\ /\ mol\ e^-) \times 1.10\ V$

$\Delta G° = -2.12 \times 10^5\ J$

2.12×10^2 kJ is the maximum work derived per mole of Zn reacted by Cu^{2+}.

The direction of electron flow

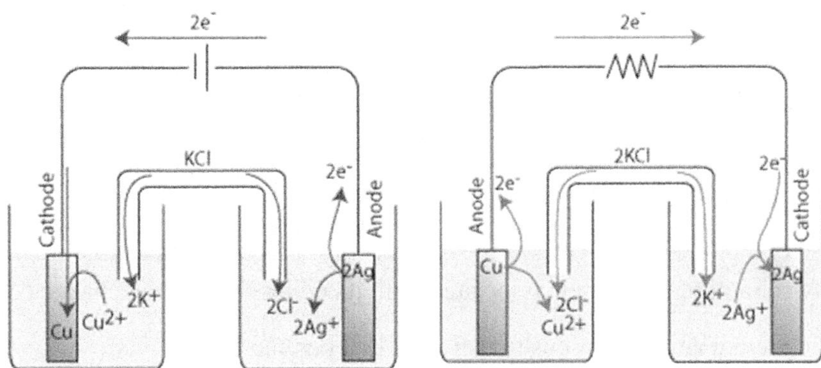

Electrolytic Charge Flow *Galvanic (Voltaic) Charge Flow*

Electrons flow from anode to cathode. Mnemonic: *A to C in alphabetical order.*

Alternatively, AC power as *the A comes first and stands for Anode.*

Oxidation (at the anode) produces electrons (and cations) and releases electrons toward the cathode. The cathode receives those electrons and uses them for reduction.

The species with the highest oxidation potential (lowest reduction potential) is the anode, and the species with the highest reduction potential is the cathode.

In the diagram above, the galvanic (voltaic) cell shows a natural flow because Cu (higher oxidation potential/lower reduction potential) is the anode, and Ag (higher reduction potential) is the cathode.

The electrolytic cell shows exactly the opposite. To force Cu to be the cathode and Ag to be the anode, a battery is used to drive the reaction.

Electrons flow in wires and electrodes, while ions flow in the electrolyte solution, thus creating a completed circuit.

Concentration Cells

A *concentration cell* is an electrochemical cell in which both half-cells are of the same type, but with different electrolyte concentrations.

The following cell notations are examples of concentration cells:

$$Cu|Cu^{2+}(aq, 0.0010\ M)||Cu^{2+}(aq, 1.0\ M)|Cu$$

$$Ag|Ag^{+}(aq, 0.0010\ M)\ ||Ag^{+}(aq, 0.10\ M)|Ag$$

In concentration cells, the half-cell with the lower electrolyte concentration serves as an anode half-cell, and one with the higher electrolyte concentration is the cathode half-cell. At the anode half-cell, an oxidation reaction occurs to increase the electrolyte concentration, and at the cathode half-cell, a reduction reaction occurs to decrease its electrolyte concentration. The oxidation-reduction reaction continues until the electrolyte concentrations in both half-cells become equal.

At anode half-cell: $Cu\ (s) \rightarrow Cu^{2+}\ (aq) + 2\ e^-$ (in 0.0010 M Cu^{2+})

At cathode half-cell: $Cu^{2+}\ (aq) + 2\ e^- \rightarrow Cu\ (s)$ (in 0.50 M Cu^{2+})

Batteries

Batteries are galvanic cells (or a group of galvanic cells) connected in series, where the total battery potential is equal to the sum of the potentials of the cells.

Three types of batteries: primary batteries, secondary batteries, and fuel cells.

Primary batteries are not rechargeable.

Secondary batteries are rechargeable.

Fuel cells last if there is an ample supply of fuel to provide the energy.

Dry Cells: *Primary* batteries include common (acidic) dry batteries, alkaline batteries, and mercury batteries.

The *acidic dry battery* consists of Zinc casing as container, anode, and the reducing agent; graphite rod as an (inert) cathode; aqueous NH_4Cl paste as an electrolyte and MnO_2 powder as the oxidizing agent.

$$Zn|Zn^{2+},NH_4^+,NH_3\ (aq)||Mn_2O_3,MnO_2|C\ (s)$$

Anode reaction:

$$Zn\ (s) \rightarrow Zn^{2+}\ (aq) + 2\ e^-$$

Cathodic reaction:

$$2\ MnO_2\ (s) + 2\ NH_4^+\ (aq) + 2\ e^- \rightarrow Mn_2O_3\ (s) + 2\ NH_3\ (aq) + H_2O\ (l)$$

Net reaction:

$$Zn\ (s) + 2\ MnO_2\ (s) + 2\ NH_4^+\ (aq) \rightarrow Zn^{2+}\ (aq) + Mn_2O_3\ (s) + 2\ NH_3\ (aq) + H_2O\ (l)$$

The reverse reaction is prevented by the formation of $[Zn(NH_3)_4]^{2+}$ ions.

A new dry cell battery has a potential of about 1.5 V regardless of the size, but the amount of energy that a battery can deliver does depend on its size.

A D-size battery can deliver more current (greater amperes) than an AAA-size battery. Standard dry batteries use aqueous NH_4Cl paste as an electrolyte and are referred to as acidic batteries due to the ionization:

$$NH_4^+ (aq) \rightarrow NH_3 (aq) + H^+ (aq).$$

Alkaline batteries also use zinc (as the reducing agent) and MnO_2 (as the oxidizing agent), but an aqueous paste containing KOH, instead of NH_4Cl, as the electrolyte.

The anode and cathode reactions are as follows:

At anode: $Zn\ (s) + 2\ OH^-\ (aq) \rightarrow ZnO\ (s) + H_2O\ (l) + 2\ e^-$

At cathode: $2\ MnO_2\ (s) + H_2O\ (l) + 2\ e^- \rightarrow Mn_2O_3\ (s) + 2\ OH^-\ (aq)$

Net reaction: $Zn\ (s) + 2\ MnO_2\ (s) \rightarrow ZnO\ (s) + Mn_2O_3\ (s)$

Since all reactants involved are solid, alkaline batteries can deliver a constant voltage until the limiting reactant is exhausted. They last longer because zinc metal corrodes more slowly under basic conditions.

Mercury batteries are used in calculators and watches.

The following reactions occur at the anode and cathode sections of the cell:

Anode reaction: $Zn\ (s) + 2\ OH^-\ (aq) \rightarrow ZnO\ (s) + H_2O + 2\ e^-$

Cathode reaction: $HgO\ (s) + H_2O + 2\ e^- \rightarrow Hg\ (l) + 2\ OH^-\ (aq)$

Net cell reaction: $Zn\ (s) + HgO\ (s) \rightarrow ZnO\ (s) + Hg\ (l)$

Electromotive force, Voltage

Electromotive force (emf), or cell potential (E_{cell}), is the driving force that enables electrons to flow from one electrode to the other, which has the unit volt (V). A volt is one Joule per coulomb, where coulomb is the unit of charge.

Voltage is the driving force for reactions, often called the EMF (electromotive force). The higher the voltage, the greater the driving force.

A voltmeter must have its negative terminal connected to the anode and the positive terminal to the cathode to get the correct polarity (± sign).

Voltage is always positive for a spontaneous reaction.

Lead-storage batteries

Lead-storage batteries are used in automobiles. They contain sulfuric acid as an electrolyte. Each lead-storage battery cell contains several grids of lead alloy. One set of alternating grids is packed with lead metal and the other with lead(IV) oxide, PbO_2.

Each set of grids, which are the electrodes, are connected in a parallel arrangement, which enables the cell to deliver more current; the amount of current delivered depends on the surface area of the electrode.

Each cell in a lead storage battery produces a potential of about 2.01 V.

The standard 12-V battery used in most cars contains six cells connected in series.

Spontaneous Reactions that occur in the lead storage battery are:

Anode reaction: $Pb\ (s) + HSO_4^-\ (aq) \rightarrow PbSO_4\ (s) + H^+\ (aq) + 2\ e^-$

Cathode reaction: $PbO_2\ (s) + 3\ H^+\ (aq) + HSO_4^-\ (aq) + 2\ e^- \rightarrow PbSO_4\ (s) + 2\ H_2O\ (m)$

Net reaction:

$Pb\ (s) + PbO_2\ (s) + 2\ H^+\ (aq) + 2\ HSO_4^-\ (aq) \rightarrow 2\ PbSO_4\ (s) + 2\ H_2O\ (l) \rightarrow$ discharging

---

The greatest advantage of lead-storage batteries (i.e., secondary batteries) is that they are rechargeable. Starting the engine causes the discharge reaction to occur. However, while the car is being driven, the battery obtains energy from the motor through the alternator.

The following re-charging reaction occurs:

$$2\ PbSO_4\ (s) + 2\ H_2O\ (l) \rightarrow Pb\ (s) + PbO_2\ (s) + 2\ H^+\ (aq) + 2\ HSO_4^-\ (aq)$$

Lead-storage batteries also have a longer lifetime than other batteries, and they can deliver a relatively large amount of current and electrical energy within a short time.

The major disadvantages of lead-storage batteries are:

(1) they are very heavy, bulky, and relatively non-portable

(2) lead is a toxic metal, and its disposal creates environmental problems

(3) the battery must be kept upright because H_2SO_4 is very corrosive

Nickel-cadmium batteries

Nickel-cadmium batteries are rechargeable batteries used in small appliances such as cordless phones. They contain cadmium as the anode and hydrated nickel oxide as the cathode.

The electrolyte is an aqueous KOH paste.

Anode reaction:

$$Cd\ (s) + 2\ OH^-\ (aq) \rightarrow Cd(OH)_2\ (s) + 2\ e^-$$

Cathode reaction:

$$2\ NiO(OH)\ (s) + 2\ H_2O\ (l) + 2\ e^- \rightarrow 2\ Ni(OH)_2\ (s) + 2\ OH^-\ (aq)$$

Net cell reaction:

$$Cd\ (s) + 2\ NiO(OH)\ (s) + 2\ H_2O\ (l) \rightarrow Cd(OH)_2\ (aq) + 2\ Ni(OH)_2\ (aq)$$

Standard Reduction Potentials

Reduction potential is a measure of the tendency of a chemical species to acquire electrons and become reduced (gain of electrons is a reduction).

Reduction potential is measured in volts (V) or millivolts (mV).

The more positive the E°, the higher the tendency for the species to be reduced (a strong oxidizing agent).

The more negative the E°, the higher the tendency for the species to be oxidized (a strong reducing agent).

TABLE OF STANDARD REDUCTION POTENTIALS

Standard Reduction Potentials in Aqueous Solution at 25 °C	
Acidic Solutions	**E° (V)**
F_2 (*g*) + 2 e^- → 2 F^- (*aq*)	+2.87
Co^{3+} (*aq*) + e^- → Co^{2+} (*aq*)	+1.82
Pb^{4+} (*aq*) + 2 e^- → Pb^{2+} (*aq*)	+1.8
H_2O_2 (*aq*) + 2 H^+ (*aq*) + 2 e^- → 2 H_2O	+1.77
NiO_2 (*s*) + 4 H^+ (*aq*) + 2 e^- → Ni^{2+} (*aq*) + 2 H_2O	+1.7
PbO_2 (*s*) + SO_4^{2-} (*aq*) + 4 H^+ (*aq*) + 2 e^- → $PbSO_4$ (*s*) + 2 H_2O	+1.685
Au^+ (*aq*) + e^- → Au (*s*)	+1.68
2 HClO (*aq*) + 2 H^+ (*aq*) + 2 e^- → Cl_2 (g) + 2 H_2O	+1.63
Ce^{4+} (*aq*) + e^- → Ce^{3+} (*aq*)	+1.61
$NaBiO_3$ (*s*) + 6 H^+ (*aq*) + 2 e^- → Bi^{3+} (*aq*) + Na^+ (*aq*) + 3 H_2O	+1.6
MnO_4^- (*aq*) + 8 H^+ (*aq*) + 5 e^- → Mn^{2+} (*aq*) + 4 H_2O	+1.51
Au^{3+} (*aq*) + 3 e^- → Au (*s*)	+1.5
ClO_3^- (*aq*) + 6 H^+ (*aq*) + 5 e^- → 1/2 Cl_2 (*g*) + 3 H_2O	+1.47
BrO_3^- + 6 H^+ (*aq*) + 6 e^- → Br^- (*aq*) + 3 H_2O	+1.44
Cl_2 (*g*) + 2 e^- → 2 Cl^- (*aq*)	+1.358
$Cr_2O_7^{2-}$ + 14 H^+ (*aq*) + 6 e^- → 2 Cr^{3+} (*aq*) + 7 H_2O	+1.33
$N_2H_5^+$ (*aq*) + 3 H^+ (*aq*) + 2 e^- → 2 NH_4^+ (*aq*)	+1.24
MnO_2 (*s*) + 4 H^+ (*aq*) + 2 e^- → Mn^{2+} (*aq*) + 2 H_2O	+1.23
O_2 (*g*) + 4 H^+ (*aq*) + 4 e^- → 2 H_2O	+1.229
Pt^{2+} (*aq*) + 2 e^- → Pt (*s*)	+1.2
IO_3^- (*aq*) + 6 H^+ (*aq*) + 5 e^- → ½ I_2 (*aq*) + 3 H_2O	+1.195
ClO_4^- (*aq*) + 2 H^+ (*aq*) + 2 e^- → ClO_3^- (*aq*) + H_2O	+1.19
Br_2 (*l*) + 2 e^- → 2 Br^- (*aq*)	+1.066
$AuCl_4^-$ + 3 e^- → Au (*s*) + 4 Cl^- (*aq*)	+1
Pd^{2+} (*aq*) + 2 e^- → Pd (*s*)	+0.987
NO_3^- (*aq*) + 4 H^+ (*aq*) + 3 e^- → NO (*g*) + 2 H_2O	+0.96
NO_3^- (*aq*) + 3 H^+ (*aq*) + 2 e^- → HNO_2 (*aq*) + H_2O	+0.94
2 Hg^{2+} (*aq*) + 2 e^- → Hg_2^{2+} (*aq*)	+0.92
Hg^{2+} (*aq*) + 2 e^- → Hg (*l*)	+0.855
Ag^+ (*aq*) + e^- → Ag (*s*)	+0.7994
Hg_2^{2+} (*aq*) + 2 e^- → 2 Hg (*l*)	+0.789

Fe^{3+} (*aq*) + e^- → Fe^{2+} (*aq*)	+0.771
$SbCl_6^-$ (*aq*) + 2 e^- → $SbCl_4^-$ (*aq*) + 2 Cl^- (*aq*)	+0.75
$[PtCl_4]^{2-}$ (*aq*) + 2 e^- → Pt (*s*) + 4 Cl^- (*aq*)	+0.73
O_2 (*g*) + 2 H^+ (*aq*) + 2 e^- → H_2O_2 (*aq*)	+0.682
$[PtCl_6]^{2-}$ (*aq*) + 2 e^- → $[PtCl_4]^{2-}$ (*aq*) + 2 Cl^- (*aq*)	+0.68
H_3AsO_4 (*aq*) + 2 H^+ (*aq*) + 2 e^- → H_3AsO_3 (*aq*) + H_2O	+0.58
I_2 (*s*) + 2 e^- → 2 I^- (*aq*)	+0.535
TeO_2 (*s*) + 4 H^+ (*aq*) + 4 e^- → Te (*s*) + 2 H_2O	+0.529
Cu^+ (*aq*) + e^- → Cu (*s*)	+0.521
$[RhCl_6]^{3-}$ (*aq*) + 3 e^- → Rh (*s*) + 6 Cl^- (*aq*)	+0.44
Cu^{2+} (*aq*) + 2 e^- → Cu (*s*)	+0.337
$HgCl_2$ (*s*) + 2 e^- → 2 Hg (*l*) + 2 Cl^- (*aq*)	+0.27
AgCl (*s*) + e^- → Ag (*s*) + Cl^- (*aq*)	+0.222
SO_4^{2-} (*aq*) + 4 H^+ (*aq*) + 2 e^- → SO_2 (*g*) + 2 H_2O	+0.2
SO_4^{2-} (*aq*) + 4 H^+ (*aq*) + 2 e^- → H_2SO_3 (*g*) + H_2O	+0.17
Cu^{2+} (*aq*) + e^- → Cu^+ (*aq*)	+0.153
Sn^{4+} (*aq*) + 2 e^- → Sn^{2+} (*aq*)	+0.15
S (*s*) + 2 H^+ (*aq*) + 2 e^- → H_2S (*aq*)	+0.14
AgBr (*s*) + e^- → Ag (*s*) + Br^- (*aq*)	+0.0713
2 H^+ (*aq*) + 2 e^- → H_2 (*g*) (reference electrode)	0.0
N_2O (*g*) + 6 H^+ (*aq*) + H_2O + 4 e^- → 2 NH_3OH^+ (*aq*)	−0.05
Pb^{2+} (*aq*) + 2 e^- → Pb (*s*)	−0.126
Sn^{2+} (*aq*) + 2 e^- → Sn (*s*)	−0.14
AgI (*s*) + e^- → Ag (*s*) + I^- (*aq*)	−0.15
Sn^{4+} (*aq*) + 2 e^- → Sn^{2+} (*aq*)	+0.15
S (*s*) + 2 H^+ (*aq*) + 2 e^- → H_2S (*aq*)	+0.14
AgBr (*s*) + e^- → Ag (*s*) + Br^- (*aq*)	+0.0713
2 H^+ (*aq*) + 2 e^- → H_2 (*g*) (reference electrode)	0.0
N_2O (*g*) + 6 H^+ (*aq*) + H_2O + 4 e^- → 2 NH_3OH^+ (*aq*)	−0.05
Pb^{2+} (*aq*) + 2 e^- → Pb (*s*)	−0.126
Sn^{2+} (*aq*) + 2 e^- → Sn (*s*)	−0.14
AgI (*s*) + e^- → Ag (*s*) + I^- (*aq*)	−0.15
$[SnF_6]^{2-}$ (*aq*) + 4 e^- → Sn (*s*) + 6 F^- (*aq*)	−0.25
Ni^{2+} (*aq*) + 2 e^- → Ni (*s*)	−0.25

Co^{2+} (*aq*) + 2 e^- → Co (*s*)	−0.28
Tl^+ (*aq*) + e^- → Tl (*s*)	−0.34
$PbSO_4$ (*s*) + 2 e^- → Pb (*s*) + SO_4^{2-} (*aq*)	−0.356
Se (*s*) + 2 H^+ (*aq*) + 2 e^- → H_2Se (*aq*)	−0.4
Cd^{2+} (*aq*) + 2 e^- → Cd (*s*)	−0.403
Cr^{3+} (*aq*) + e^- → Cr^{2+} (*aq*)	−0.41
Fe^{2+} (*aq*) + 2 e^- → Fe (*s*)	−0.44
2 CO_2 (*g*) + 2 H^+ (*aq*) + 2 e^- → $(COOH)_2$ (*aq*)	−0.49
Ga^{3+} (*aq*) + 3 e^- → Ga (*s*)	−0.53
HgS (*s*) + 2 H^+ (*aq*) + 2 e^- → Hg (*l*) + H_2S (*g*)	−0.72
Cr^{3+} (*aq*) + 3 e^- → Cr (*s*)	−0.74
Zn^{2+} (*aq*) + 2 e^- → Zn (*s*)	−0.763
2 H_2O (*l*) + 2 e^- → H_2 (*g*) + 2 OH^- (*aq*)	−0.8277
Cr^{2+} (*aq*) + 2 e^- → Cr (*s*)	−0.91
Mn^{2+} (*aq*) + 2 e^- → Mn (*s*)	−1.18
V^{2+} (*aq*) + 2 e^- → V (*s*)	−1.18
Zr^{4+} (*aq*) + 4 e^- → Zr (*s*)	−1.53
Al^{3+} (*aq*) + 3 e^- → Al (*s*)	−1.66
H_2 (*g*) + 2 e^- → 2 H^- (*aq*)	−2.25
Mg^{2+} (*aq*) + 2 e^- → Mg (*s*)	−2.37
Na^+ (*aq*) + e^- → Na (*s*)	−2.714
Ca^{2+} (*aq*) + 2 e^- → Ca (*s*)	−2.87
Sr^{2+} (*aq*) + 2 e^- → Sr (*s*)	−2.89
Ba^{2+} (*aq*) + 2 e^- → Ba (*s*)	−2.9
Rb^+ (*aq*) + e^- → Rb (*s*)	−2.925
K^+ (*aq*) + e^- → K (*s*)	−2.925
Li^+ (*aq*) + e^- → Li (*s*)	−3.045
Basic Solutions	**E° (V)**
ClO^- (*aq*) + H_2O + 2 e^- → Cl^- (*aq*) + 2 OH^- (*aq*)	+0.89
OOH^- (*aq*) + H_2O + 2 e^- → 3 OH^- (*aq*)	+0.88
2 NH_2OH (*aq*) + 2 e^- → N_2H_4 (*aq*) + 2 OH^- (*aq*)	+0.74
ClO_3^- (*aq*) + 3 H_2O + 6 e^- → Cl^- (*aq*) + 6 OH^- (*aq*)	+0.62
MnO_4^- (*aq*) + 2 H_2O + 3 e^- → MnO_2 (*s*) + 4 OH^- (*aq*)	+0.588
MnO_4^- (*aq*) + e^- → MnO_4^{2-} (*aq*)	+0.564

NiO_2 (*s*) + 2 H_2O + 2 e^- → $Ni(OH)_2$ (s) + 2 OH^- (*aq*)	+0.49
Ag_2CrO_4 (*s*) + 2 e^- → 2 Ag (s) + CrO_4^{2-} (*aq*)	+0.446
O_2 (*g*) + 2 H_2O + 4 e^- → 4 OH^- (*aq*)	+0.4
ClO_4^- (*aq*) + H_2O + 2 e^- → ClO_3^- (*aq*) + 2 OH^- (*aq*)	+0.36
Ag_2O (*s*) + H_2O + 2 e^- → 2 Ag (*s*) + 2 OH^- (*aq*)	+0.34
2 NO_2^- (*aq*) + 3 H_2O + 4 e^- → N_2O (*g*) + 6 OH^- (*aq*)	+0.15
N_2H_4 (*aq*) + 2 H_2O + 2 e^- → 2 NH_3 (*aq*) + 2 OH^- (*aq*)	+0.1
$[Co(NH_3)_6]^{3+}$ (*aq*) + e^- → $[Co(NH_3)_6]^{2+}$ (*aq*)	+0.1
HgO (*s*) + H_2O + 2 e^- → Hg (*l*) + 2 OH^- (*aq*)	+0.0984
O_2 (*g*) + H_2O + 2 e^- → OOH^- (*aq*) + OH^- (*aq*)	+0.076
NO_3^- (*aq*) + H_2O + 2 e^- → NO_2^- (*aq*) + 2 OH^- (*aq*)	+0.01
MnO_2 (s) + 2 H_2O + 2 e^- → $Mn(OH)_2$ (*s*) + 2 OH^- (*aq*)	−0.05
CrO_4^{2-} (*aq*) + 4 H_2O + 3 e^- → $Cr(OH)_3$ (*s*) + 5 OH^- (*aq*)	−0.12
$Cu(OH)_2$ (*s*) + 2 e^- → Cu (*s*) + 2 OH^- (*aq*)	−0.36
$Fe(OH)_3$ (*s*) + e^- → $Fe(OH)_2$ (*s*) + OH^- (*aq*)	−0.56
2 H_2O + 2 e^- → H_2 (*g*) + 2 OH^- (*aq*)	−0.8277
2 NO_3^- (*aq*) + 2 H_2O + 2 e^- → N_2O_4 (g) + 4 OH^- (*aq*)	−0.85
$Fe(OH)_2$ (*s*) + 2 e^- → Fe (*s*) + 2 OH^- (*aq*)	−0.877
SO_4^{2-} (*aq*) + H_2O + 2 e^- → SO_3^{2-} (*aq*) + 2 OH^- (*aq*)	−0.93
N_2 (*g*) + 4 H_2O + 4 e^- → N_2H_4 (*aq*) + 4 OH^- (*aq*)	−1.15
$[Zn(OH)_4]^{2-}$ (*aq*) + 2 e^- → Zn (*s*) + 4 OH^- (*aq*)	−1.22
$Zn(OH)_2$ (*s*) + 2 e^- → Zn (*s*) + 2 OH^- (*aq*)	−1.245
$[Zn(CN)_4]^{2-}$ (*aq*) + 2 e^- → Zn (*s*) + 4 CN^- (*aq*)	−1.26
$Cr(OH)_3$ (*s*) + 3 e^- → Cr (*s*) + 3 OH^- (*aq*)	−1.3
SiO_3^{2-} (*aq*) + 3 H_2O + 4 e^- → Si (*s*) + 6 OH^- (*aq*)	−1.7

Practice Questions

1. Nickel-cadmium batteries are used in rechargeable electronic calculators. Given the following reaction for a discharging NiCad battery, what substance is being reduced?

$$Cd\ (s) + NiO_2\ (s) + 2H_2O\ (l) \rightarrow Cd(OH)_2\ (s) + Ni(OH)_2\ (s)$$

A. $Cd(OH)_2$ **B.** H_2O **C.** NiO_2 **D.** Cd **E.** $Ni(OH)_2$

2. How long must a current of 2 amps run to liberate 3 moles of H_2O when electrolyzed in the following reaction? (Use Faraday's constant = 96,500C/mol and the conversion factor of 1 amp = 1 coulomb/sec)

$$H_2O_2 + 2\ H^+ + 2\ e^- \rightleftarrows 2\ H_2O$$

A. 4.83×10^4 sec
B. 4.83×10^5 sec
C. 1.45×10^5 sec
D. 2.41×10^5 sec
E. 9.65×10^3 sec

3. Which is NOT valid for the following redox reaction occurring in an electrolytic cell?

$$Cd\ (s) + Zn(NO_3)_2\ (aq) \xrightarrow{\text{Electricity}} Zn\ (s) + Cd(NO_3)_2\ (aq)$$

A. Cd^{2+} is produced at the anode
B. Zn metal is produced at the anode
C. Oxidation half-reaction: $Cd \rightarrow Cd^{2+} + 2\ e^-$
D. Reduction half-reaction: $Zn^{2+} + 2\ e^- \rightarrow Zn$
E. Zn metal is produced at the cathode

4. Which of the statements is true regarding the following redox reaction occurring in a nonspontaneous electrolytic cell?

$$Br_2\ (l) + 2\ NaCl\ (aq) \xrightarrow{\text{Electricity}} Cl_2\ (g) + 2\ NaBr\ (aq)$$

A. Br_2 liquid is produced at the cathode
B. Cl_2 gas is produced at the cathode
C. Reduction half-reaction: $2\ Cl^- \rightarrow Cl_2 + 2\ e^-$
D. Oxidation half-reaction: $Br_2 + 2\ e^- \rightarrow 2\ Br^-$
E. Cl_2 gas is produced at the anode

5. Which of the following is true regarding the redox reaction occurring in a spontaneous electrochemical cell?

$$Sn\ (s) + Cu^{2+}\ (aq) \rightarrow Cu\ (s) + Sn^{2+}\ (aq)$$

A. Electrons flow from the Sn electrode to the Cu electrode
B. Anions in the salt bridge flow from the Sn half-cell to the Cu half-cell
C. Sn is oxidized at the cathode
D. Cu^{2+} is reduced at the anode
E. None of the above

6. Which of the following is a unit of electrical charge?

A. joule **B.** volt **C.** coulomb **D.** ampere **E.** watt

7. If a galvanic cell has two electrodes, which statement is correct?

A. Oxidation occurs at the negatively charged cathode
B. Oxidation occurs at the positively charged cathode
C. Oxidation occurs at the negatively charged anode
D. Oxidation occurs at the positively charged anode
E. Oxidation occurs at the uncharged dynode

8. A battery operates by:

A. oxidation
B. reduction
C. both oxidation and reduction
D. neither oxidation nor reduction
E. none of the above

9. What is the term for the electrode in an electrochemical cell where oxidation occurs?

A. oxidation electrode
B. reduction electrode
C. anode
D. cathode
E. none of the above

10. What is the general term for an apparatus that contains two solutions with electrodes in separate compartments that are connected by a wire and salt bridge?

A. electrolytic cell
B. voltaic cell
C. battery
D. electrochemical cell
E. dry cell

11. The electrode with the standard reduction potential of 0 V is assigned as the standard reference electrode and uses the half-reaction:

A. $2\ NH_4^+\ (aq) + 2\ e^- \leftrightarrows H_2\ (g) + 2\ NH_3\ (g)$
B. $Ag^+\ (aq) + e^- \leftrightarrows Ag\ (s)$
C. $Cu^{2+}\ (aq) + 2\ e^- \leftrightarrows Cu\ (s)$
D. $Zn^{2+}\ (aq) + 2\ e^- \leftrightarrows Zn\ (s)$
E. $2\ H^+\ (aq) + 2\ e^- \leftrightarrows H_2\ (g)$

12. In which type of cell does the following reaction occur when electrons are forced into a system by applying an external voltage?

$$Fe^{2+} + 2\ e^- \rightarrow Fe\ (s) \qquad E° = -0.44\ V$$

A. concentration cell
B. battery
C. electrochemical cell
D. galvanic cell
E. electrolytic cell

13. What is the term for a reaction that represents separate oxidation or reduction processes?

A. reduction reaction
B. redox reaction
C. oxidation reaction
D. half-reaction
E. none of the above

14. What is the term for the providing of electricity to a nonspontaneous redox process to cause a reaction?

A. hydrolysis
B. protolysis
C. electrochemistry
D. electrolysis
E. none of the above

15. How is electrolysis different from the chemical process inside a battery?

A. Pure compounds cannot be generated *via* electrolysis
B. Electrolysis only uses electrons from a cathode
C. Electrolysis does not use electrons
D. They are the same process in reverse
E. Chemical changes do not occur in electrolysis

Detailed Explanations

1. C is correct.

Determine the oxidation number of each species.

The reduced substance has a decrease in the oxidation number.

In NiO_2, the oxidation number of Ni = +4.

In $Ni(OH)_2$, the oxidation number of Ni = +2.

2. C is correct.

Three moles of H_2O need 3 moles of e^-.

Calculate time needed for 3 moles of H_2O:

$$(3 \text{ moles } e^-) \times (96{,}500 \text{ coulomb} / 1 \text{ mole } e^-) \times (1 \text{ sec} / 2 \text{ coulomb})$$

$$\text{time (for 3 moles of } H_2O) = 1.45 \times 10^5 \text{ sec}$$

3. B is correct.

An electrolytic cell is a nonspontaneous electrochemical cell that requires electrical energy to be supplied (e.g., battery) to initiate the reaction.

The anode is positive, and the cathode is the negative electrode.

For both electrolytic and galvanic cells, oxidation occurs at the anode, while reduction occurs at the cathode.

Zn metal is produced at the cathode because it is a reduction product (from an oxidation number of +3 on the left to 0 on the right). Zn will not be produced at the anode.

4. E is correct.

In electrochemical (i.e., galvanic) cells, oxidation always occurs at the anode, and reduction occurs at the cathode.

Cl_2 gas is produced at the anode because it is an oxidation product (from an oxidation number of –1 on the left to 0 on the right).

5. A is correct.

A spontaneous electrochemical cell is also known as a galvanic cell.

6. C is correct.

A coulomb is a unit of electrical charge.

7. C is correct.

Oxidation (i.e., loss of electrons) occurs at the anode.

8. C is correct.

A battery is an electrochemical (i.e., galvanic) cell, which has both reduction and oxidation reactions happening within them to generate electricity.

9. C is correct.

Oxidation (i.e., loss of electrons) occurs at the anode.

10. D is correct.

The salt bridge should be the most obvious indication that it is an electrochemical cell – electrolytic cells do not have salt bridges.

11. E is correct.

By convention, the reference standard for potential is always hydrogen reduction.

12. E is correct.

Electrolytic cells do not occur spontaneously; the reaction only occurs with the addition of external electrical energy.

13. D is correct.

A redox reaction, or oxidation-reduction reaction, involves the transfer of electrons between two reacting substances.

An oxidation reaction refers explicitly to the substance that is losing electrons.

A reduction reaction refers explicitly to the substance that is gaining reactions.

The oxidation and reduction reactions alone are half-reactions because they always occur together to form a whole reaction.

Therefore, half-reaction is the answer because it can represent either a separate oxidation process or a separate reduction process.

14. D is correct.

Electrolysis is a method of using a direct electric current to provide electricity to a nonspontaneous redox process to drive the reaction.

The direct electric current must be passed through an ionic substance or solution that contains electrolytes.

15. D is correct.

Electrolysis is the same process in reverse for the chemical process inside a battery.

Electrolysis is often used to separate elements.

Appendix

Periodic Table of the Elements

Key: Atomic Number | Valence | **Symbol** | Name | Atomic Mass

1 IA 1A	2 IIA 2A	3 IIIB 3B	4 IVB 4B	5 VB 5B	6 VIB 6B	7 VIIB 7B	8	9 VIII 8	10	11 IB 1B	12 IIB 2B	13 IIIA 3A	14 IVA 4A	15 VA 5A	16 VIA 6A	17 VIIA 7A	18 VIIIA 8A
1 -1,+1 **H** Hydrogen 1.008																	2 0 **He** Helium 4.003
3 +1 **Li** Lithium 6.941	4 +2 **Be** Beryllium 9.012											5 +3 **B** Boron 10.811	6 +4,+3,+2,+1,-4,-3,-2,-1 **C** Carbon 12.011	7 +5,+3,-3 **N** Nitrogen 14.007	8 -2 **O** Oxygen 15.999	9 -1 **F** Fluorine 18.998	10 0 **Ne** Neon 20.180
11 +1 **Na** Sodium 22.990	12 +2 **Mg** Magnesium 24.305											13 +3 **Al** Aluminum 26.982	14 +4,-4 **Si** Silicon 28.086	15 +5,+3,-3 **P** Phosphorus 30.974	16 +6,+4,+2,-2 **S** Sulfur 32.066	17 +7,+5,+3,+1,-1 **Cl** Chlorine 35.453	18 0 **Ar** Argon 39.948
19 +1 **K** Potassium 39.098	20 +2 **Ca** Calcium 40.078	21 +3 **Sc** Scandium 44.956	22 +4 **Ti** Titanium 47.88	23 +5 **V** Vanadium 50.942	24 +6,+3 **Cr** Chromium 51.996	25 +7,+4,+2 **Mn** Manganese 54.938	26 +6,+3,+2 **Fe** Iron 55.845	27 +3,+2 **Co** Cobalt 58.933	28 +2 **Ni** Nickel 58.693	29 +2 **Cu** Copper 63.546	30 +2 **Zn** Zinc 65.38	31 +3 **Ga** Gallium 69.723	32 +4,+2,-4 **Ge** Germanium 72.631	33 +5,+3,-3 **As** Arsenic 74.922	34 +6,+4,+2,-2 **Se** Selenium 78.971	35 +5,+3,+1,-1 **Br** Bromine 79.904	36 +2,0 **Kr** Krypton 84.798
37 +1 **Rb** Rubidium 85.468	38 +2 **Sr** Strontium 87.62	39 +3 **Y** Yttrium 88.906	40 +4 **Zr** Zirconium 91.224	41 +5 **Nb** Niobium 92.906	42 +6,+4 **Mo** Molybdenum 95.95	43 +7,+4 **Tc** Technetium 98.907	44 +4,+3 **Ru** Ruthenium 101.07	45 +3 **Rh** Rhodium 102.906	46 +4,+2 **Pd** Palladium 106.42	47 +1 **Ag** Silver 107.868	48 +2 **Cd** Cadmium 112.414	49 +3 **In** Indium 114.818	50 +4,+2,-4 **Sn** Tin 118.711	51 +5,+3,-3 **Sb** Antimony 121.760	52 +6,+4,+2,-2 **Te** Tellurium 127.6	53 +7,+5,+3,+1,-1 **I** Iodine 126.904	54 +6,+4,+2,0 **Xe** Xenon 131.294
55 +1 **Cs** Cesium 132.905	56 +2 **Ba** Barium 137.328	57-71	72 +4 **Hf** Hafnium 178.49	73 +5 **Ta** Tantalum 180.948	74 +6,+4 **W** Tungsten 183.85	75 +4 **Re** Rhenium 186.207	76 +4 **Os** Osnium 190.23	77 +4,+3 **Ir** Iridium 192.22	78 +4,+2 **Pt** Platinum 195.08	79 +3 **Au** Gold 196.967	80 +2,+1 **Hg** Mercury 200.59	81 +3,+1 **Tl** Thallium 204.383	82 +4,+2 **Pb** Lead 207.2	83 +3 **Bi** Bismuth 208.980	84 +4,+2,-2 **Po** Polonium [208.982]	85 +1,-1 **At** Astatine 209.987	86 +2,0 **Rn** Radon 222.018
87 +1 **Fr** Francium 223.020	88 +2 **Ra** Radium 226.025	89-103	104 +4 **Rf** Rutherfordium [261]	105 +5 **Db** Dubnium [262]	106 +6 **Sg** Seaborgium [266]	107 +7 **Bh** Bohrium [264]	108 +8 **Hs** Hassium [269]	109 unknown **Mt** Meitnerium [278]	110 unknown **Ds** Darmstadtium [281]	111 unknown **Rg** Roentgenium [280]	112 +2 **Cn** Copernicium [285]	113 unknown **Nh** Nihonium [286]	114 unknown **Fl** Flerovium [289]	115 unknown **Mc** Moscovium [289]	116 unknown **Lv** Livermorium [293]	117 unknown **Ts** Tennessine [294]	118 unknown **Og** Oganesson [294]

Series															
Lanthanide Series	57 +3 **La** Lanthanum 138.905	58 +4,+3 **Ce** Cerium 140.116	59 +3 **Pr** Praseodymium 140.908	60 +3 **Nd** Neodymium 144.243	61 +3 **Pm** Promethium 144.913	62 +3 **Sm** Samarium 150.36	63 +3,+2 **Eu** Europium 151.964	64 +3 **Gd** Gadolinium 157.25	65 +3 **Tb** Terbium 158.925	66 +3 **Dy** Dysprosium 162.500	67 +3 **Ho** Holmium 164.930	68 +3 **Er** Erbium 167.259	69 +3 **Tm** Thulium 168.934	70 +3 **Yb** Ytterbium 173.055	71 +3 **Lu** Lutetium 174.967
Actinide Series	89 +3 **Ac** Actinium 227.028	90 +4 **Th** Thorium 232.038	91 +5 **Pa** Protactinium 231.036	92 +6 **U** Uranium 238.029	93 +5 **Np** Neptunium 237.048	94 +4 **Pu** Plutonium 244.064	95 +3 **Am** Americium 243.061	96 +3 **Cm** Curium 247.070	97 +3 **Bk** Berkelium 247.070	98 +3 **Cf** Californium 251.080	99 +3 **Es** Einsteinium [254]	100 +3 **Fm** Fermium 257.095	101 +3 **Md** Mendelevium 258.1	102 +2 **No** Nobelium 259.101	103 +3 **Lr** Lawrencium [262]

Common Chemistry Equations

L, mL	= liter(s), milliliter(s)	mm Hg	= millimeters of mercury
g	= gram(s)	J, kJ	= joule(s), kilojoule(s)
nm	= nanometer(s)	V	= volt(s)
atm	= atmosphere(s)	mol	= mole(s)

ATOMIC STRUCTURE

$E = h\nu$

$c = \lambda\nu$

E = energy

ν = frequency

λ = wavelength

Planck's constant, $h = 6.626 \times 10^{-34}$ J s

Speed of light, $c = 2.998 \times 10^{8}$ m s^{-1}

Avogadro's number $= 6.022 \times 10^{23}$ mol^{-1}

Electron charge, $e = -1.602 \times 10^{-19}$ coulomb

EQUILIBRIUM

$$K_c = \frac{[C]^c[D]^d}{[A]^a[B]^b}, \text{ where } a\,A + b\,B \rightleftarrows c\,C + d\,D$$

$$K_p = \frac{(P_C)^c(P_D)^d}{(P_A)^a(P_B)^b}$$

$$K_a = \frac{[H^+][A^-]}{[HA]}$$

$$K_b = \frac{[OH^-][HB^+]}{[B]}$$

$$K_w = [H^+][OH^-] = 1.0 \times 10^{-14} \text{ at } 25°C$$

$$= K_a \times K_b$$

$$pH = -\log[H^+],\ pOH = -\log[OH^-]$$

$$14 = pH + pOH$$

$$pH = pK_a + \log\frac{[A^-]}{[HA]}$$

$$pK_a = -\log K_a,\ pK_b = -\log K_b$$

Equilibrium Constants

K_c (molar concentrations)

K_p (gas pressures)

K_a (weak acid)

K_b (weak base)

K_w (water)

KINETICS

$$\ln[A]_t - \ln[A]_0 = -kt$$

$$\frac{1}{[A]_t} - \frac{1}{[A]_0} = kt$$

$$t_{1/2} = \frac{0.693}{k}$$

k = rate constant

t = time

$t_{1/2}$ = half-life

GASES, LIQUIDS, AND SOLUTIONS

$PV = nRT$

$P_A = P_{total} \times X_A$, where $X_A = \frac{\text{moles A}}{\text{total moles}}$

$P_{total} = P_A + P_B + P_C + \ldots$

$n = \frac{m}{M}$

$K = °C + 273$

$D = \frac{m}{V}$

KE per molecule $= \frac{1}{2}mv^2$

Molarity, M = moles of solute per liter of solution

$A = abc$

P = pressure
V = volume
T = temperature
n = number of moles
m = mass
M = molar mass
D = density
KE = kinetic energy
v = velocity
A = absorbance
a = molar absorptivity
b = path length
c = concentration

Gas constant, $R = 8.314\ \text{J mol}^{-1}\ \text{K}^{-1}$
$= 0.08206\ \text{L atm mol}^{-1}\ \text{K}^{-1}$
$= 62.36\ \text{L torr mol}^{-1}\ \text{K}^{-1}$
1 atm = 760 mm Hg
= 760 torr
STP = 0.00 °C and 1.000 atm

THERMOCHEMISTRY/ ELECTROCHEMISTRY

$q = mc\Delta T$

$\Delta S° = \sum S°$ products $- \sum S°$ reactants

$\Delta H° = \sum \Delta H_f°$ products $- \sum \Delta H_f°$ reactants

$\Delta G° = \sum \Delta G_f°$ products $- \sum \Delta G_f°$ reactants

$\Delta G° = \Delta H° - T\Delta S°$
$= -RT \ln K$
$= -nFE°$

$I = \frac{q}{t}$

q = heat
m = mass
c = specific heat capacity
T = temperature
$S°$ = standard entropy
$H°$ = standard enthalpy
$G°$ = standard free energy
n = number of moles
$E°$ = standard reduction potential
I = current (amperes)
q = charge (coulombs)
t = time (seconds)

Faraday's constant, F = 96,485 coulombs per mole of electrons

1 volt = $\frac{\text{1 joule}}{\text{1 coulomb}}$

Glossary of Chemistry Terms

A

Absolute entropy (of a substance) – the increase in the entropy (ΔS) of a substance as it goes from a perfectly ordered crystalline form at 0 K (where its entropy is zero) to the temperature in question.

Absolute zero – the zero point on the absolute temperature scale; –273.15 °C or 0 Kelvin; theoretically, the temperature at which molecular motion ceases (i.e., the system does not emit or absorb energy, and all atoms are at rest).

Absorption spectrum – spectrum associated with absorption of electromagnetic radiation by atoms (or other species), resulting from transitions from lower to higher energy states.

Accuracy – how close a value is to the actual or true value; see *Precision*.

Acid – a substance that produces H^+ (*aq*) ions in aqueous solution and gives a pH of less than 7.0; strong acids ionize entirely or almost entirely in dilute aqueous solution; weak acids ionize only slightly.

Acid dissociation constant – an equilibrium constant for the dissociation of a weak acid.

Acidic salt – a salt containing an ionizable hydrogen atom; does not necessarily produce acidic solutions.

Actinides – the fifteen chemical elements that are between actinium (89) and lawrencium (103).

Activated complex – a structure that forms because of a collision between molecules while new bonds are formed.

Activation energy (E_a) – the amount of energy that must be absorbed by reactants in their ground states to reach the transition state needed for a reaction can occur.

Active metal – a metal with low ionization energy that loses electrons readily to form cations.

Activity (of a component of an ideal mixture) – a dimensionless quantity whose magnitude is equal to the molar concentration in an ideal solution; equal to partial pressure in an ideal gas mixture; 1 for pure solids or liquids.

Activity series – a listing of metals (and hydrogen) in order of decreasing activity.

Actual yield – the amount of a specified pure product obtained from a given reaction; see *Theoretical Yield.*

Addition reaction – a reaction in which two atoms or groups of atoms are added to a molecule, one on each side of a double or triple bond.

Adhesive forces – forces of attraction between a liquid and another surface.

Adsorption – the adhesion of a species onto the surfaces of particles.

Aeration – the mixing of air into a liquid or a solid.

Alcohol – hydrocarbon derivative containing a hydroxyl (~OH) group attached to a carbon atom, not in an aromatic ring.

Alkali metals – metals of Group 1 on the periodic table (Na, K, Rb).

Alkaline battery – a dry cell in which the electrolyte contains KOH.

Alkaline earth metals – Group 2 metals on the periodic table; see *Earth metals*.

Allomer – a substance that has a different composition than another but the same crystalline structure.

Allotropes – elements that can have different structures (and therefore different forms), such as carbon (e.g., diamonds, graphite, and fullerene).

Allotropic modifications (allotropes) – different forms of the same element in the same physical state.

Alloying – mixing of metal with other substances (usually other metals) to modify its properties.

Alpha (α) particle – a helium nucleus; helium ion with 2+ charge; an assembly of two protons and two neutrons.

Amorphous solid – a non-crystalline solid with no well-defined ordered structure.

Ampere – unit of electrical current; one ampere equals one coulomb per second.

Amphiprotic – the ability of a substance to exhibit amphiprotism by accepting donated protons.

Amphoterism – the ability to react with both acids and bases; the ability of a substance to act as either an acid or a base.

Amplitude – the maximum distance that the particles of the medium carrying the wave move away from their rest position.

Anion – a negative ion; an atom or group of atoms that has gained one or more electrons.

Anode – in a cathode ray tube, the positive electrode (electrode at which oxidation occurs); the positive side of a dry cell battery or a cell.

Antibonding orbital – a molecular orbital higher in energy than any of the atomic orbitals from which it is derived; lends instability to a molecule or ion when populated with electrons; denoted with a star (*) superscript or symbol.

Artificial transmutation – an artificially induced nuclear reaction caused by the bombardment of a nucleus with subatomic particles or small nuclei.

Associated ions – short-lived species formed by the collision of dissolved ions of opposite charges.

Atmosphere – a unit of pressure; the pressure that will support a column of mercury 760 mm high at 0 °C.

Atom – a chemical element in its smallest form; made up of neutrons and protons within the nucleus and electrons circling the nucleus.

Atomic mass unit (amu) – one-twelfth of the mass of an atom of the carbon (^{12}C) isotope; used for stating atomic and formula weights; also known as a dalton.

Atomic number (*Z*) – the number representing an element which corresponds with the number of protons within the nucleus.

Atomic orbital – a region or volume in space in which the probability of finding electrons is highest.

Atomic radius – radius of an atom.

Atomic weight – weighted average of the masses of the constituent isotopes of an element; the relative masses of atoms of different elements.

Aufbau ("building up") principle – describes the order in which electrons fill orbitals in atoms.

Autoionization – an ionization reaction between identical molecules.

Avogadro's Law – at the same temperature and the same pressure, equal volumes of all gases will contain the same number of molecules.

Avogadro's number (*N*) – the number (6.022×10^{23}) of atoms, molecules, or particles found in precisely 1 mole of a substance.

B

Background radiation – radiation extraneous to an experiment; usually, the low-level natural radiation from cosmic rays and trace radioactive substances present in our environment.

Band – a series of very closely spaced, nearly continuous molecular orbitals that belong to the crystal as a whole.

Band of stability – band containing nonradioactive nuclides in a plot of the number of neutrons versus their atomic number.

Band theory of metals – a theory that accounts for the bonding and properties of metallic solids.

Barometer – a device used to measure the pressure in the atmosphere.

Base – a substance that produces ^{-}OH (*aq*) ions in aqueous solution; accepts a proton and has a high pH; strongly soluble bases are soluble in water and are entirely dissociated; weak bases ionize only slightly; a typical example of a base is sodium hydroxide – NaOH.

Basic anhydride – the oxide of a metal that reacts with water to form a base.

Basic salt – a salt containing an ionizable ~OH group.

***Beta* (*β*) particle** – an electron emitted from the nucleus when a neutron decays to a proton and an electron.

Binary acid – a binary compound in which H is bonded to one or more of the more electronegative nonmetals.

Binary compound – a compound consisting of two elements; it may be ionic or covalent.

Binding energy (nuclear binding energy) – the energy equivalent ($E - mc^2$) of the mass deficiency of an atom (where E is the energy in Joules, m is the mass in kilograms and c is the speed of light in m/s^2).

Boiling – the phase transition of liquid vaporizing.

Boiling point – the temperature at which the vapor pressure of a liquid is equal to the applied pressure; also, the condensation point.

Boiling point elevation – the increase in the boiling point of a solvent caused by the dissolution of a nonvolatile solute.

Bomb calorimeter – a device used to measure the heat transfer between a system and its surroundings at constant volume.

Bond – the attraction and repulsion between atoms and molecules that is a cornerstone of chemistry.

Bond energy – the amount of energy necessary to break one mole of bonds in a substance, dissociating the substance in its gaseous state into atoms of its elements in the gaseous state.

Bond order – half the number of electrons in bonding orbitals minus half the number of electrons in antibonding orbitals.

Bonding orbital – a molecular orbit lower in energy than any of the atomic orbitals from which it is derived; lends stability to a molecule or ion when populated with electrons.

Bonding pair – pair of electrons involved in a covalent bond.

Boron hydrides – binary compounds of boron and hydrogen.

Born-Haber cycle – a series of reactions (and the accompanying enthalpy changes) which, when summed, represents the hypothetical one-step reaction by which elements in their standard states are converted into crystals of ionic compounds (and the accompanying enthalpy changes).

Boyle's Law – at a constant temperature, the volume occupied by a definite mass of a gas is inversely proportional to the applied pressure ($P \propto 1 / V$) or $P_1V_1 = P_2V_2$

Breeder reactor – a nuclear reactor that produces more fissionable nuclear fuel than it consumes.

Brønsted-Lowry acid – a chemical species that donates a proton.

Brønsted-Lowry base – a chemical species that accepts a proton.

Buffer solution – resists change in pH; contains either a weak acid and a soluble ionic salt of the acid or a weak base and a soluble ionic salt of the base.

Buret (burette)– a piece of volumetric glassware, usually graduated in 0.1 mL intervals, used to deliver solutions to be used in titrations in a quantitative (drop-like) manner.

C

Calorie – the amount of heat required to raise the temperature of one gram of water from 14.5 °C to 15.5 °C; 1 calorie = 4.184 joule.

Calorimeter – a device used to measure the heat transfer between a system and its surroundings.

Canal ray – stream of positively charged particles (cations) that moves toward the negative electrode in cathode ray tubes; observed to pass through canals in the negative electrode.

Capillary – a tube having an exceedingly small inside diameter.

Capillary action – the drawing of a liquid up the inside of a small-bore tube when adhesive forces exceed cohesive forces; the depression of the surface of the liquid when cohesive forces exceed the adhesive forces.

Catalyst – a chemical compound used to change the rate (to speed it or slow it) of a reaction that is regenerated (i.e., not consumed) at the end of the reaction.

Catenation – bonding of atoms of the same element into chains or rings (i.e., the ability of an element to bond with itself).

Cathode – the electrode at which reduction occurs; in a cathode ray tube, the negative electrode.

Cathodic protection – protection of a metal (making a cathode) against corrosion by attaching it to a sacrificial anode of more easily oxidized metal.

Cathode ray tube – a closed glass tube containing gas under low pressure, with electrodes near the ends and a luminescent screen at the end near the positive electrode; produces cathode rays when high voltage is applied.

Cation – a positive ion; an atom or group of atoms that have lost one or more electrons.

Cell potential – the potential difference, E_{cell}, between oxidation and reduction half-cells under nonstandard conditions; the force in a galvanic cell that pulls electrons through a reducing agent to an oxidizing agent.

Central atom – an atom in a molecule or polyatomic ion that is bonded to more than one other atom.

Chain reaction – a reaction that, once initiated, sustains itself and expands; a reaction in which reactive species, such as radicals, are produced in more than one step; these reactive species propagate the chain reaction.

Charles' Law – at constant pressure, the volume occupied by a definite mass of gas is directly proportional to its absolute temperature. $V \propto T$ or $V_1 / T_1 = V_2 / T_2$

Chemical bonds – the attractive forces that hold atoms together in elements or compounds.

Chemical change – a change in which one or more new substances are formed.

Chemical equation – description of a chemical reaction by placing the formulas of the reactants on the left of an arrow and the formulas of the products on the right.

Chemical equilibrium – a state of dynamic balance in which the rates of forward and reverse reactions are equal; there is no net change in concentrations of reactants or products while a system is at equilibrium.

Chemical kinetics – the study of rates and mechanisms of chemical reactions and of the factors on which they depend.

Chemical periodicity – the variations in properties of elements with their position in the periodic table.

Chemical reaction – the change of one or more substances into another or multiple substances.

Cloud chamber – a device for observing the paths of speeding particles as vapor molecules condense on them to form fog-like tracks.

Coefficient of expansion – the ratio of the change in the length or the volume of a body to the original length or volume for a unit change in temperature.

Cohesive forces – all the forces of attraction among particles of a liquid.

Colligative properties – physical properties of solutions that depend upon the number but not the kind of solute particles present.

Collision theory – theory of reaction rates that states that effective collisions between reactant molecules must occur for the reaction to occur.

Colloid – a heterogeneous mixture in which solute-like particles do not settle out (e.g., many kinds of milk).

Combination reaction – reaction in which two substances (elements or compounds) combine to form one compound.

Combustible – classification of liquid substances that will burn based on flashpoints; a combustible liquid means a liquid having a flashpoint at or above 37.8 °C (100 °F) but below 93.3 °C (200 °F), except any mixture having components with flashpoints of 93.3 °C (200 °F) or higher, the total of which makes up 99% or more of the total volume of the mixture.

Combustion – an exothermic reaction between an oxidant and fuel with heat and often light.

Common ion effect – suppression of ionization of a weak electrolyte by the presence in the same solution of a strong electrolyte containing one of the same ions as the weak electrolyte.

Complexions – ions resulting from the formation of coordinate covalent bonds between simple ions and other ions or molecules.

Composition stoichiometry – describes the quantitative (mass) relationships among elements in compounds.

Compound – a substance of two or more chemically bonded elements in fixed proportions; can be decomposed into their constituent elements.

Compressed gas – a gas or mixture of gases having (in a container) an absolute pressure exceeding 40 psi at 21.1 °C (70 °F).

Compression – an area in a longitudinal wave where the particles are closer and pushed in.

Concentration – the amount of solute per unit volume, the mass of solvent, or solution.

Condensation – the phase change from gas to liquid.

Condensed phases – the liquid and solid phases; phases in which particles interact strongly.

Condensed states – the solid and liquid states.

Conduction band – a partially filled band or a band of vacant energy levels just higher in energy than a filled band; a band within which, or into which, electrons must be promoted to allow electrical conduction to occur in a solid.

Conductor – material that allows electric flow more freely.

Conjugate acid-base pair – in Brønsted-Lowry terminology, a reactant, and a product that differs by a proton (H^+).

Conformations – structures of a compound that differ by the extent of their rotation about a single bond.

Continuous spectrum – spectrum that contains all wavelengths in a specified region of the electromagnetic spectrum.

Control rods – rods of materials such as cadmium or boron steel that act as neutron absorbers (not merely moderators), used in nuclear reactors to control neutron fluxes and therefore rates of fission.

Conjugated double bonds – double bonds that are separated from each other by one single bond –C=C–C=C–.

Contact process – an industrial process by which sulfur trioxide and sulfuric acid are produced from sulfur dioxide.

Coordinate covalent bond – a covalent bond in which the same species furnish both shared electrons; a bond between a Lewis acid and a Lewis base.

Coordination compound or complex – a compound containing coordinate covalent bonds.

Coordination number – the number of donor atoms coordinated to a metal; in describing crystals, the number of nearest neighbors of an atom or ion.

Coordination sphere – the metal ion and its coordinating ligands but not any uncoordinated counter-ions.

Corrosion – oxidation of metals in the presence of air and moisture.

Coulomb – the SI unit of electrical charge; unit symbol – C, 1 coulomb = 6.242×10^{18} electrons.

Covalent bond – a chemical bond formed by the sharing of one or more electron pairs between two atoms.

Covalent compounds – compounds made of two or more nonmetal atoms that are bonded by sharing valence electrons.

Critical mass – the minimum mass of a fissionable nuclide in a given volume required to sustain a nuclear chain reaction.

Critical point – the combination of critical temperature and critical pressure of a substance.

Critical pressure – the pressure required to liquefy a gas (vapor) at its *critical temperature*.

Critical temperature – the temperature above which a gas cannot be liquefied; the temperature above which a substance cannot exhibit distinct gas and liquid phases.

Crystal – a solid that is packed with ions, molecules, or atoms in an orderly lattice structure.

Crystal field stabilization energy – a measure of the net energy of stabilization gained by a metal ion's nonbonding *d* electrons because of complex formation.

Crystal field theory – theory of bonding in transition metal complexes in which ligands and metal ions are treated as point charges; a purely ionic model; ligand point charges represent the crystal (electrical) field perturbing the metal's *d* orbitals containing nonbonding electrons.

Crystal lattice – a pattern of arrangement of particles in a crystal.

Crystal lattice energy – the amount of energy that holds a crystal together; the energy change when a mole of solid is formed from its constituent molecules or ions (for ionic compounds) in their gaseous state (always negative).

Crystalline solid – a solid characterized by a regular, ordered arrangement of particles.

Curie (Ci) – the basic unit used to describe the intensity of radioactivity in a sample of material; one curie equals 37 billion disintegrations per second or approximately the amount of radioactivity given off by 1 gram of radium.

Cuvette – glassware used in spectroscopic experiments; usually made of plastic, glass, or quartz, and should be as clean and transparent as possible.

Cyclotron – a device for accelerating charged particles along a spiral path.

D

Daughter nuclide – nuclide that is produced in nuclear decay.

Debye (*D*)– the unit used to express dipole moments.

Degenerate – in orbitals, describes orbitals of the same energy.

Deionization – the removal of ions; in the case of water, mineral ions such as sodium, iron, and calcium.

Deliquescence – substances that absorb water from the atmosphere to form liquid solutions.

Delocalization – in reference to electrons, bonding electrons that are distributed among more than two atoms that are bonded together; occurs in species that exhibit resonance.

Density (φ) – mass per unit volume; φ = mV.

Deposition – settling of particles within a solution or mixture; the direct solidification of vapor by cooling; see *Sublimation*.

Derivative – a compound that can be imagined arising from a parent compound by replacement of one atom with another atom or group of atoms; used extensively in organic chemistry to assist in identifying compounds.

Detergent – a soap-like emulsifier that contains a sulfate, SO_3, or a phosphate group instead of a carboxylate group.

Deuterium – an isotope of hydrogen whose atoms are twice as massive as ordinary hydrogen; deuterium atoms contain both a proton and a neutron in the nucleus.

Dextrorotatory – refers to an optically active substance that rotates the plane of plane-polarized light clockwise, also known as "*dextro*" or (+).

Diagonal similarities – refers to chemical similarities in the Periodic Table of Elements of Period 2 to elements of Period 3 one group to the right, especially evident toward the left of the periodic table.

Diamagnetism – weak repulsion by a magnetic field.

Differential Scanning Calorimetry (DSC) – a technique for measuring the temperature, direction, and magnitude of thermal transitions in a sample material by heating/cooling and comparing the amount of energy required to maintain its rate of temperature increase or decrease with an inert reference material under similar conditions.

Differential Thermal Analysis (DTA) – a technique for observing the temperature, direction, and magnitude of thermally induced transitions in a material by heating/cooling a sample and comparing its temperature with that of inert reference material under similar conditions.

Differential thermometer – a thermometer used for accurate measurement of minimal changes in temperature.

Dilution – the process of reducing the concentration of a solute in a solution, usually only by mixing it with more solvent.

Dimer – molecule formed by the combination of two smaller (identical) molecules.

Dipole – electric or magnetic separation of charge; the separation of charge between two covalently bonded atoms.

Dipole-dipole interactions – attractive interactions between polar molecules (i.e., between molecules with permanent dipoles).

Dipole moment – the product of the distance separating opposite charges of the equal magnitude of the charge; a measure of the polarity of a bond or molecule; a measured dipole moment refers to the dipole moment of an entire molecule.

Dispersing medium – the solvent-like phase in a colloid.

Dispersed phase – the solute-like species in a colloid.

Displacement reactions – reactions in which one element displaces another from a compound.

Disproportionation reactions – redox reactions in which the oxidizing agent and the reducing agent are the same species.

Dissociation – in an aqueous solution, the process by which a solid ionic compound separates into its ions.

Dissociation constant – equilibrium constant that applies to the dissociation of a complexion into a simple ion and coordinating species (ligands).

Dissolution or solvation – the spread of ions in a monosaccharide.

Distilland – the material in a distillation apparatus that is to be distilled.

Distillate – the material in a distillation apparatus that is collected in the receiver.

Distillation – the separation of a liquid mixture into its components based on differences in boiling points; the process in which components of a mixture are separated by boiling away the more volatile liquid.

Domain – a cluster of atoms in a ferromagnetic substance, which will all align in the same direction in the presence of an external magnetic field.

Donor atom – a ligand atom whose electrons are shared with a Lewis acid.

***d* orbitals** – beginning in the third energy level, a set of five degenerate orbitals per energy level, higher in energy than *s* and *p* orbitals of the same energy level.

Dosimeter – a small, calibrated electroscope worn by laboratory personnel, designed to detect and measure incident ionizing radiation or chemical exposure.

Double bond – covalent bond resulting from the sharing of four electrons (two pairs) between two atoms.

Double salt – solid consisting of two co-crystallized salts.

Doublet – two peaks or bands of about equal intensity appearing close together on a spectrogram.

Downs cell – electrolytic cell for the commercial electrolysis of molten sodium chloride.

DP number – the degree of polymerization; the average number of monomer units per polymer unit.

Dry cells – ordinary batteries (voltaic cells) for flashlights, radios, etc.

Dumas method – a method used to determine the molecular weights of volatile liquids.

Dynamic equilibrium – an equilibrium in which the processes occur continuously with no net change.

E

Earth metal – highly reactive elements in Group IIA of the periodic table (includes beryllium, magnesium, calcium, strontium, barium, and radium); see *Alkaline earth metal*.

Effective collisions – a collision between molecules resulting in a reaction; one in which the molecules collide with proper relative orientations and enough energy to react.

Effective molality – the sum of the molalities of all solute particles in a solution.

Effective nuclear charge – the nuclear charge experienced by the outermost electrons of an atom; the actual nuclear charge minus the effects of shielding due to inner-shell electrons (e.g., a set of dx_2-y_2 and dz_2 orbitals); those *d* orbitals within a set with lobes directed along the *x*, *y,* and *z*-axes.

Electrical conductivity – the measure of how easily an electric current can flow through a substance.

Electric charge – a measured property (coulombs) that determines electromagnetic interaction.

Electrochemical cell – using a chemical reaction's current; electromotive force is made.

Electrochemistry – the study of chemical changes produced by electrical current and the production of electricity by chemical reactions.

Electrodes – surfaces upon which oxidation and reduction half-reactions occur in electrochemical cells.

Electrode potentials – potentials, *E*, of half-reactions as reductions versus the standard hydrogen electrode.

Electrolysis – a process that occurs in electrolytic cells; chemical decomposition that occurs by the passing of an electric current through a solution containing ions.

Electrolyte – a solution that conducts a certain amount of current and can be split categorically as weak and strong electrolytes.

Electrolytic cells – electrochemical cells in which electrical energy causes nonspontaneous redox reactions to occur (i.e., forced to occur by the application of an outside source of electrical energy).

Electrolytic conduction – conduction of electrical current by ions through a solution or pure liquid.

Electromagnetic radiation – energy that is propagated using electric and magnetic fields that oscillate in directions perpendicular to the direction of travel of the energy; a type of wave that can go through vacuums as well as material; classified as a "self-propagating wave."

Electromagnetism – fields that have an electric charge and electric properties that change the way that particles move and interact.

Electromotive force – a device that gains energy as electric charges pass through it.

Electromotive series – the relative order of tendencies for elements and their simple ions to act as oxidizing or reducing agents, also known as the "activity series."

Electron – a subatomic particle having a mass of 0.00054858 amu and a charge of –1.

Electron affinity – the amount of energy absorbed in the process in which an electron is added to a neutral isolated gaseous atom to form a gaseous ion with a 1– charge; has a negative value if energy is released.

Electron configuration – the specific distribution of electrons in atomic orbitals of atoms or ions.

Electron-deficient compounds – compounds that contain at least one atom (other than H) that shares fewer than eight electrons.

Electron shells – an orbital around the atom's nucleus that has a fixed number of electrons (usually two or eight).

Electronic transition – the transfer of an electron from one energy level to another.

Electronegativity – a measure of the relative tendency of an atom to attract electrons to itself when chemically combined with another atom.

Electronic geometry – the geometric arrangement of orbitals containing the shared and unshared electron pairs surrounding the central atom of a molecule or polyatomic ion.

Electrophile – positively charged or electron-deficient.

Electrophoresis – a technique for the separation of ions by their rate of migration and direction of migration in an electric field.

Electroplating – plating a metal onto a (cathodic) surface by electrolysis.

Element – a substance that cannot be decomposed into simpler substances by chemical means; defined by its *Atomic number*.

Eluant or eluent – the solvent used in the process of elution, as in liquid chromatography.

Eluate – a solvent (or mobile phase) which passes through a chromatographic column and removes the sample components from the stationary phase.

Emission spectrum – spectrum associated with emission of electromagnetic radiation by atoms (or other species) resulting from electronic transitions from higher to lower energy states.

Empirical formula – gives the simplest whole-number ratio of atoms of each element present in a compound; also known as the simplest formula.

Emulsifying agent – a substance that coats the particles of the dispersed phase and prevents coagulation of colloidal particles; an emulsifier.

Emulsion – colloidal suspension of a liquid in a liquid.

Endergonic (+ΔG) – energy is absorbed by the system: nonspontaneous. Products have more energy than reactants.

Endothermic (+ΔH) – describes processes that absorb heat energy.

Endothermicity – the absorption of heat by a system as the process occurs.

Endpoint – the point at which an indicator changes color, and a titration is stopped.

Energy – a system's ability to do work.

Enthalpy (*H*) – the heat content of a specific amount of substance; E= PV.

Entropy (*S*) – a thermodynamic state or property that measures the degree of disorder or randomness of a system; the amount of energy not available for work in a closed thermodynamic system.

Enzyme – a protein that acts as a catalyst in biological systems.

Equation of state – an equation that describes the behavior of matter in a given state; the van der Waals equation describes the behavior of the gaseous state.

Equilibrium or chemical equilibrium – a state of dynamic balance in which the rates of forward and reverse reactions are equal; the state of a system when neither the forward nor reverse reaction is thermodynamically favored.

Equilibrium constant (*K*)– a quantity that characterizes the position of equilibrium for a reversible reaction; its magnitude is equal to the mass action expression at equilibrium; equilibrium *K* varies with temperature.

Equivalence point – the point at which chemically equivalent amounts of reactants have reacted.

Equivalent weight – an oxidizing or reducing agent whose mass gains (oxidizing agents) or loses (reducing agents) 6.022×10^{23} electrons in a redox reaction.

Evaporation – vaporization of a liquid below its boiling point.

Evaporation rate – the rate at which a substance will vaporize (evaporate) when compared to the rate of a known substance such as ethyl ether, especially useful for health and fire-hazard considerations.

Excited state – any state other than the ground state of an atom or molecule; see *Ground state*.

Exergonic ($-\Delta G$) – a positive flow of energy from the system to surroundings: spontaneous. Products have less energy than reactants.

Exothermic – describes processes that release heat energy.

Exothermicity – the release of heat by a system as a process occurs.

Explosive – a chemical or compound that causes a sudden, almost instantaneous release of pressure, gas, heat, and light when subjected to sudden shock, pressure, high temperature or applied potential.

Explosive limits – the range of concentrations over which a flammable vapor mixed with the proper ratios of air will ignite or explode if a source of ignition is provided.

Extensive property – a property that depends upon the amount of material in a sample.

Extrapolate – to estimate the value of a result outside the range of a series of known values; a technique used in standard additions calibration procedure.

F

Faraday constant (*F*) – a unit of electrical charge widely used in electrochemistry and equal to ~ 96,500 coulombs; represents 1 mole of electrons, or the Avogadro number of electrons: 6.022×10^{23} electrons.

Faraday's Law of Electrolysis – a two-part law that Michael Faraday published about electrolysis: (a) the mass of a substance altered at an electrode during electrolysis is directly proportional to the quantity of electricity transferred at that electrode; (b) the mass of an elemental material altered at an electrode is directly proportional to the element's equivalent weight; one equivalent weight of a substance is produced at each electrode during the passage of 96,487 coulombs of charge through an electrolytic cell.

Fast neutron – a neutron ejected at high kinetic energy in a nuclear reaction.

Ferromagnetism – the ability of a substance to become permanently magnetized by exposure to an external magnetic field.

Flashpoint – the temperature at which a liquid will yield enough flammable vapor to ignite; there are various recognized industrial testing methods; therefore, the method used must be stated.

Fluorescence – absorption of high energy radiation by a substance and subsequent emission of visible light.

First Law of Thermodynamics – the total amount of energy in the universe is constant (i.e., energy is neither created nor destroyed in ordinary chemical reactions and physical changes); also known as the Law of Conservation of Energy.

Fluids – substances that flow freely; gases and liquids.

Flux – a substance added to react with the charge, or a product of its reduction; in metallurgy, usually added to lower a melting point.

Foam – colloidal suspension of a gas in a liquid.

Formal charge – a method of counting electrons in a covalently bonded molecule or ion; it counts bonding electrons as though they were equally shared between the two atoms.

Formula – a combination of symbols that indicates the chemical composition of a substance.

Formula unit – the smallest repeating unit of a substance; the molecule for nonionic substances.

Formula weight – the mass of one formula unit of a substance in atomic mass units.

Fractional distillation – the process in which a fractioning column is used in a distillation apparatus to separate the components of a liquid mixture that have different boiling points.

Fractional precipitation – removal of some ions from a solution by precipitation while leaving other ions with similar properties in the solution.

Free energy change – the indicator of the spontaneity of a process at constant temperature (T) and pressure (P); e.g., if ΔG is negative, the process is spontaneous.

Free radical – a highly reactive chemical species carrying no charge and having a single unpaired electron in an orbital.

Freezing – phase transition from liquid to solid.

Freezing point depression – the decrease in the freezing point of a solvent caused by the presence of a solute.

Frequency – the number of repeating corresponding points on a wave that pass a given observation point per unit time; the unit is 1 hertz = 1 cycle per 1 second.

Fuel cells – a voltaic cell that converts the chemical energy of a fuel and an oxidizing agent directly into electrical energy continuously.

G

Gamma (γ) ray – a highly penetrating type of nuclear radiation similar to x-ray radiation, except that it comes from within the nucleus of an atom and has higher energy; energy-wisc, very similar to cosmic rays except that cosmic rays originate from outer space.

Galvanic cell – battery made up of electrochemical with two different metals connected by a salt bridge.

Galvanizing – placing a thin layer of zinc on a ferrous material to protect the underlying surface from corrosion.

Gangue – sand, rock and other impurities surrounding the mineral of interest in an ore.

Gas – a state of matter in which the particles have no definite shape or volume, though they do fill their container.

Gay-Lussac's Law – the expression Gay-Lussac's Law, used for each of the two relationships named after the French chemist Joseph Louis Gay-Lussac and which concern the properties of gases; more usually applied to his law of combining volumes.

Geiger counter – a gas-filled tube which discharges electrically when ionizing radiation passes through it.

Gel – colloidal suspension of a solid dispersed in a liquid; a semi-rigid solid.

Gibbs Free Energy (Δ*G*) – the thermodynamic state function of a system that indicates the amount of energy available for the system to do useful work at constant temperature (T) and pressure (P); value that indicates the spontaneity of a reaction.

Graham's Law – the rates of effusion of gases are inversely proportional to the square roots of their molecular weights or densities.

Ground state – the lowest energy state or most stable state of an atom, molecule, or ion; see *Excited state*.

Group – a vertical column in the periodic table; also known as a family.

H

Haber process – a process for the catalyzed industrial production of ammonia from N_2 and H_2 at high temperature and pressure.

Half-cell – the compartment in which the oxidation or reduction half-reaction occurs in a voltaic cell.

Half-life – the time required for half of a reactant to be converted into product(s); the time required for half of a given sample to undergo radioactive decay.

Half-reaction – either the oxidation part or the reduction part of a redox reaction.

Halogens – Group VIIA elements: F, Cl, Br, I; all halogens are non-metals.

Heat – a form of energy that flows between two samples of matter because of their differences in temperature.

Heat capacity – the amount of heat required to raise the temperature of a body (of any mass) one degree Celsius (1 °C).

Heat of condensation – the amount of heat that must be removed from one gram of vapor at its condensation point to condense the vapor with no change in temperature.

Heat of crystallization – the amount of heat that must be removed from one gram of a liquid at its freezing point to freeze it with no change in temperature.

Heat of fusion – the amount of heat required to melt one gram of a solid at its melting point with no change in temperature; usually expressed in J/g; the molar heat of fusion is the amount of heat required to melt one mole of a solid at its melting point with no change in temperature and is usually expressed in kJ/mol.

Heat of solution – the amount of heat absorbed in the formation of a solution that contains one mole of the solute; the value is positive if heat is absorbed (endothermic) and negative if heat is released (exothermic).

Heat of vaporization – the amount of heat required to vaporize one gram of a liquid at its boiling point with no change in temperature; usually expressed in J/g; the molar heat of vaporization is the amount of heat required to vaporize one mole of liquid at its boiling point with no change in temperature and is usually expressed as ion kJ/mol.

Heisenberg's uncertainty principle – states that it is impossible to accurately determine both the momentum (p) and the position (x) of an electron simultaneously.

Henry's Law – the pressure of the gas above a solution is proportional to the concentration of the gas in the solution.

Hess' Law of Heat Summation – the enthalpy change for a reaction is the same, whether it occurs in one step or a series of steps.

Heterogeneous catalyst – a catalyst that exists in a different phase (solid, liquid or gas) from the reactants; a contact catalyst.

Heterogeneous equilibria – equilibria involving species in more than one phase.

Heterogeneous mixture – a mixture that does not have uniform composition and properties throughout.

Heteronuclear – consisting of different elements.

High spin complex – crystal field designation for an outer orbital complex; all t_{2g} and e_g orbitals are singly occupied before any pairing occurs.

Homogeneous catalyst – a catalyst that exists in the same phase (solid, liquid or gas) as the reactants.

Homogeneous equilibria – when all *Reagents* and products are of the same phase (i.e., all gases, all liquids, or all solids).

Homogeneous mixture – a mixture which has uniform composition and properties throughout.

Homologous series – a series of compounds in which each member differs from the next by a specific number and kind of atoms.

Homonuclear – consisting of only one element.

Hund's rule – all orbitals of a given sublevel must be occupied by single electrons before pairing begins; see *Aufbau* (*"building up"*) *principle*.

Hybridization – mixing a set of atomic orbitals to form a new set of atomic orbitals with the same total electron capacity and with properties and energies intermediate between those of the original unhybridized orbitals.

Hydrate – a solid compound that contains a definite percentage of bound water.

Hydrate isomers – isomers of crystalline complexes that differ in whether water is present inside or outside the coordination sphere.

Hydration – the reaction of a substance with water.

Hydration energy – the energy change accompanying the hydration of a mole of gas and ions.

Hydride – a binary compound of hydrogen.

Hydrocarbons – compounds that contain only carbon and hydrogen (e.g., methane or octane).

Hydrogen bond – a strong dipole-dipole interaction (but still considerably weaker than the covalent or ionic bonds) between molecules containing hydrogen directly bonded to a small, highly electronegative atom, such as N, O or F.

Hydrogenation – the reaction in which hydrogen adds across a double or triple bond.

Hydrogen-oxygen fuel cell – a fuel cell in which hydrogen is the fuel (reducing agent) and oxygen is the oxidizing agent.

Hydrolysis – the reaction of a substance with water or its ions.

Hydrolysis constant – an equilibrium constant for a hydrolysis reaction.

Hydrometer – a device used to measure the densities of liquids and solutions.

Hydrophilic colloids – colloidal particles that repel water molecules.

I

Ideal gas – a hypothetical gas that obeys exactly all postulates of the kinetic-molecular theory.

Ideal Gas Law – the product of pressure and the volume of an ideal gas is directly proportional to the number of moles of the gas and the absolute temperature ($PV = nRT$).

Ideal solution – a solution that obeys Raoult's Law.

Indicators – for acid-base titrations, organic compounds that exhibit different colors in solutions of different acidities; used to determine the point at which reaction between two solutes is complete.

Inert pair effect – characteristic of the post-transition minerals; the tendency of the electrons in the outermost atomic *s* orbital to remain un-ionized or unshared in compounds of post-transition metals.

Inhibitory catalyst – an inhibitor; a catalyst that decreases the rate of reaction.

Inner orbital complex – valence bond designation for a complex in which the metal ion utilizes *d* orbitals for one shell inside the outermost occupied shell in its hybridization.

Inorganic chemistry – a part of chemistry concerned with inorganic (non-carbon-based) compounds.

Insulator – a material that resists the flow of electric current or transfer of heat.

Insoluble compound – a substance that will not dissolve in a solvent, even after mixing.

Integrated rate equation – an equation giving the concentration of a reactant remaining after a specified time; has a different mathematical form for different orders of reactants.

Intermolecular forces – forces between individual particles (atoms, molecules, ions) of a substance.

Ion – a molecule that has gained or lost one or more electrons; an atom or a group of atoms that carries an electric charge.

Ion product for water – equilibrium constant for the ionization of water; $K_w = [H_3O^+]\cdot[OH^-] = 1.00 \times 10^{-14}$ at 25 °C.

Ionic bond – the electrostatic attraction between oppositely charged ions.

Ionic bonding – chemical bonding resulting from the transfer of one or more electrons from one atom or group of atoms to another.

Ionic compounds – compounds containing predominantly ionic bonding.

Ionic geometry – the arrangement of atoms (not lone pairs of electrons) about the central atom of a polyatomic ion.

Ionization – the breaking up of a compound into separate ions; in aqueous solution, the process by which a molecular compound reacts with water and forms ions.

Ionization constant – equilibrium constant for the ionization of a weak electrolyte.

Ionization energy – the minimum amount of energy required to remove the most loosely held electron of an isolated gaseous atom or ion.

Ionization isomers – isomers that result from the interchange of ions inside and outside the coordination sphere.

Isoelectric – having the same electronic configurations.

Isomers – different substances that have the same molecular formula.

Isomorphous – refers to crystals having the same atomic arrangement.

Isotopes – two or more forms of atoms of the same element with different masses; atoms containing the same number of protons but different numbers of neutrons.

IUPAC – acronym for "International Union of Pure and Applied Chemistry."

J

Joule – a unit of energy in the SI system; one joule is 1 $kg{\cdot}m^2/s^2$, which is also 0.2390 calorie.

K

K capture – absorption of a K shell (n = 1) electron by a proton converted to a neutron.

Kelvin (K) – a unit of measure for temperature based upon an absolute scale.

Kinetics – a sub-field of chemistry specializing in reaction rates.

Kinetic energy (*KE*) – energy that matter processes by its motion.

Kinetic-molecular theory – a theory that attempts to explain macroscopic observations on gases in microscopic or molecular terms.

L

Lanthanides – elements 57 (lanthanum) through 71 (lutetium); grouped because of their similar behavior in chemical reactions.

Lanthanide contraction – a decrease in the radii of the elements following the lanthanides compared to what would be expected if there were no f-transition metals.

Lattice – unique arrangement of atoms or molecules in a crystalline liquid or solid.

Law of combining volumes (Gay-Lussac's Law) – at constant temperature and pressure, the volumes of reacting gases (and any gaseous products) can be expressed as ratios of small whole numbers.

Law of conservation of energy – energy cannot be created or destroyed; it can only be changed from one form to another.

Law of conservation of matter – there is no detectable change in the quantity of matter during an ordinary chemical reaction.

Law of conservation of matter and energy – the total amount of matter and energy available in the universe is fixed.

Law of definite proportions (law of constant composition) – different samples of a pure compound will always contain the same elements in the same proportions by mass.

Law of partial pressures (Dalton's Law) – the total pressure exerted by a mixture of gases is the sum of the partial pressures of the individual gases.

Laws of thermodynamics – physical laws that define quantities of thermodynamic systems describe how they behave and (by extension) set certain limitations such as perpetual motion.

Lead storage battery – secondary voltaic cell used in most automobiles.

Leclanche cell – a common type of *Dry cell.*

Le Châtelier's principle – states that a system at equilibrium, or striving to attain an equilibrium, responds in such a way as to counteract any stress placed upon it; if stress (change of conditions) is applied to a system at equilibrium, the system will shift in the direction that reduces stress.

Leveling effect – effect by which all acids stronger than the acid that is characteristic of the solvent react with the solvent to produce that acid; a similar statement applies to bases. The strongest acid (base) that can exist in a given solvent is the acid (base) characteristic of the solvent.

Levorotatory – refers to an optically active substance that rotates the plane of plane-polarized light counterclockwise, also known as a "*levo*" or (–).

Lewis acid – any species that can accept a share in an electron pair.

Lewis base – any species that can make available a share in an electron pair.

Lewis dot formula (electron dot formula) – representation of a molecule, ion or formula unit by showing atomic symbols and only outer shell electrons.

Ligand – a Lewis base in a coordination compound.

Light – that portion of the electromagnetic spectrum visible to the naked eye; also known as "visible light."

Limiting reactant – a substance that stoichiometrically limits the amount of product(s) that can be formed.

Linear accelerator – a device used for accelerating charged particles along a straight-line path.

Line spectrum – an atomic emission or absorption spectrum.

Linkage isomers – isomers in which a ligand bonds to a metal ion through different donor atoms.

Liquid – a state of matter which takes the shape of its container.

Liquid aerosol – colloidal suspension of a liquid in a gas.

London dispersion forces – very weak and very short-range attractive forces between short-lived temporary (induced) dipoles; also known as "dispersion forces."

Lone pair – pair of electrons residing on one atom and not shared by other atoms; unshared pair.

Low spin complex – crystal field designation for an inner orbital complex; contains electrons paired t_{2g} orbitals before e_g orbitals are occupied in octahedral complexes.

M

Magnetic quantum number (mc) – quantum mechanical solution to a wave equation that designates the orbital within a given set (*s*, *p*, *d*, *f*) in which an electron resides.

Manometer – a two-armed barometer.

Mass – a measure of the amount of matter in an object; mass is usually measured in grams or kilograms.

Mass action expression – for a reversible reaction, aA + bB cC + dD; the product of the concentrations of the products (species on the right), each raised to the power that corresponds to its coefficient in the balanced chemical equation, divided by the product of the concentrations of reactants (species on the left), each raised to the power that corresponds to its coefficient in the balanced chemical equation; at equilibrium the mass action expression equals K.

Mass deficiency – the amount of matter that would be converted into energy if an atom were formed from constituent particles.

Mass number (*A*) – the sum of the numbers of protons and neutrons in an atom; always an integer.

Mass spectrometer – an instrument that measures the charge-to-mass ratio of charged particles.

Matter – anything that has mass and occupies space.

Mechanism – the sequence of steps by which reactants are converted into products.

Melting point – the temperature at which liquid and solid coexist in equilibrium.

Meniscus – the shape assumed by the surface of a liquid in a cylindrical container.

Melting – the phase change from a solid to a liquid.

Metal – a chemical element that is a good conductor of both electricity and heat and forms cations and ionic bonds with non-metals; elements below and to the left of the stepwise division (metalloids) in the upper right corner of the periodic table; about 80% of known elements are metals.

Metallic bonding – bonding within metals due to the electrical attraction of positively charged metal ions for mobile electrons that belong to the crystal as a whole.

Metallic conduction – conduction of electrical current through metal or along a metallic surface.

Metalloid – a substance possessing both the properties of metals and non-metals (B, Al, Si, Ge, As, Sb, Te, Po, and At).

Metathesis reactions – reactions in which two compounds react to form two new compounds, with no changes in oxidation number; reactions in which the ions of two compounds exchange partners.

Method of initial rates – method of determining the rate-law expression by carrying out a reaction with different initial concentrations and analyzing the resultant changes in initial rates.

Methylene blue – a heterocyclic aromatic chemical compound with the molecular formula $C_{16}H_{18}N_3SCl$.

Miscibility – the ability of one liquid to mix with (dissolve in) another liquid.

Mixture – a sample of matter composed of two or more substances, each of which retains its identity and properties.

Moderator – a substance, such as hydrogen, deuterium, oxygen, or paraffin, capable of slowing fast neutrons upon collision.

Molality (m) – a concentration expressed as the number of moles of solute per kilogram of solvent.

Molarity (*M*) – the number of moles of solute per liter of solution.

Molar solubility – the number of moles of a solute that dissolves to produce a liter of a saturated solution.

Mole – a measurement of an amount of substance; a single mole contains approximately 6.022×10^{23} units or entities; abbreviated mol.

Molecule – a chemically bonded number of electrically neutral atoms.

Molecular equation – an equation for a chemical reaction in which all formulas are written as if all substances existed as molecules; only complete formulas are used.

Molecular formula – a formula that indicates the actual number of atoms present in a molecule of a molecular substance.

Molecular geometry – the arrangement of atoms (not lone pairs of electrons) around a central atom of a molecule or polyatomic ion.

Molecular orbital (MO) – an orbit resulting from the overlap and mixing of atomic orbitals on different atoms (i.e., a region where an electron can be found in a molecule, as opposed to an atom); a MO belongs to the molecule as a whole.

Molecular orbital theory – a theory of chemical bonding based upon the postulated existence of molecular orbitals.

Molecular weight (MW) – the mass of one molecule of a nonionic substance in atomic mass units.

Molecule – the smallest particle of a compound capable of a stable, independent existence.

Mole fraction (Χ) – the number of moles of a component of a mixture divided by the total number of moles in the mixture.

Monoprotic acid – an acid that can form only one hydronium ion per molecule; may be strong or weak.

Mother nuclide – nuclide that undergoes nuclear decay.

N

Native state – refers to the occurrence of an element in an uncombined or free state in nature.

Natural radioactivity – spontaneous decomposition of an atom.

Neat – conditions with a liquid reagent or gas performed with no added solvent or co-solvent.

Nernst equation – corrects standard electrode potentials for nonstandard conditions.

Net ionic equation – an equation that results from canceling spectator ions and eliminating brackets from a total ionic equation.

Neutralization – the reaction of an acid with a base to form a salt and water; usually, the reaction of hydrogen ions with hydrogen ions to form water molecules.

Neutrino – a particle that can travel at speeds close to the speed of light; created because of radioactive decay.

Neutron – a neutral unit or subatomic particle that has no net charge and a mass of 1.0087 amu.

Nickel-cadmium cell (NiCd battery) – a dry cell in which the anode is Cd, the cathode is NiO2, and the electrolyte is basic.

Nitrogen cycle – the complex series of reactions by which nitrogen is slowly but continually recycled in the atmosphere, lithosphere, and hydrosphere.

Noble gases – elements of the periodic Group 0; He, Ne, Ar, Kr, Xe, Rn; also known as "*rare gases*;" formerly called "*inert gases*."

Nodal plane – a region in which the probability of finding an electron is zero.

Nonbonding orbital – a molecular orbital derived only from an atomic orbital of one atom; lends neither stability nor instability to a molecule or ion when populated with electrons.

Nonelectrolyte – a substance whose aqueous solutions do not conduct electricity.

Non-metal – an element that is not metallic.

Nonpolar bond – a covalent bond in which electron density is symmetrically distributed.

Nuclear – of or about the atomic nucleus.

Nuclear binding energy – the energy equivalent of the mass deficiency; the energy released in the formation of an atom from the subatomic particles.

Nuclear fission – the process in which a heavy nucleus splits into nuclei of intermediate masses, and one or more protons are emitted.

Nuclear magnetic resonance spectroscopy – a technique that exploits the magnetic properties of specific nuclei; useful for identifying unknown compounds.

Nuclear reaction – involves a change in the composition of a nucleus and can emit or absorb an extraordinarily large amount of energy.

Nuclear reactor – a system in which controlled nuclear fission reactions generate heat energy on a large scale, which is subsequently converted into electrical energy.

Nucleons – particles comprising the nucleus; protons and neutrons.

Nucleus – the exceedingly small and dense, positively charged center of an atom containing protons and neutrons, as well as other subatomic particles; the net charge is positive.

Nuclides – refers to different atomic forms of all elements; in contrast to isotopes, which refer only to different atomic forms of a single element.

Nuclide symbol – symbol for an atom A/Z E, in which E is the symbol of an element, Z is its atomic number, and A is its mass number.

Number density – a measure of the concentration of countable objects (e.g., atoms, molecules, etc.) in a space; the number per volume.

O

Octahedral – a term used to describe molecules and polyatomic ions that have one atom in the center and six atoms at the corners of an octahedron.

Octane number – a number that indicates how smoothly a gasoline burns.

Octet rule – many representative elements attain at least a share of eight electrons in their valence shells when they form molecular or ionic compounds; there are some limitations.

Open sextet – refers to species that have only six electrons in the highest energy level of the central element (many Lewis acids).

Orbital – may refer to either an atomic orbital or a molecular orbital.

Organic chemistry – the chemistry of substances that contain carbon-hydrogen bonds.

Organic compound – compounds that contain carbon.

Osmosis – the process by which solvent molecules pass through a semi-permeable membrane from a dilute solution into a more concentrated solution.

Osmotic pressure – the hydrostatic pressure produced on the surface of a semi-permeable membrane by osmosis.

Outer orbital complex – valence bond designation for a complex in which the metal ion utilizes *d* orbitals in the outermost (occupied) shell in hybridization.

Overlap – the interaction of orbitals on different atoms in the same region of space.

Oxidation – an algebraic increase in the oxidation number; may correspond to a loss of electrons.

Oxidation numbers – arbitrary numbers that can be used as mechanical aids in writing formulas and balancing equations; for single-atom ions, they correspond to the charge on the ion; more electronegative atoms are assigned negative oxidation numbers; also known as "oxidation states."

Oxidation-reduction reactions – reactions in which oxidation and reduction occur; also known as "redox reactions."

Oxide – a binary compound of oxygen.

Oxidizing agent – the substance that oxidizes another substance and is reduced.

P

Pairing – a favorable interaction of two electrons with opposite *m* values in the same orbital.

Pairing energy – the energy required to pair two electrons in the same orbital.

Paramagnetism – attraction toward a magnetic field, stronger than diamagnetism but still weak compared to ferromagnetism.

Partial pressure – the pressure exerted by one gas in a mixture of gases.

Particulate matter – fine, divided solid particles suspended in polluted air.

Pauli exclusion principle – no two electrons in the same atom may have identical sets of four quantum numbers.

Percentage ionization – the percentage of the weak electrolyte that ionizes in a solution of a given concentration.

Percent by mass – 100% times the actual yield divided by the theoretical yield.

Percent composition – the mass percent of each element in a compound.

Percent purity – the percent of a specified compound or element in an impure sample.

Period – the elements in a horizontal row of the periodic table.

Periodicity – regular periodic variations of properties of elements with their atomic number (and position in the periodic table).

Periodic law – the properties of the elements are periodic functions of their atomic numbers.

Periodic table – an arrangement of elements in order of increasing atomic numbers that also emphasizes periodicity.

Peroxide – a compound containing oxygen in the –1 oxidation state; metal peroxides contain the peroxide ion, O_2^{2-}.

pH – the measure of acidity (or basicity) of a solution; negative logarithm of the concentration (mol/L) of the H_3O^+ $[H^+]$ ion; scale is commonly used over a range 0 to 14, $pH = -\log[H^+]$.

Phase diagram – a diagram that shows equilibrium temperature-pressure relationships for different phases of a substance.

Photoelectric effect – emission of an electron from the surface of a metal caused by impinging electromagnetic radiation of specific minimum energy; the current increases with increasing intensity of radiation.

Photon – a carrier of electromagnetic radiation of all wavelengths, such as gamma rays and radio waves; also known as "quantum of light."

Physical change – in which a substance changes from one physical state to another, but no substances with different composition are formed; physical change may involve a phase change (e.g., melting, freezing, etc.) or other physical change such as crushing a crystal or separating one volume of liquid into different containers; never produces a new substance.

Plasma – a physical state of matter which exists at extremely high temperatures in which all molecules are dissociated, and most atoms are ionized.

Polar bond – a covalent bond in which there is an unsymmetrical distribution of electron density.

Polarimeter – a device used to measure optical activity.

Polarization – the buildup of a product of oxidation or a reduction of an electrode, preventing further reaction.

Polydentate – refers to ligands with more than one donor atom.

Polyene – a compound that contains more than one double bond per molecule.

Polymerization – the combination of many small molecules to form large molecules.

Polymer – a large molecule consisting of chains or rings of linked monomer units, usually characterized by high melting and boiling points.

Polymorphous – refers to substances that can crystallize in more than one crystalline arrangement.

Polyprotic acid – an acid that can form two or more hydronium ions per molecule; often at least one step of ionization is weak.

Positron – a nuclear particle with the mass of an electron but opposite charge (positive).

Potential energy – energy stored in a body or a system due to its position in a force field or due to its configuration.

Precipitate – an insoluble solid formed by mixing in solution the constituent ions of a slightly soluble solution.

Precision – how close the results of multiple experimental trials are; see *Accuracy*.

Primary standard – a substance of a known high degree of purity that undergoes one invariable reaction with the other reactant of interest.

Primary voltaic cells – voltaic cells that cannot be recharged; no further chemical reaction is possible once the reactants are consumed.

Proton – a subatomic particle having a mass of 1.0073 amu and a charge of +1, found in the nuclei of atoms.

Protonation – the addition of a proton (H^+) to an atom, molecule, or ion.

Pseudobinaryionic compounds – compounds that contain more than two elements but are named like binary compounds.

Q

Quanta – the minimum amount of energy emitted by radiation.

Quantum mechanics – the study of how atoms, molecules, subatomic particles behave and are structured; a mathematical method of treating particles based on quantum theory, which assumes that energy (of small particles) is not infinitely divisible.

Quantum numbers – numbers that describe the energies of electrons in atoms; derived from quantum mechanical treatment.

Quarks – elementary particle and a fundamental constituent of matter; they combine to form hadrons (protons and neutrons).

R

Radiation – high energy particles or rays emitted during the nuclear decay processes.

Radical – an atom or group of atoms that contains one or more unpaired electrons; usually a very reactive species.

Radioactive dating – method of dating ancient objects by determining the ratio of amounts of mother and daughter nuclides present in an object and relating the ratio to the object's age via half-life calculations.

Radioactive tracer – a small amount of radioisotope replacing a nonradioactive isotope of the element in a compound whose path (e.g., in the body) or whose decomposition products are to be monitored by detection of radioactivity; also known as a "radioactive label."

Radioactivity – the spontaneous disintegration of atomic nuclei.

Raoult's Law – the vapor pressure of a solvent in an ideal solution decreases as its mole fraction decreases.

Rate-determining step – the slowest step in a mechanism; the step that determines the overall rate of reaction.

Rate-law expression – equation relating the rate of a reaction to the concentrations of the reactants and the specific rate of the constant.

Rate of reaction – the change in the concentration of a reactant or product per unit time.

Reactants – substances consumed in a chemical reaction.

Reaction quotient – the mass action expression under any set of conditions (not necessarily equilibrium); its magnitude relative to K determines the direction in which the reaction must occur to establish equilibrium.

Reaction ratio – the relative amounts of reactants and products involved in a reaction; may be the ratio of moles, millimoles, or masses.

Reaction stoichiometry – description of the quantitative relationships among substances as they participate in chemical reactions.

Reactivity series (or activity series) – an empirical, calculated and structurally analytical progression of a series of metals, arranged by their "reactivity" from highest to lowest; used to summarize information about the reactions of metals with acids and water, double displacement reactions and the extraction of metals from their ores.

Reagent – a substance or compound added to a system to cause a chemical reaction or to see if a reaction occurs; the terms reactant and reagent are often used interchangeably; however, a reactant is more specifically a substance consumed in the course of a chemical reaction.

Reducing agent – a substance that reduces another substance and is itself oxidized.

Resonance – the concept in which two or more equivalent dot formulas for the same arrangement of atoms (resonance structures) are necessary to describe the bonding in a molecule or ion.

Reverse osmosis – forcing solvent molecules to flow through a semi-permeable membrane from a concentrated solution into a dilute solution by the application of greater hydrostatic pressure on the concentrated side than the osmotic pressure opposing it.

Reversible reaction – reactions that do not go to completion and occur in both the forward and reverse direction.

S

Saline solution – a general term for NaCl in water.

Salts – ionic compounds composed of anions and cations.

Salt bridge – a U-shaped tube containing electrolyte, which connects the two half-cells of a voltaic cell.

Saturated solution – solution in which no more solute will dissolve.

***s*-block elements** – Group 1 and 2 elements (alkali and alkaline metals), which include Hydrogen and Helium.

Schrödinger equation – quantum state equation which represents the behavior of an electron around an atom; describes the wave function of a physical system evolving.

Second law of thermodynamics – the universe tends toward a state of greater disorder in spontaneous processes.

Secondary standard – a solution that has been titrated against a primary standard; a standard solution.

Secondary voltaic cells – voltaic cells that can be recharged; original reactants can be regenerated by reversing the direction of the current flow.

Semiconductor – a substance that does not conduct electricity at low temperatures but will do so at higher temperatures.

Semi-permeable membrane – a thin partition between two solutions through which specific molecules can pass, but others cannot.

Shielding effect – electrons in filled sets of *s*, *p* orbitals between the nucleus and outer shell electrons shield the outer shell electrons somewhat from the effect of protons in the nucleus; also known as the "screening effect."

***Sigma* (σ) bonds** – bonds resulting from the head-on overlap of atomic orbitals, in which the region of electron sharing is along and (cylindrically) symmetrical to the intermolecular axis connecting the bonded atoms.

***Sigma* orbital** – molecular orbital resulting from the head-on overlap of two atomic orbitals.

Single bond – covalent bond resulting from the sharing of two electrons (one pair) between two atoms.

Sol. – a suspension of solid particles in liquid; artificial examples include sol-gels.

Solid – one of the states of matter, where the molecules are packed close together, and there is a resistance to movement/deformation and volume change.

Solubility product constant (K_{sp}) – equilibrium constant that applies to the dissolution of a slightly soluble compound.

Solubility product principle – the solubility product constant expression for a slightly soluble compound is the product of the concentrations of the constituent ions, each raised to the power that corresponds to the number of ions in one formula unit.

Solute – the dispersed (dissolved) phase of a solution; the part of the solution that is mixed into the solvent (e.g., NaCl in saline water).

Solution – a homogeneous mixture made up of multiple substances; made up of solutes and solvents.

Solvation – the process by which solvent molecules surround and interact with solute ions or molecules.

Solvent – the dispersing medium of a solution (e.g., H_2O in saline water).

Solvolysis – the reaction of a substance with the solvent in which it is dissolved.

***s* orbital** – a spherically symmetrical atomic orbital; one per energy level.

Specific gravity – the ratio of the density (φ) of a substance to the density of water.

Specific heat – the amount of heat required to raise the temperature of one gram of substance one degree Celsius.

Specific rate constant – an experimentally determined (proportionality) constant, which is different for different reactions and which changes only with temperature; *k* in the rate-law expression: rate = k [A] × [B].

Spectator ions – ions in a solution that do not participate in a chemical reaction.

Spectral line – any of several lines corresponding to definite wavelengths of an atomic emission or absorption spectrum; marks the energy difference between two energy levels.

Spectrochemical series – arrangement of ligands in order of increasing ligand field strength.

Spectroscopy – the study of radiation and matter, such as X-ray absorption and emission spectroscopy.

Spectrum – display of component wavelengths (colors) of electromagnetic radiation.

Speed of light – the speed at which radiation travels through a vacuum ($c = 299{,}792{,}458$ m/sec).

Square planar – a term used to describe molecules and polyatomic ions that have one atom in the center and four atoms at the corners of a square.

Square planar complex – complex in which the metal is in the center of a square plane, with ligand donor atoms at each of the four corners.

Standard conditions for temperature and pressure (STP) – a standardization used to compare experimental results at 273.15 K (0 °C, 32 °F) and 10^5 Pa (100.0 kPa, 1 bar).

Standard electrodes – half-cells in which the oxidized and reduced forms of a species are present at the unit activity (1.0 M solutions of dissolved ions, 1.0 atm partial pressure of gases, pure solids, and liquids).

Standard electrode potential – by convention, potential ($E°$) of a half-reaction as a reduction relative to the standard hydrogen electrode when all species are present at unit activity.

Standard entropy (*S*) – the absolute entropy of a substance in its standard state at 298 K.

Standard molar enthalpy of formation ($H°_f$) – the amount of heat absorbed in the formation of one mole of a substance in a specified state from its elements in their standard states.

Standard molar volume – the volume occupied by one mole of an ideal gas under standard conditions; 22.4 L.

Standard reaction – a reaction in which the numbers of moles of reactants shown in the balanced equation, all in their standard states, are completely converted to the numbers of moles of products shown in the balanced equation, also all at their standard state.

State of matter – matter having a homogeneous, macroscopic phase (e.g., as a gas, plasma, liquid, or solid in increasing concentration).

Stoichiometry – description of the quantitative relationships among elements and compounds as they undergo chemical changes.

Strong electrolyte – a substance that conducts electricity well in a dilute aqueous solution.

Strong field ligand – ligand that exerts a strong crystal or ligand electrical field and generally forms low spin complexes with metal ions when possible.

Structural isomers – compounds that contain the same number of the same kinds of atoms in different geometric arrangements.

Subatomic particles – particles that comprise an atom (e.g., protons, neutrons, and electrons).

Sublimation – the direct vaporization of a solid by heating without passing through the liquid state; a phase transition from solid to limewater fuel or gas.

Substance – any kind of matter, all specimens of which have the same chemical composition and physical properties.

Substitution reaction – a reaction in which another atom or group of atoms replaces an atom or a group of atoms.

Supercooled liquids – liquids that, when cooled, apparently solidify but continue to flow very slowly under the influence of gravity.

Supercritical fluid – a substance at a temperature above its critical temperature.

Supersaturated solution – a solution that contains a higher than saturation concentration of solute; slight disturbance or seeding causes crystallization of excess solute.

Suspension – a heterogeneous mixture in which solute-like particles settle out of the solvent-like phase sometime after their introduction.

T

Talc – a mineral representing the one on the Mohs Scale and composed of hydrated magnesium silicate with the chemical formula $H_2Mg_3(SiO_3)_4$ or $Mg_3Si_4O_{10}(OH)_2$.

Temperature – a measure of the kinetic energy of particles.

Ternary acid – a ternary compound containing H, O, and another element, often a nonmetal.

Ternary compound – a compound consisting of three elements; may be ionic or covalent.

Tetrahedral – a term used to describe molecules and polyatomic ions that have one atom in the center and four atoms at the corners of a tetrahedron; ideal bond angle equals 109.5°.

Theoretical yield – the maximum amount of a specified product that could be obtained from specified amounts of reactants, assuming complete consumption of the limiting reactant according to only one reaction and complete recovery of the product; see *Actual yield*.

Theory – an established model describing the nature of a phenomenon.

Thermal conductivity – a property of a material to conduct heat (often noted as k).

Thermal cracking – decomposition by heating a substance in the presence of a catalyst and in the absence of air.

Thermochemistry – the study of absorption/release of heat within a chemical reaction.

Thermodynamics – the study of the effects of changing temperature, volume, or pressure (or work, heat, and energy) on a macroscopic scale.

Thermodynamic stability – when a system is in its lowest energy state with its environment (equilibrium).

Thermometer – a device that measures the average energy of a system.

Thermonuclear energy – energy from nuclear fusion reactions.

Third Law of Thermodynamics – the entropy of a hypothetical pure, perfect crystalline substance at absolute zero temperature is zero.

Titration – a procedure in which one solution is added to another solution until the chemical reaction between the two solutes is complete; the concentration of one solution is known, and that of the other is unknown.

Torr – a unit of pressure; 1 Torr is equivalent to 133.322 Pa or 1.3158×10^{-3} atm.

Total ionic equation – an equation for a chemical reaction written to show the predominant form of all species in aqueous solution or contact with water.

Transition elements (metals) – B Group elements except IIB in the periodic table; sometimes called transition elements, elements that have incomplete d sub-shells; may also be referred to as "the *d*-block elements."

Transition state theory – theory of reaction rates that states that reactants pass through high-energy transition states before forming products.

Transuranic element – an element with an atomic number greater than 92; none of the transuranic elements are stable.

Triple bond – the sharing of three pairs of electrons within a covalent bond (e.g., N_2 as $N{\equiv}N$).

Triple point – the place where the temperature and pressure of three phases are the same; water has a distinct phase diagram.

Tyndall effect – the effect of light scattering by colloidal particles (a mixture where one substance is dispersed evenly throughout another) or by suspended particles.

U

Uncertainty – the characteristic that any measurement that involves the estimation of any amount cannot be exactly reproducible.

Uncertainty principle – knowing the location of a particle makes the momentum uncertain, while knowing the momentum of a particle makes the location uncertain.

Unit cell – the smallest repeating unit of a lattice.

Unit factor – statements used in converting between units.

Universal or ideal gas constant – proportionality constant in the ideal gas law (0.08206 L·atm/(K·mol)).

UN number – a four-digit code used to note hazardous and flammable substances.

Unsaturated hydrocarbons – hydrocarbons that contain double or triple carbon-carbon bonds.

V

Valence bond theory – assumes that covalent bonds are formed when atomic orbitals on different atoms overlap, and the electrons are shared.

Valence electrons – outermost electrons of atoms; usually those electrons involved in bonding.

Valence shell electron pair repulsion theory – assumes that electron pairs are arranged around the central element of a molecule or polyatomic ion so that there is maximum separation (and minimum repulsion) among regions of high electron density.

Van der Waals' equation – an equation of a state that extends the ideal gas law to real gases by the inclusion of two empirically determined parameters, which are different for different gases.

Van der Waals force – one of the forces (attraction/repulsion) between molecules.

Van't Hoff factor – the ratio of moles of particles in solution to moles of solute dissolved.

Vapor – when a substance is below the critical temperature while in the gas phase.

Vaporization – the phase change from liquid to gas.

Vapor pressure – the particle pressure of vapor at the surface of its parent liquid.

Viscosity – the resistance of a liquid to flow (e.g., oil has a higher viscosity than water).

Volt – one joule of work per coulomb; the unit of electrical potential transferred.

Voltage – the potential difference between two electrodes; measure of chemical potential for a redox reaction.

Voltaic cells – electrochemical cells in which spontaneous chemical reactions produce electricity; also known as "galvanic cells."

Voltmeter – an instrument that measures the cell potential.

Volumetric analysis – measuring the volume of a solution (of known concentration) to determine the concentration of the substance being measured within said solution; see *Titration*.

W

Water equivalent – the amount of water that would absorb the same amount of heat as the calorimeter per degree of temperature increase.

Weak electrolyte – a substance that conducts electricity poorly in a dilute aqueous solution.

Weak field ligand – a ligand that exerts a weak crystal or ligand field and generally forms high spin complexes with metals.

X

X-ray – electromagnetic radiation between gamma and ultraviolet (UV) rays.

X-ray diffraction – a method for establishing structures of crystalline solids using single wavelength X-rays and studying the diffraction pattern.

X-ray photoelectron spectroscopy – a spectroscopic technique used to measure the composition of a material.

Y

Yield – the amount of product produced during a chemical reaction.

Z

Zone melting – a way to remove impurities from an element by melting it and slowly traveling it down an ingot (cast).

Zone refining – a method of purifying a bar of metal by passing it through an induction heater; this causes impurities to move along a melted portion.

Zwitterion (formerly known as a dipolar ion) – a neutral molecule with both a positive and a negative electrical charge; multiple positive and negative charges can be present; distinct from dipoles at different locations within that molecule; also known as "inner salts."

We want to hear from you

Your feedback is important to us because we strive to provide the highest quality prep materials. Email us any questions or comments, so we can incorporate your feedback into future editions.

Customer Satisfaction Guarantee

If you have any concerns about this book, including printing issues, contact us and we will resolve any issues to your satisfaction.

info@sterling–prep.com

*We reply to all emails – **check your spam folder***

Thank you for choosing our products to achieve your educational goals!

Please, leave your Customer Review on Amazon

Made in the USA
Coppell, TX
13 January 2022